IRMA Lectures in Mathematics and Theoretical Physics 1

Edited by Vladimir G. Turaev

Deformation Quantization

Proceedings of the meeting
of theoretical physicists and mathematicians
Strasbourg, May 31 – June 2, 2001

Rencontre entre physiciens théoriciens
et mathématiciens
Strasbourg, 31 mai – 2 juin 2001

Editor

Gilles Halbout

Walter de Gruyter · Berlin · New York 2002

Editor

Gilles Halbout

Institut de Recherche Mathématique Avancée (IRMA), Université Louis Pasteur – C.N.R.S., 7, rue René Descartes, 67084 Strasbourg Cedex, France, e-mail: halbout@math.u-strasbg.fr

Series Editor

Vladimir G. Turaev

Institut de Recherche Mathématique Avancée (IRMA), Université Louis Pasteur – C.N.R.S., 7, rue René Descartes, 67084 Strasbourg Cedex, France, e-mail: turaev@math.u-strasbg.fr

Mathematics Subject Classification:

53-06; 53D055

Key words:

deformation quantization, Poisson manifold, quantum physics, representation theory, non commutative geometry

♾ Printed on acid-free paper which falls within the guidelines of the ANSI to ensure permanence and durability.

Library of Congress – Cataloging-in-Publication Data

Deformation quantization : proceedings of the meeting between mathematicians and theoretical physicists, Institut de recherche mathématiques avancée de Strasbourg May 31–June 2, 2001 / editor G. Halbout.

p. cm. – (IRMA-lectures in mathematics and theoretical physics ; 1)

ISBN 3 11 017247 X (alk. paper)

1. Geometric quantization – Congresses. 2. Mathematical physics – Congresses. I. Halbout, G. (Gilles) II. Institut de recherche mathématique avancée. III. Series.

QC174.17.G46 D44 2002

530.14'3–dc21

2002017376

Die Deutsche Bibliothek – Cataloging-in-Publication Data

Deformation quantization : proceedings of the Meeting of Theoretical Physicists and Mathematicians, Strasbourg, May 31–June 2, 2001 / ed. Gilles Halbout. – Berlin ; New York : de Gruyter, 2002

(IRMA lectures in mathematics and theoretical physics ; 1)

ISBN 3-11-017247-X

Printed in Germany.

Cover design: I. Zimmermann, Freiburg.

Depicted on the cover is the Strasbourg Cathedral.

Typeset using the authors' TEX files: I. Zimmermann, Freiburg.

Printing and binding: Hubert & Co. GmbH & Co. KG, Göttingen.

Preface

This volume presents the proceedings of the 68th meeting between theoretical physicists and mathematicians in Strasbourg. The subject of this meeting was deformation quantization. This topic was introduced in the 1970s to clarify mathematically the relationships between classical and quantum mechanics. It has recently witnessed a spectacular development due to the works of Drinfeld, Etingof, Kazhdan and Kontsevich. A short overview of the subject is given in the editor's paper which also presents other contributions.

The papers in this volume cover a number of topics including star-products on Poisson manifolds, quantization of Hopf algebras, index theorems, globalization and cohomological problems. Both mathematical and physical approaches ranging from operads and prop theory to the asymptotic quantum electrodynamics are presented. In the contribution of Dito and Sternheimer, historical remarks and surveys set the results in the proper perspective.

Strasbourg, February 2002 *Gilles Halbout*

Table of Contents

Deformation quantization, methods and applications

Gilles Halbout

Département de Mathématiques
Université Louis Pasteur
7, rue R. Descartes, 67 084 Strasbourg, France
e-mail: `halbout@math.u-strasbg.fr`

Abstract. This paper is a presentation of various methods in the theory of deformation quantization. After recalling the definitions of deformation for Poisson manifolds and Lie bialgebras, we discuss different approaches: cohomological, explicit constructions, globalization. We will also present other papers of these proceedings and show how they are related to the various approaches. Finally, we give two open problems and we indicate possible ways of solving them using these methods.

2000 Mathematics Subject Classification: 53D55 (17B37, 16E40)

1 Introduction

To clarify mathematically the analogy between classical and quantum mechanics, Bayen, Flato, Frønsdal, Lichnerowicz and Sternheimer (*cf.* [BFFLS1] and [BFFLS2]) expressed quantum mechanics as a "deformation" of classical mechanics. The classical phase space is seen as a differential manifold M. Instead of considering new objects such as operators, we consider smooth functions on M and change the algebraic structure of this algebra $A = C^\infty(M)$. To express the evolution equations, we suppose that $(M, \{.,.\})$ is a Poisson manifold, that is to say a manifold endowed with a Poisson bracket $\{.,.\} : A \times A \to A$ which is a bilinear, antisymmetric form satisfying

$$-\forall f, g, h \in A,\ \{f, \{g, h\}\} + \{g, \{h, f\}\} + \{h, \{f, g\}\} = 0 \quad \text{(the Jacobi identity)}$$

and

$$-\forall f, g_1, g_2 \in A,\ \{f, g_1 g_2\} = \{f, g_1\} g_2 + g_1\{f, g_2\} \quad \text{(the Leibniz identity).}$$

Bayen, Flato, Frønsdal, Lichnerowicz and Sternheimer expressed the notion of deformation by introducing star-products, *i.e.*, $\mathbb{R}[[\hbar]]$-linear and associative maps

$$\star : A[[\hbar]] \times A[[\hbar]] \to A[[\hbar]],$$
$$(f, g) \mapsto f \star g = fg + \hbar P_1(f, g) + \hbar^2 P_2(f, g) + \cdots,$$

where $P_i : A \times A \to A$ are $\mathbb{R}[[\hbar]]$-bilinear bidifferential maps that vanish on constant functions. More generally, we will denote $C^k(A, A)$ the space of $\mathbb{R}[[\hbar]]$-linear differential maps $A^{\otimes k} \to A$ that vanish on constant functions. If A is endowed with a Poisson structure $\{.,.\}$, one says that the star-product $\star$ is a deformation of the algebra $(A[[\hbar]], \{.,.\})$ if it satisfies: $\frac{1}{2}(P_1(f, g) - P_1(g, f)) = \{f, g\}$. One often supposes, without loss of generality, that $P_1 = \{.,.\}$.

A deformation $\star$ of a Poisson manifold allows one to define a "quantum" bracket $[.,.]_\hbar : \mathbb{A} \times \mathbb{A} \to \mathbb{A}$,

$$[f, g]_\hbar = \frac{1}{2\hbar}(f \star g - g \star f) = \{f, g\}\hbar + \cdots.$$

In this way, one has constructed a one parameter family of algebras (with parameter $\hbar$) having the same underlying vector space A. Equations of quantum mechanics correspond to a non zero value of $\hbar$ whereas the equations of classical mechanics correspond to $\hbar = 0$.

An extension of this approach to partial differential equations is given in the contribution of G. Dito on asymptotic quantum electrodynamics.

In many physical problems, the manifold M has a Lie group structure compatible with the Poisson structure. In this case, one usually denotes the manifold by G. We denote by e its neutral element, and we say that G is a Poisson–Lie group. The group G can be seen as the set of Dirac measures on itself and is then naturally embedded in $\mathcal{E}$, the set of compactly supported distributions. The space $\mathcal{E}$ has the structure of a topological Hopf algebra $(\mathcal{E}, \mu, \Delta, \eta, \epsilon, S)$ (*cf.*, for example, [BFGP] and [BFP]) where μ is the usual convolution product,

$$\Delta : D \mapsto \Big(f \times g \mapsto \int_G f(x)g(x)D\Big), \quad \eta(1)(f) = f(e),$$

$$\epsilon : D \mapsto \int_G D, \quad \text{and} \quad S : D \mapsto \Big(f \mapsto \int_G f(x^{-1})D\Big).$$

We usually restrict ourselves to the Hopf subalgebra $\mathcal{E}' \subset \mathcal{E}$ of distributions whose support is included in $\{e\}$. This subalgebra can be algebraically identified with the universal enveloping algebra $U(\mathfrak{g})$ generated by the complexified Lie algebra $\mathfrak{g}$ of G. Its Hopf algebra structure, $(U(\mathfrak{g}), \cdot, \Delta_0, \eta_0, \epsilon_0, S_0)$, can be defined by

$$\forall X, Y \in \mathfrak{g}, \ X \cdot Y - Y \cdot X = [X, Y], \quad \Delta_0(X) = 1 \otimes X + X \otimes 1,$$
$$\eta_0(1) = 1, \quad \epsilon_0(X) = 0, \quad S_0(X) = -X.$$

When $\mathfrak{g}$ is finite dimensional, the topological linear dual $U^*(\mathfrak{g})$ of $U(\mathfrak{g})$ is still a Hopf algebra. This commutative algebra can be identified with the ring $\mathbb{C}[[\mathfrak{g}]]$ of formal power series over $\mathfrak{g}$ and can be seen as the algebra of functions over a space $\bar{G}$. This space $\bar{G}$ is called the formal group associated to G ($Fun(\bar{G}) = \mathbb{C}[[\mathfrak{g}]]$).

When G is a Poisson–Lie group, one obtains, at the infinitesimal level, a Lie bialgebra structure on its Lie algebra $\mathfrak{g}$: the Poisson bracket on G induces a map $\delta : \mathfrak{g} \to \mathfrak{g} \wedge \mathfrak{g}$ which endows $\mathfrak{g}$ with a Lie bialgebra structure. One can extend δ to $U(\mathfrak{g})$ by setting $\delta(ab) = \delta(a) \cdot \Delta_0(b) + \Delta_0(a) \cdot \delta(b)$, for a, b in $U(\mathfrak{g})$. We then obtain on $(U(\mathfrak{g}), \cdot, \Delta_0, \delta)$ a Hopf co-Poisson algebra structure (*cf.*, for example, [EK]). Considering its linear dual and denoting $\{.,.\}$ the dual map of δ, one verifies that the Hopf algebra $(Fun(\bar{G}), \{.,.\})$, is a Hopf–Poisson algebra. This allows us to view $\bar{G}$ as a "formal" Poisson group.

Following Drinfeld (*cf.* [D1,D2]), deforming a Poisson–Lie group G comes down to deforming its associated formal Poisson group $(\bar{G}, \{.,.\})$ or, by duality, to deforming the Hopf co-Poisson algebra $U(\mathfrak{g})$: if $(H, \cdot, \Delta, \delta)$ is a Hopf co-Poisson cocommutative algebra, a Hopf algebra $(H_\hbar, \cdot_\hbar, \Delta_\hbar)$ is a quantization of H if:

- $H_\hbar = H[[\hbar]]$ as a $\mathbb{C}[[\hbar]]$-module,
- $H_\hbar/\hbar H_\hbar$ is isomorphic to H as a Hopf algebra,
- for all x_0 in H and x in $H_\hbar$ such as $x \equiv x_0 \bmod \hbar$, one has: $\frac{\Delta_\hbar(x)-\tau\Delta_\hbar(x)}{\hbar} \equiv \delta(x_0) \bmod \hbar$, where $\tau : H_\hbar \otimes H_\hbar \to H_\hbar \otimes H_\hbar,\ x \otimes y \mapsto y \otimes x$.

Let G be a Poisson–Lie group and $\bar{G}$ the formal Poisson group associated to it. Suppose we have constructed $U_\hbar(\mathfrak{g})$, a deformation of $U(\mathfrak{g})$. The product $\star_\hbar$ of the Hopf algebra $Fun(\bar{G})[[\hbar]]$ (the linear dual of $U_\hbar(\mathfrak{g})$), satisfies for all f and g in $Fun(\bar{G})$:

$$f \star_\hbar g = fg + \frac{\hbar}{2}\{f, g\} + O(\hbar).$$

This product looks like a star-product but the terms involved in $\star_\hbar$ are no longer supposed to be differential operators.

As in the classical case, it is often very interesting to look at the representation theory of the "deformed" objects. Results on representation theory and deformation quantization will be found in the contribution of S. Waldmann.

2 Methods

In their contribution, G. Dito and D. Sternheimer give an overview of the history and of the different results of deformation quantization theory. In this section, we summarize different approaches to deformation. For each general method we will give examples of the construction of star-products over a Poisson manifold and of quantizations of Lie bialgebras. In the particular case of unimodular fields, C. Roger gives in his contribution a more complete description of these methods.

2.1 Explicit formulas

The basic examples of Poisson manifolds are symplectic manifolds (M, ω). For such manifolds, by the Darboux theorem, we know that locally the manifold can be endowed with a constant symplectic form, say ω_0. In that case, we know how to construct a star-product, the Moyal–Weyl product: for any two functions f and g we set $f \star g = \sum_{k,i_1,\dots,i_k,j_1,\dots,j_k} \frac{1}{k!}\left(\frac{\hbar}{2}\right)^k \pi_{\omega_0}^{i_1 j_1} \cdots \pi_{\omega_0}^{i_k j_k} \partial_{y_{i_1}} \cdots \partial_{y_{i_k}} f \partial_{y_{j_1}} \cdots \partial_{y_{j_k}} g$ where π_ω is the constant Poisson tensor field, dual of ω and x_i, y_j are the local coordinates on M.

One had to wait for Kontsevich [K] to see explicit formulas for the construction of a star-product in the general case of Poisson manifolds. In his work, Kontsevich gave explicit calculations of the coefficients in the power series of the star-product in terms of convergent integrals.

For a manifold M, Kontsevich deduced the existence of star-products from the formality conjecture, that is the existence of a L_∞-homomorphism between the two graded Lie algebras

$$\mathfrak{g}_1 = (\Gamma(M, \wedge^{\cdot} TM), [.,.]_S) \quad \text{and} \quad \mathfrak{g}_2 = (C^{\cdot}(A, A) = \bigotimes_k C^k(A, A), [.,.]_G),$$

where $[.,.]_S$ and $[.,.]_G$ are respectively the Schouten and the Gerstenhaber bracket (*cf.* [K]) which give the two spaces a Lie algebra stucture. Recall that if $\mathfrak{g}_1^{\cdot}$ and $\mathfrak{g}_2^{\cdot}$ are two $\mathbb{N}$-graded Lie algebras, a L_∞-homomorphism is given by maps $\{\rho_i : \wedge^i \mathfrak{g}_1 \to \mathfrak{g}_2\}_{i\in\mathbb{N}}$ of degree $1 - i$ satisfying

$$[\rho, \rho] = \rho \circ \partial^{\mathrm{Lie}},$$

where $\rho = \sum_i \rho_i$, ∂^{Lie} is the Lie differential, and $[\rho, \rho]$ is defined by

$$[\rho, \rho](g_1 \wedge \cdots \wedge g_l) = \sum_{i=0}^{l} \sum_{\sigma \in S_l} \mathrm{sign}(\sigma)[\rho_i(g_{\sigma(1)} \wedge \cdots \wedge g_{\sigma(i)}), \rho_{l-i}(g_{\sigma(i+1)} \wedge \cdots \wedge g_{\sigma(j)})]$$

($\mathrm{sign}(\sigma) = \pm 1$ is equal to 1 for even σ).

Kontsevich proved the formality conjecture. Then, If $(M, \{.,.\})$ is a Poisson manifold, π the associated tensor field and $(K_i)_{i\in\mathbb{N}}$ the Kontsevich L_∞-homomorphism, the bilinear map

$$\star := \sum_l \frac{1}{l!} K_l(\hbar\pi \wedge \cdots \wedge \hbar\pi) \in C^2(A, A)[[\hbar]]$$

is a deformation of $(M, \{.,.\})$.

Shoikhet ([S]) extended the work of Kontsevich to chains: if we denote by $C_k(A)$ the space $A^{\otimes k}$, this space is the "non-commutative analogue" of the space $\Omega^{\cdot}(M)$ of differential forms on M (in the same way the space $C^{\cdot}(A, A)$ is the "non-commutative analogue" of the space $\Gamma(M, \wedge^{\cdot} TM)$ of antisymmetric tensor fields on M). Shoikhet gave an explicit L_∞-homomorphism between the two graded Lie algebras

$$(\Gamma(M, \wedge^{\cdot} TM) \oplus \Omega^{\cdot}(M), [.,.]'_S) \quad \text{and} \quad (C^{\cdot}(A, A) \oplus C_{\cdot}(A), [.,.]'_G),$$

proving explicitly the "formality conjecture for chains" (here the bracket $[.,.]'_S$ and $[.,.]'_G$ are natural extensions of the brackets $[.,.]_S$ and $[.,.]_G$).

In the case of quantization of Lie bialgebras, the first explicit constructions were given in the case of semi-simple Lie algebras. One can find a good description in [L]. An explicit constructions in the general case were given by Etingof and Kazhdan (*cf.* [EK]). Their construction uses the existence of Drinfeld associators (whose construction is far from being explicit).

In their contribution Arnaudon, Avan, Frappat and Ragoucy use explicit constructions to emphasize the relations between the deformed double Yangian type algebras and the quantum affine elliptic algebras.

2.2 Cohomological methods

Any deformation problem has a corresponding cohomological approach. For a star-product, the only problem in constructing the product step by step is ensuring its associativity. An easy computation shows that the obstructions lie in the third group of Hochschild cohomology. In the case where this group is not trivial, one has tried to impose extra conditions on the products so that the new cohomological setting gives a trivial third cohomology group. This method was used followed by De Wilde and Lecomte [DWL1] who first proved the existence of a star-products on a general symplectic manifold by simultaneously constructing a product and a corresponding derivation.

The existence of star-products in the general case was proved by Tamarkin [T]. He succeeded in proving the formality conjecture by considering morphisms up to homotopy not only between Lie algebras but also between "homotopy" Gerstenhaber algebras. This extra condition ensures the triviality of the corresponding cohomology group containing the obstructions to deformation.

In his contribution, Tamarkin uses the same method to prove the existence of a deformation on any cocommutative Hopf algebra.

In the same spirit, Frønsdal studies Abelian deformations of ordinary algebras of functions and relates the problem to Harrison cohomology.

Cohomological methods usually give only the existence of deformations. To get explicit formulas one has to construct explicit homotopy formulas. For the Hochschild cohomology and homology, some formulas were given in *cf.* [DWL1] and [H].

2.3 Globalization

The existence of star-products on a Poisson manifold M, is often proved locally. This is the case when M is symplectic because the symplectic form can be chosen standard locally. The problem is then to "glue" together the different local star-products.

The first solution to the globalization problem was given by Fedosov in 1994 ([F1]). Fedosov used a "flat" connection on the Weyl bundle constructed over the symplectic manifold M to identify horizontal sections of that connection with the algebra of smooth functions on M. The connection was constructed using cohomological methods.

In the general case, Kontsevich proposed explicit local formulas for deformation. Using Fedosov techniques, Cataneo, Felder and Tomassini describe, in their contribution a method to globalize the Kontsevich star-product.

Fedosov's construction on symplectic manifolds allowed him to prove the index theorem for "operators on a symplectic manifold" (*cf.* [F2]). To do this he had to construct a trace on the deformed algebra and a corresponding trace density (which is the analogue of the formality map for chains). In his contribution to these proceedings, Fedosov explains how to compute explicitly this trace density map.

Using the Fedosov connection, Boutet de Monvel investigates the relation between Toeplitz algebras and semi-classical algebras (star-products) arising for example in geometric prequantization.

3 Open problems

We end this paper with two open problems.

The first problem is the quantization of Poisson morphisms: let (M_1, π_1) and (M_2, π_2) be two Poisson manifolds endowed with star-products $\star_1$ and $\star_2$, and let $\varphi : M_1 \to M_2$ be a Poisson morphism. Can one find a quantization of φ that is to say a morphism of deformed algebras giving φ for $\hbar = 0$? Bordemann has shown ([B]) that this problem is related to the symplectic restriction of star-products. There is no general answer yet but it seems that a good possible cohomological approach would involve a suitable equivariant BRST cohomology.

For Lie bialgebras a similar question is the quantization problem for braidings. Starting from a quasi-triangular Lie bialgebra one can always construct a braiding structure on the dual Poisson formal group (*cf.* [GH]). The proof of [GH] used the quantization of quasi-triangular Lie bialgebras. Conversely, can we always quantize a braiding? A possible solution to this question could rely on the Tamarkin approach developed in this volume.

Acknowledgements. I would like to thank C. Kassel and V. Turaev for their help. The conference and these proceedings would not exist without the precious cooperation of the staff of IRMA (Institut de Recherche Mathématique Avancée de Strasbourg) and of dc Gruytcr's mathcmatics editors.

References

[BFFLS1] Bayen, F., Flato, M., Frønsdal, C., Lichnerowicz, A., Sternheimer, D., Quantum mechanics as a deformation of classical mechanics, *Lett. Math. Phys.* 1 (1975/77), no. 6, 521–530.

[BFFLS2] Bayen, F., Flato, M., Frønsdal, C., Lichnerowicz, A., Sternheimer D., Deformation theory and quantization, I and II, *Ann. Phys.* 111 (1977), 61–151.

[BFGP] Bonneau, P., Flato, M., Gerstenhaber, M., Pinczon, G., The hidden group structure of quantum groups: strong duality, rigidity and preferred deformations, *Comm. Math. Phys.* 161 (1994), no. 1, 125–156.

[BFP] Bonneau, P., Flato, M., Pinczon, G., A natural and rigid model of quantum groups, *Lett. Math. Phys.* 25 (1992), no. 1, 75–84.

[B] Bordemann, M., The deformation quantization of certain super-Poisson brackets and BRST cohomology, in *Quantization, Deformations, and Symmetries* (Dito, G., Sternheimer, D., eds.), Conférence Moshé Flato 1999, Math. Phys. Stud. 22, Kluwer Academic Publishers, Dordrecht, Boston, London 2000, 45–68.

[D1] Drinfeld, V., Quantum groups, in *Proceedings of the International Congress of Mathematicians*, vol. 1, Amer. Math. Soc., Providence, RI, 1987 , 798–820.

[D2] Drinfeld, V., Hamiltonian structures on Lie groups, Lie bialgebras and the geometric meaning of classical Yang-Baxter equations, *Dokl. Akad. Nauk SSSR* 268 (1983), no. 2, 285–287.

[DWL1] De Wilde, M., Lecomte, P., Existence of star-product and of formal deformation of the Poisson Lie algebra of arbitrary symplectic manifold, *Lett. Math. Phys.* 7 (1983), no.6, 487–496.

[DWL2] De Wilde, M., Lecomte, P., An homotopy formula for the Hochschild cohomology, *Compositio Math.* 96 (1995), no.1, 99–109.

[EK] Etingof, P., Kazhdan, D., Quantization of Lie bialgebras, I, *Selecta Math.* (N. S.) 2 (1996), 1–41.

[F1] Fedosov, B.V., A simple geometrical construction of deformation quantization, *J. Differential Geom.* 40 (1994), no.2, 213–238.

[F2] Fedosov, B., *Deformation Quantization and Index Theory*, Math. Top. 9, Akademie Verlag, Berlin 1996.

[GH] Gavarini, F., Halbout, G., Tressages des groupes de Poisson formels à dual quasitriangulaire, *J. Pure Appl. Algebra* 161 (2001), no. 3, 295–307.

[H] Halbout, G., Formule d'homotopie entre les complexes de Hochschild et de de Rham, *Compositio Math.* 126 (2001), no. 2, 123–145.

[K] Kontsevich, M., Deformation quantization of Poisson varieties I, q-alg/9709040; Operads and motives in deformation quantization, *Lett. Math. Phys.* 48 (1999), 35–72; Deformation quantization of algebraic varieties, *Lett. Math. Phys.* 56 (3) (2001), in press, math.AG/0106006.

[L] Lusztig, G., *Introduction to quantum groups*, Progr. Math. 110, Birkhäuser, Boston 1993.

[R] Reshetikhin, N., Quasitriangularity of quantum groups at roots of 1, *Comm. Math. Phys.* 170 (1995), 79–99.

[S] Shoikhet, B., Tsygan formality and Duflo formula, *QA/0012066.*

[T] Tamarkin, D., Another proof of M. Kontsevich formality theorem, math. QA/9803025 v.4; Tamarkin, D. and Tsygan, B., Noncommutative differential calculus, homotopy BV algebras and formality conjectures, *Methods Funct. Anal. Topology* 6 (2000), 85–100, math.KT/0002116.

Deformation quantization: genesis, developments and metamorphoses*

Giuseppe Dito and Daniel Sternheimer

Laboratoire Gevrey de Mathématique Physique, CNRS UMR 5029
Département de Mathématiques
Université de Bourgogne
BP 47870, 21078 Dijon Cedex, France
e-mail: `Giuseppe.Dito@u-bourgogne.fr`
e-mail: `Daniel.Sternheimer@u-bourgogne.fr`

Abstract. We start with a short exposition of developments in physics and mathematics that preceded, formed the basis for, or accompanied, the birth of deformation quantization in the seventies. We indicate how the latter is at least a viable alternative, autonomous and conceptually more satisfactory, to conventional quantum mechanics and mention related questions, including covariance and star representations of Lie groups. We sketch Fedosov's geometric presentation, based on ideas coming from index theorems, which provided a beautiful frame for developing existence and classification of star-products on symplectic manifolds. We present Kontsevich's formality, a major metamorphosis of deformation quantization which implies existence and classification of star-products on general Poisson manifolds and has numerous ramifications. Its alternate proof using operads gave a new metamorphosis which in particular showed that the proper context is that of deformations of algebras over operads, while still another is provided by the extension from differential to algebraic geometry. In this panorama some important aspects are highlighted by a more detailed account.

2000 Mathematics Subject Classification: 53D55 (53-02,81S10,81T70,53D17,18D50,22Exx)

1 Introduction: the background and around

The passage from commutative to noncommutative structures is a staple in frontier physics and mathematics. The advent of quantum mechanics is of course an important example but it is now becoming increasingly clear that the appearance, 25 years ago [FLSq, BFFLS], of deformation quantization, followed by numerous developments and metamorphoses, has been a major factor in the present trend. The epistemological background can be traced back [Fl, SD] to the deformation philosophy which Moshé Flato developed in the early seventies, motivated by deep (decades old) physical ideas and (then recent) mathematical developments. Since that time, inspiring many, he has

*This is an extended version of the presentation by D.S. on May 30, 2001, at the RCP in Strasbourg

consistently pursued and promoted physical and mathematical consequences of that idea in several directions, of which deformation quantization is presently the most widely recognized.

Sir Michael Atiyah said recently, after Oscar Wilde, that mathematics and physics are two communities separated by a common language. Sir Michael added that the two communities managed to communicate rather well until the beginning of the twentieth century, then became so separated that, half a century ago, Eugene Wigner marveled at what he called the unreasonable effectiveness of mathematics in physics. In the past decade however, in spite of the ever increasing "Babel tower effect" in Science, some form of communication was developed, with a converse phenomenon: what may seem an unreasonable effectiveness of physics in mathematics, including in such abstract fields as algebraic geometry. Now, if we remember that mathematics arose as an abstraction of our understanding of the physical world, neither effectiveness should be unreasonable. We shall see here that deformation quantization is a perfect example of both of them.

Physical theories have their domain of applicability defined by the relevant distances, velocities, energies, etc. concerned. But the passage from one domain (of distances, etc.) to another does not happen in an uncontrolled way. Rather, experimental phenomena appear that cause a paradox and contradict accepted theories. Eventually a new fundamental constant enters and the formalism is modified. Then the attached structures (symmetries, observables, states, etc.) *deform* the initial structure. Namely, we have a new structure which in the limit, when the new parameter goes to zero, coincides with the previous formalism. The only question is, in which category do we seek for deformations? Usually physics is rather conservative and if we start e.g. with the category of associative or Lie algebras, we tend to deform in the same category, but there are important examples of generalizations of this principle (e.g. quantum groups are deformations of Hopf algebras).

The discovery of the non-flat nature of Earth may be the first example of this phenomenon. Closer to us, the paradox coming from the Michelson and Morley experiment (1887) was resolved in 1905 by Einstein with the special theory of relativity: there the Galilean geometrical symmetry group of Newtonian mechanics is deformed to the Poincaré group, the new fundamental constant being c^{-1} where c is the velocity of light in vacuum. At around the same time the Riemann surface theory can be considered as a first mathematical example of deformations, even if deformations became systematically studied in the mathematical literature only at the end of the fifties with the profound works of Kodaira and Spencer [KS] on deformations of complex analytic structures. Now, when one has an action on a geometrical structure, it is natural to try and "linearize" it by inducing from it an action on an algebra of functions on that structure. This is implicitly what Gerstenhaber did in 1963 [Ge] with his definition and thorough study of deformations of rings and algebras. It is in the Gerstenhaber sense that the Galileo group is deformed to the Poincaré group; that operation is the inverse of the notion of group contraction introduced ten years before, empirically, by Ínönü and Wigner [WI]. This fact triggered strong interest for deformation theory in France among a number of theoretical physicists, including Flato who had just arrived from the Racah school and knew well the effectiveness of symmetry in physical problems.

In 1900, as a last resort to explain the black body radiation, Planck proposed the quantum hypothesis: the energy of light is not emitted continuously but in quanta proportional to its frequency. He wrote h for the proportionality constant which bears his name. This paradoxical situation got a beginning of a theoretical basis when, again in 1905, Einstein came with the theory of the photoelectric effect. Around 1920, an "agrégé d'histoire", Prince Louis de Broglie, was introduced to the photoelectric effect, together with the Planck–Einstein relations and the theory of relativity, in the laboratory of his much older brother, Maurice duc de Broglie. This led him, in 1923, to his discovery of the duality of waves and particles, which he described in his celebrated Thesis published in 1925, and to what he called 'mécanique ondulatoire'. German and Austrian physicists, in particular, Hermann Weyl, Werner Heisenberg and Erwin Schrödinger, transformed it into the quantum mechanics that we know, where the observables are operators in Hilbert spaces of wave functions.

Intuitively, classical mechanics is the limit of quantum mechanics when $\hbar = \frac{h}{2\pi}$ goes to zero. But how can this be realized when in classical mechanics the observables are functions over phase space (a Poisson manifold) and not operators? The deformation philosophy promoted by Flato showed the way: one has to look for deformations of algebras of functions over Poisson manifolds endowed with the Poisson bracket and realize, in an *autonomous* manner, quantum mechanics there. This required, as a preliminary, a detailed study of the corresponding cohomology spaces. As a first step, in 1974, the cochains were assumed [FLS1] to be 1-differentiable (given by bidifferential operators of order $(1, 1)$). This fell short of the solution but inspired Vey [Ve] who was able in 1975 (for symplectic manifolds with vanishing third Betti number) to show the existence of such differentiable deformations. Doing so, he rediscovered a formula for the deformed bracket (the sine of the Poisson bracket) that he did not know had been obtained in an entirely different context by Moyal [Mo] in 1949. The technical obstacle (a solution of which, in an algebraic context, could later be traced back to a result hidden in a fundamental paper [HKR] from 1962) was lifted and deformation quantization could be developed [FLSq, BFFLS] in 1976-78. In this approach, as we shall see in the next section, *quantization is a deformation* of the associative (and commutative) product of classical observables (functions on phase space) driven by the Poisson bracket, namely a star-product.

In a context which then seemed unrelated to deformation theory, pseudodifferential operators were introduced also at the end of the fifties and became a very hot subject in mathematics thanks to the publication in 1963 of the first index theorems by Atiyah and Singer [AS], which express an analytically defined index in topological terms. The composition of symbols of pseudodifferential operators, an important ingredient in the proof, is a nontrivial example of a star-product, but this fact was noticed only years later, after deformation quantization was introduced.

There have been many generalizations of the original index theorems, including algebraic versions developed in particular by Connes [Co] in the context of noncommutative geometry, a natural continuation of his important works of the seventies on operator algebras that were motivated by physical problems. It was developed shortly after the appearance of deformation quantization, using cyclic rather than plain Hochschild cohomology. If one adds to star-products the notion of trace (see Section 3.2.2), which in this case comes from integration over the manifold on which the functions are defined, one gets closed star-products [CFS] (classified by cyclic cohomology) that provide [Co] other examples of algebras falling in the framework of noncommutative geometry. Now the Gelfand isomorphism provides a realization of a commutative algebra as an algebra of functions over a manifold, i.e., its spectrum. A natural but difficult question is then to develop a theory of noncommutative manifolds. Truly nontrivial examples (if we exclude the group case, cf. below) are appearing only recently (see e.g. [Ma, CDV]).

Simple Lie groups and algebras are rigid for the Gerstenhaber notion of deformation but if one goes to the category of Hopf algebras, they can be deformed. This is what Drinfeld [Dr] realized with a notion to which he gave the spectacular (albeit somewhat misleading) name of quantum group, by considering star-products deforming the product in a Hopf algebra of functions on Lie groups having a compatible Poisson structure. The dual approach of deforming the coproduct in a closure of enveloping algebras, generalizing the $\mathfrak{su}(2)$ example discovered empirically [KR] in 1981 in relation with quantum inverse scattering problems, was also taken independently by Jimbo [Ji]. The domain has since known an extensive development, at the beginning mostly in Russia [FRT]. Numerous applications to physics were developed in many fields and several metamorphoses occurred both in the mathematical methods used and in the concept itself. The subject now covers thousands of references, a fraction of which can be found in [ES, Maj, SSt], which are recent complementary expositions ([SSt] has 1264 references!) We shall not attempt to develop further this important avatar of deformation quantization.

2 Genesis and first developments

2.1 A few preliminaries

In this part we recall some very classical notions, definitions and properties, for the sake of completeness and the benefit of theoretical physicists who might not be familiar with them but would be willing to take up the mathematical jargon required without having to read an extensive literature. It may be skipped by many readers. As for all of this paper, the interested reader is strongly encouraged to study the references, and the references quoted in references (etc.), at least in the topics that are closest to his interests: the paper is meant to be mostly a starting point for newcomers in the domains covered.

2.1.1 Hochschild and Chevalley–Eilenberg cohomologies. When A is an *associative* algebra (over some commutative ring $\mathbb{K}$), we can consider it as a module over itself with the adjoint action (algebra multiplication) and we shall here define cohomology in that context. The generalization to cohomology valued in a general module is straightforward. A Hochschild *p-cochain* is a p-linear map C from A^p to A and its *coboundary* bC is a $(p+1)$-cochain given by

$$\begin{aligned} bC(u_0, \dots, u_p) = {} & u_0 C(u_1, \dots, u_p) - C(u_0 u_1, u_2, \dots, u_p) + \cdots \\ & + (-1)^p C(u_0, u_1, \dots, u_{p-1} u_p) + (-1)^{p+1} C(u_0, \dots, u_{p-1}) u_p. \end{aligned} \tag{2.1}$$

One checks that we have here what is called a complex, i.e. $b^2 = 0$. We say that a p-cochain C is a *p-cocycle* if $bC = 0$. We denote by $Z^p(A, A)$ the space of p-cocycles and by $B^p(A, A)$ the space of those p-cocycles which are coboundaries (of a $(p-1)$-cochain). The pth *Hochschild cohomology* space (of A valued in A seen as a bimodule) is defined as $H^p(A, A) = Z^p(A, A)/B^p(A, A)$. *Cyclic cohomology* is defined using a bicomplex which includes the Hochschild complex. We refer to [Co] for a detailed treatment.

For *Lie algebras* (with bracket $\{\cdot\,, \cdot\}$) one has a similar definition, due to Chevalley and Eilenberg [CE]. The (ad-valued) p-cochains are here skew-symmetric, i.e. linear maps $B : \wedge^p A \longrightarrow A$, and the Chevalley coboundary operator ∂ is defined on a p-cochain B by (where $\hat{u}_j$ means that u_j has to be omitted):

$$\begin{aligned} \partial C(u_0, \dots, u_p) = {} & \sum_{j=0}^{p} (-1)^j \{u_j, C(u_0, \dots, \hat{u}_j, \dots, u_p)\} \\ & + \sum_{i<j} (-1)^{i+j} C(\{u_i, u_j\}, u_0, \dots, \hat{u}_i, \dots, \hat{u}_j, \dots, u_p). \end{aligned} \tag{2.2}$$

Again one has a complex ($\partial^2 = 0$), cocycles and coboundaries spaces $\mathcal{Z}^p$ and $\mathcal{B}^p$ (resp.) and by quotient the *Chevalley cohomology* spaces $\mathcal{H}^p(A, A)$, or in short $\mathcal{H}^p(A)$; the collection of all cohomology spaces will be denoted here $\mathcal{H}^\bullet$, or $H^\bullet$ for the Hochschild cohomology.

2.1.2 Gerstenhaber theory of deformations of algebras. Let A be an algebra. By this we mean an *associative, Lie* or *Hopf* algebra, or a *bialgebra* [an associative algebra A where one has in addition a coproduct $\Delta : A \longrightarrow A \otimes A$ and the obvious compatibility relations, see e.g. [ES] for a precise definition]. Whenever needed we assume it is also a *topological* algebra, i.e., endowed with a locally convex topology for which all needed algebraic laws are continuous. For simplicity we may think that the base (commutative) ring $\mathbb{K}$ is the field of complex numbers $\mathbb{C}$ or that of the real numbers $\mathbb{R}$. Extending it to the ring $\mathbb{K}[[\lambda]]$ of formal series in some parameter λ gives the module $\tilde{A} = A[[\lambda]]$, on which we can consider the preceding various algebraic (and topological) structures. In a number of instances we also need to look at $A[\lambda^{-1}, \lambda]]$, formal series in λ and polynomials in λ^{-1} (considered at first as an independent parameter).

2.1.2.1. *Deformations and cohomologies.* A concise formulation of a Gerstenhaber deformation of an algebra (which we shall call in short a *DrG-deformation* whenever a confusion may arise with more general deformations) is [Ge, GS, BFGP]:

Definition 2.1. A *deformation* of such an algebra A is a $\mathbb{K}[[\lambda]]$-algebra $\tilde{A}$ such that $\tilde{A}/\lambda\tilde{A} \approx A$. Two deformations $\tilde{A}$ and $\tilde{A}'$ are said *equivalent* if they are isomorphic over $\mathbb{K}[[\lambda]]$ and $\tilde{A}$ is said *trivial* if it is isomorphic to the original algebra A considered by base field extension as a $\mathbb{K}[[\lambda]]$-algebra.

Whenever we consider a topology on A, $\tilde{A}$ is supposed to be topologically free. For associative (resp. Lie) algebras, Definition 2.1 tells us that there exists a new product $\star$ (resp. bracket $[\cdot\,,\cdot]$) such that the new (deformed) algebra is again associative (resp. Lie). Denoting the original composition laws by ordinary product (resp. $\{\cdot\,,\cdot\}$) this means that, for $u, v \in A$ we have:

$$u \star v = uv + \sum_{r=1}^{\infty} \lambda^r C_r(u, v) \tag{2.3}$$

$$[u, v] = \{u, v\} + \sum_{r=1}^{\infty} \lambda^r B_r(u, v) \tag{2.4}$$

the C_r being Hochschild 2-cochains and the B_r (skew-symmetric) Chevalley 2-cochains, such that for $u, v, w \in A$ we have $(u\star v)\star w = u\star(v\star w)$ and $\mathcal{S}[[u, v], w] = 0$, where $\mathcal{S}$ denotes summation over cyclic permutations (we extend (2.3) and (2.4) to $A[[\lambda]]$ by $\mathbb{K}[[\lambda]]$-linearity). At each level r we therefore need to fulfill the equations $(j, k \geq 1)$:

$$D_r(u, v, w) \equiv \sum_{j+k=r} (C_j(C_k(u, v), w) - C_j(u, C_k(v, w))) = bC_r(u, v, w) \tag{2.5}$$

$$E_r(u, v, w) \equiv \sum_{j+k=r} \mathcal{S} B_j(B_k(u, v), w) = \partial B_r(u, v, w) \tag{2.6}$$

where b and ∂ denote (respectively) the Hochschild and Chevalley coboundary operators. In particular we see that for $r = 1$ the driver C_1 (resp. B_1) must be a 2-cocycle. Furthermore, assuming one has shown that (2.5) or (2.6) are satisfied up to some order $r = t$, a simple calculation shows that the left-hand sides for $r = t + 1$ are then 3-cocycles, depending only on the cochains C_k (resp. B_k) of order $k \leq t$. If we want to extend the deformation up to order $r = t + 1$ (i.e. to find the required 2-cochains C_{t+1} or B_{t+1}), this cocycle has to be a coboundary (the coboundary of the required cochain): *The obstructions to extend a deformation from one step to the next lie in the* 3-*cohomology*. In particular (and this was Vey's trick) if one can manage to pass always through the null class in the 3-cohomology, a cocycle can be the driver of a full-fledged (formal) deformation.

For a (topological) *bialgebra*, denoting by $\otimes_\lambda$ the tensor product of $\mathbb{K}[[\lambda]]$-modules, we can identify $\tilde{A} \hat{\otimes}_\lambda \tilde{A}$ with $(A \hat{\otimes} A)[[\lambda]]$, where $\hat{\otimes}$ denotes the algebraic tensor product completed with respect to some operator topology (e.g. projective for Fréchet nuclear topology), we similarly have a deformed coproduct $\tilde{\Delta} = \Delta + \sum_{r=1}^{\infty} \lambda^r D_r$, $D_r \in \mathcal{L}(A, A\hat{\otimes}A)$ and in this context appropriate cohomologies can be introduced. As we have said we shall not elaborate on these, nor on the additional requirements for Hopf algebras, referring for more details to original papers and books.

2.1.2.2. *Equivalence* means that there is an isomorphism $T_\lambda = I + \sum_{r=1}^{\infty} \lambda^r T_r$, $T_r \in \mathcal{L}(A, A)$ so that $T_\lambda(u \star' v) = (T_\lambda u \star T_\lambda v)$ in the associative case, denoting by $\star$ (resp. $\star'$) the deformed laws in $\tilde{A}$ (resp. $\tilde{A}'$); and similarly in the Lie case. In particular we see (for $r = 1$) that a deformation is trivial at order 1 if it starts with a 2-cocycle which is a 2-coboundary. More generally, exactly as above, we can show [BFFLS] that if two deformations are equivalent up to some order t, the condition to extend the equivalence one step further is that a 2-cocycle (defined using the T_k, $k \leq t$) is the coboundary of the required T_{t+1} and therefore *the obstructions to equivalence lie in the 2-cohomology*. In particular, if that space is null, all deformations are trivial.

An important property is that a *deformation of an associative algebra with unit* (what is called a unital algebra) is again unital, and *equivalent to a deformation with the same unit*. This follows from a more general result of Gerstenhaber (for deformations leaving unchanged a subalgebra); a proof can be found in [GS].

In the case of (topological) *bialgebras* or *Hopf* algebras, *equivalence* of deformations has to be understood as an isomorphism of (topological) $\mathbb{K}[[\lambda]]$-algebras, the isomorphism starting with the identity for the degree 0 in λ. A deformation is again said *trivial* if it is equivalent to that obtained by base field extension. For Hopf algebras the deformed algebras may be taken (by equivalence) to have the same unit and counit, but in general not the same antipode.

2.1.2.3. *Homotopy of deformations.* Recently Gerstenhaber has considered (for reasons that are related to the so-called Donald–Flanigan conjecture, see [GG, GGS]) the question of (formal) *compatibility* of deformations, a kind of homotopy in the variety of algebras between two deformations (2.3) with parameters λ and λ' and cochains C_r and C'_r. By this he means a 2-parameter deformation of the form

$$u \tilde{\star} v = uv + \lambda C_1(u, v) + \lambda' C'_1(u, v) + \sum_{r=2}^{\infty} \Phi_r(u, v; \lambda, \lambda') \tag{2.7}$$

where each Φ_r is a polynomial of total degree r in λ and λ', which reduces to the first one-parameter deformation when $\lambda' = 0$ and to the second when $\lambda = 0$. At the first order the condition for this to hold (e.g. for associative algebras) is that the Gerstenhaber bracket [Ge] $[C_1, C'_1]_G$ is a 3-coboundary, and here also there are higher obstructions. As an example, it follows from [HKR] that the Weyl algebra and the quantum plane are formally (but non analytically [GZ]) compatible nonequivalent deformations of the polynomial algebra $\mathbb{C}[x, y]$. In (2.2.3.4) below we shall see another appearance of such a 2-parameter deformation in a physical context.

2.1.3 The differentiable case, Poisson manifolds. Consider the algebra $N = C^\infty(X)$ of functions on a differentiable manifold X. When we look at it as an associative algebra acting on itself by pointwise multiplication, we can define the corresponding Hochschild cohomologies. Now let Λ be a skew-symmetric contravariant two-tensor (possibly degenerate) defined on X, satisfying $[\Lambda, \Lambda]_{SN} = 0$ in the sense of the Schouten–Nijenhuis bracket (a definition of which, both intrinsic and in terms of local coordinates, can be found in [BFFLS, FLS1]; see also (4.5) below).

Then the inner product of Λ with the 2-form $du \wedge dv$, $P(u, v) = i(\Lambda)(du \wedge dv)$, $u, v \in N$, defines a *Poisson bracket* P: it is obviously skew-symmetric, satisfies the Jacobi identity because $[\Lambda, \Lambda]_{SN} = 0$ and the Leibniz rule $P(uv, w) = P(u, w)v + uP(v, w)$. It is a bidifferential 2-cocycle for the (general or differentiable) Hochschild cohomology of N, skewsymmetric of order $(1, 1)$, therefore nontrivial [BFFLS] and thus defines an infinitesimal deformation of the pointwise product on N. (X, P) is called a *Poisson manifold* [BFFLS, LiP].

When Λ is everywhere nondegenerate (X is then necessarily of even dimension 2ℓ), its inverse ω is a closed everywhere nondegenerate 2-form ($d\omega = 0$ is then equivalent to $[\Lambda, \Lambda]_{SN} = 0$) and we say that (X, ω) is *symplectic*; ω^ℓ is a volume element on X. Then one can in a consistent manner work with differentiable cocycles [BFFLS, FLS1] and the differentiable Hochschild p-cohomology space $H^p(N)$ is that of all skew-symmetric contravariant p-tensor fields [HKR, Ve], therefore is infinite-dimensional. Thus, except when X is of dimension 2 (because then necessarily $H^3(N) = 0$), the obstructions belong to an infinite-dimensional space where they may be difficult to trace. On the other hand, when $2\ell = 2$, any 2-cocycle can be the driver of a deformation of the associative algebra N: "anything goes" in this case; some examples for $\mathbb{R}^2$ can be found already in [Ve].

Now endow N with a Poisson bracket: we get a Lie algebra and can look at its Chevalley cohomology spaces. Note that P is bidifferential of order $(1, 1)$ so it is important to check whether the Gerstenhaber theory is *consistent* when restricted to *differentiable* cochains (in both cases, of arbitrary order and order at most 1). The answer is positive; in brief, if a coboundary is differentiable, it is the coboundary of a differentiable cochain. Later it was found [Co, Na] that assuming only continuity gives the same type of results.

Since P is of order (1,1), it was natural to study first the 1-differentiable cohomologies. When the cochains are restricted to be of order $(1, 1)$ with no constant term (then they annihilate constant functions, which we write "n.c." for "null on constants") it was found [Li1] that the Chevalley cohomology $\mathcal{H}^\bullet_{1-\mathrm{diff,n.c.}}(N)$ of the Lie algebra N (acting on itself with the adjoint representation) is exactly the de Rham cohomology $H^\bullet_{dR}(X)$. Thus $\dim \mathcal{H}^p_{1-\mathrm{diff,n.c.}}(N) = b_p(X)$, the pth Betti number of the manifold X. Without the n.c. condition one gets a slightly more complicated formula [Li1]; in particular if X is symplectic with an exact 2-form $\omega = d\alpha$, one has here $\mathcal{H}^p_{1-\mathrm{diff}}(N) = H^p_{dR}(X) \oplus H^{p-1}_{dR}(X)$.

Then the "three musketeers" [FLS1] could, in 1974, study 1-*differentiable deformations* of the Poisson bracket Lie algebra N and develop some applications at the level of classical mechanics. In particular it was noticed that the "pure" order (1, 1) deformations correspond to a deformation of the 2-tensor Λ; allowing constant terms and taking the deformed bracket in Hamilton equations instead of the original Poisson bracket gave a kind of friction term.

Shortly afterwards, triggered by that work, a "fourth musketeer" J. Vey [Ve] noticed that in fact $\dim \mathcal{H}^2_{\text{diff,n.c.}}(N) = 1 + \dim \mathcal{H}^2_{1-\text{diff,n.c.}}(N)$ and that $\mathcal{H}^3_{\text{diff,n.c.}}(N)$ is also finite-dimensional. He could then study more general differentiable deformations, rediscovering in the $\mathbb{R}^{2\ell}$ case the Moyal bracket. The latter was at that time rather "exotic" and few authors (except for a number of physicists, see e.g. [AW]) paid any attention to it. In *Mathematical Reviews* this bracket, for which [Mo] is nowadays often quoted, is not even mentioned in the review! Vey's work permitted to tackle the problem with differentiable deformations, and deformation quantization was born [FLSq].

2.2 The founding papers and some follow-up

2.2.1 Classical mechanics and quantization. Classical mechanics, in Lagrangean or Hamiltonian form, assumed at first implicitly a "flat" phase space $\mathbb{R}^{2\ell}$, or at least considered only an open connected set thereof. Eventually more general configurations were needed and so the mathematical notion of manifold, on which mechanics imposed some structure, was used. This has lead in particular to using the notions of symplectic and later of Poisson manifolds, which were introduced also for purely mathematical reasons. One of these reasons has to do with families of infinite-dimensional Lie algebras, which date back to works by Élie Cartan at the beginning of last century and regained a lot of popularity (including in physics) in the past 30 years.

In 1927, Weyl came out with his quantization rule [We]. If we start with a classical observable $u(p,q)$, some function on (flat) phase space $\mathbb{R}^{2\ell}$ (with $p, q \in \mathbb{R}^\ell$), one can associate to it an operator (the corresponding quantum observable) $\Omega(u)$ in the Hilbert space $L^2(\mathbb{R}^\ell)$ by the following general recipe:

$$u \mapsto \Omega_w(u) = \int_{\mathbb{R}^{2\ell}} \tilde{u}(\xi,\eta)\exp(i(P.\xi + Q.\eta)/\hbar)w(\xi,\eta)\, d^\ell\xi d^\ell\eta \qquad (2.8)$$

where $\tilde{u}$ is the inverse Fourier transform of u, P_α and Q_α are operators satisfying the canonical commutation relations $[P_\alpha, Q_\beta] = i\hbar\delta_{\alpha\beta}$ $(\alpha, \beta = 1, \dots, \ell)$, w is a weight function and the integral is taken in the weak operator topology. What is now called normal ordering corresponds to choosing the weight $w(\xi,\eta) = \exp(-\frac{1}{4}(\xi^2 \pm \eta^2))$, standard ordering (the case of the usual pseudodifferential operators in mathematics) to the weight $w(\xi,\eta) = \exp(-\frac{i}{2}\xi\eta)$ and the original Weyl (symmetric) ordering to $w = 1$. An inverse formula was found shortly afterwards by Eugene Wigner [Wi] and maps an operator into what mathematicians call its symbol by a kind of trace formula. For example Ω_1 defines an isomorphism of Hilbert spaces between $L^2(\mathbb{R}^{2\ell})$ and Hilbert–Schmidt operators on $L^2(\mathbb{R}^\ell)$ with inverse given by

$$u = (2\pi\hbar)^{-\ell}\,\mathrm{Tr}[\Omega_1(u)\exp((\xi.P + \eta.Q)/i\hbar)] \qquad (2.9)$$

and if $\Omega_1(u)$ is of trace class one has $\mathrm{Tr}(\Omega_1(u)) = (2\pi\hbar)^{-\ell}\int u\,\omega^\ell \equiv \mathrm{Tr}_\mathrm{M}(u)$, the "Moyal trace", where ω^ℓ is the (symplectic) volume dx on $\mathbb{R}^{2\ell}$. Numerous developments followed in the direction of phase-space methods, many of which can be found described in [AW]. Of particular interest to us here is the question of finding a physical interpretation to the classical function u, symbol of the quantum operator $\Omega_1(u)$; this was the problem posed (around 15 years after [Wi]) by Blackett to his student Moyal. The (somewhat naive) idea to interpret it as a probability density had of course to be rejected (because u has no reason to be positive) but, looking for a direct expression for the symbol of a quantum commutator, Moyal found [Mo] what is now called the Moyal bracket:

$$M(u,v) = \lambda^{-1}\sinh(\lambda P)(u,v) = P(u,v) + \sum_{r=1}^{\infty}\lambda^{2r}P^{2r+1}(u,v) \tag{2.10}$$

where $2\lambda = i\hbar$, $P^r(u,v) = \Lambda^{i_1j_1}\ldots\Lambda^{i_rj_r}(\partial_{i_1\ldots i_r}u)(\partial_{j_1\ldots j_r}v)$ is the rth power ($r \geq 1$) of the Poisson bracket bidifferential operator P, $i_k, j_k = 1,\ldots,2\ell$, $k = 1,\ldots,r$ and $\Lambda = \left(\begin{smallmatrix}0 & -I\\ I & 0\end{smallmatrix}\right)$. To fix ideas we may assume here $u, v \in C^\infty(\mathbb{R}^{2\ell})$ and the sum taken as a formal series (the definition and convergence for various families of functions u and v was also studied, including in [BFFLS]). A similar formula for the symbol of a product $\Omega_1(u)\Omega_1(v)$ had been found a little earlier [Gr] and can now be written more clearly as a (Moyal) *star-product*:

$$u \star_M v = \exp(\lambda P)(u,v) = uv + \sum_{r=1}^{\infty}\lambda^r P^r(u,v). \tag{2.11}$$

Several integral formulas for the star-product have been introduced (we shall come back later to this question). In great part after deformation quantization was developed, the Wigner image of various families of operators (including all bounded operators on $L^2(\mathbb{R}^\ell)$) was studied and an adaptation to Weyl ordering of the mathematical notion of pseudodifferential operators (ordered, like differential operators, "first q, then p") had been developed. We shall not give here more details on those aspects, but for completeness mention a related development. Starting from field theory, where normal (Wick) ordering is essential (the role of q and p above is played by $q \pm ip$), Berezin [Be, BS] developed in the mid-seventies an extensive study of what he called "quantization", based on the correspondence principle and Wick symbols. It is essentially based on Kähler manifolds and related to pseudodifferential operators in the complex domain [BG]. However in his approach, as in the studies of various orderings [AW], the important concepts of *deformation* and *autonomous* formulation of quantum mechanics in general phase space are absent.

Quantization involving more general phase spaces was treated, in a somewhat systematic manner, only with Dirac constraints [Di]: second class Dirac constraints restrict phase space from some $\mathbb{R}^{2\ell}$ to a symplectic manifold W imbedded in it (with induced symplectic form), while first class constraints further restrict to a Poisson manifold with symplectic foliation (see e.g. [FLSq]). The question of quantization on such manifolds was certainly treated by many authors (including [Di]) but did not go beyond giving some (often useful) recipes and hoping for the best.

A first systematic attempt started around 1970 with what was called soon afterwards *geometric quantization* [Ko], a by-product of Lie group representations theory where it gave significant results. It turns out that it is geometric all right, but its scope as far as quantization is concerned has been rather limited since few classical observables could be quantized, except in situations which amount essentially to the Weyl case considered above. In a nutshell one considers phase-spaces W which are coadjoint orbits of some Lie groups (the Weyl case corresponds to the Heisenberg group with the canonical commutation relations and $\mathfrak{h}_\ell$ as Lie algebra); there one defines a "prequantization" on the Hilbert space $L^2(W)$ and tries to halve the number of degrees of freedom by using polarizations (often complex ones, which is not an innocent operation as far as physics is concerned) to get a Lagrangean submanifold $\mathcal{L}$ of dimension half of that of W and quantized observables as operators in $L^2(\mathcal{L})$. A recent exposition can be found in [Wo].

2.2.2 Star-products. Star-products as a deformation of the usual product of functions on a phase-space meant for an understanding of quantum mechanics were introduced in [FLSq] and their relevance examplified by the founding papers [BFFLS], which included significant applications. Their general definition is as follows.

Let X be a differentiable manifold (of finite, or possibly infinite, dimension). We assume given on X a *Poisson structure* (a Poisson bracket P).

Definition 2.2. A *star-product* on X is a deformation of the associative algebra of functions $N = C^\infty(X)$ of the form $\star = \sum_{n=0}^{\infty} \lambda^n C_n$ where $C_0(u, v) = uv$, $C_1(u, v) - C_1(v, u) = 2P(u, v)$, $u, v \in N$, and the C_n are bidifferential operators (locally of finite order). We say a star-product is *strongly closed* if $\int_X (u \star v - v \star u)dx = 0$ where dx is a volume element on X.

The parameter λ of the deformation is taken to be $\lambda = \frac{i}{2}\hbar$ in physical applications.

2.2.2.1. *a.* Using equivalence one may take $C_1 = P$. That is the case of Moyal, but other orderings like standard or normal do not verify this condition (only the skew-symmetric part of C_1 is P). Again by equivalence, in view of Gerstenhaber's result on the unit (cf. 2.1.2.2), we may take cochains C_r which are without constant term (what we called n.c. or null on constants). In fact, in the original paper [BFFLS], only this case was considered and the accent was put on "Vey products", for which the cochains C_r have the same parity as r and have P^r for principal symbol in any Darboux chart, with X symplectic.

b. It is also possible to consider star-products for which the cochains C_n are allowed to be slightly more general. Allowing them to be *local* ($C_n(u, v) = 0$ on any open set where u or v vanish) gives nothing new. Note that this is not the same as requiring the whole associative product to be local (in fact the latter condition is very restrictive and, like true pseudodifferential operators, a star-product is a nonlocal operation). In some cases (e.g. for star representations of Lie groups) it may be practical to consider pseudodifferential cochains. As far as the cohomologies are concerned, as long as one requires at least continuity for the cochains, the theory is the same as in the differentiable case [Na].

Also, due to formulas like (2.9) and the relation with Lie algebras (see below), it is sometimes convenient to take $\mathbb{K}[\lambda^{-1}, \lambda]]$ (Laurent series in λ, polynomial in λ^{-1} and formal series in λ) for the ring on which the deformation is defined. Again, this will not change the theory.

c. By taking the corresponding commutator $[u, v]_\lambda = (2\lambda)^{-1}(u \star v - v \star u)$, since the skew-symmetric part of C_1 is P, we get a deformation of the Poisson bracket Lie algebra (N, P). This is a crucial point because (at least in the symplectic case) we know the needed Chevalley cohomologies and (in contradistinction with the Hochschild cohomologies) they are small [Ve, DLG]. The interplay between both structures gives existence and classification; in addition it will explain why (in the symplectic case) the classification of star-products is based on the 1-differentiable cohomologies, hence ultimately on the de Rham cohomology of the manifold.

2.2.2.2. *Invariance and covariance.* The Poisson bracket P is (by definition) invariant under all symplectomorphisms, i.e. transformations of the manifold X generated by the flows $x_u = i(\Lambda)(du)$ defined by Hamiltonians $u \in N$. But already on $\mathbb{R}^{2\ell}$ one sees easily that its powers P^r, hence also the Moyal bracket (2.10), are invariant only under flows generated by Hamiltonians u which are polynomials of maximal order 2, forming the "affine" symplectic Lie algebra $\mathfrak{sp}(\mathbb{R}^{2\ell}) \cdot \mathfrak{h}_\ell$. For other orderings the invariance is even smaller (only $\mathfrak{h}_\ell$ remains). For general Vey products the first terms of a star-product are [BFFLS] $C_2 = P_\Gamma^2 + bH$ and $C_3 = S_\Gamma^3 + T + 3\partial H$. Here H is a differential operator of maximal order 2, T a 2-tensor corresponding to a closed 2-form, ∂ the Chevalley coboundary operator. P_Γ^2 is given (in canonical coordinates) by an expression similar to P^2 in which usual derivatives are replaced by covariant derivatives with respect to a given symplectic connection Γ (a torsionless connection with totally skew-symmetric components when all indices are lowered using Λ). S_Γ^3 is a very special cochain given by an expression similar to P^3 in which the derivatives are replaced by the relevant components of the Lie derivative of Γ in the direction of the vector field associated to the function (u or v). Fedosov's algorithmic construction [Fe] shows that the symplectic connection Γ plays a role at all orders. Therefore the invariance group of a star-product is a subgroup of the *finite-dimensional* group of symplectomorphisms preserving a connection. Its Lie algebra is $\mathfrak{g}_0 = \{a \in N;\, [a, u]_\lambda = P(a, u)\, \forall u \in N\}$, the elements of which are *preferred observables*, i.e., Hamiltonians for which the classical and quantum evolutions coincide. We are thus lead to look for a weaker notion and shall call a star-product *covariant* under a Lie algebra $\mathfrak{g}$ of functions if $[a, b]_\lambda = P(a, b)\, \forall a, b \in \mathfrak{g}$. It can be shown [ACMP] that $\star$ is $\mathfrak{g}$-covariant iff there exists a representation τ of the Lie group G whose Lie algebra is $\mathfrak{g}$ into $\mathrm{Aut}(N[[\lambda]]; \star)$ such that $\tau_g u = (Id_N + \sum_{r=1}^{\infty} \lambda^r \tau_g^r)(g.u)$ where $g \in G, u \in N$, G acts on N by the natural action induced by the vector fields associated with $\mathfrak{g}$, $(g \cdot u)(x) = u(g^{-1}x)$, and where the τ_g^r are differential operators on W. Invariance of course means that the geometric action preserves the star-product: $g \cdot u \star g \cdot v = g \cdot (u \star v)$. This is the basis for the theory of star representations which we shall briefly present below, and of the relevance of star-products for important problems in group theory.

2.2.3 Quantum mechanics. Let us start with a phase space X, a symplectic (or Poisson) manifold and N an algebra of classical observables (functions, possibly including distributions if proper care is taken for the product). We shall call *star quantization* a star-product on X invariant (or sometimes only covariant) under some Lie algebra $\mathfrak{g}_0$ of "preferred observables". Invariance of the star-product ensures that the classical and quantum evolutions of observables under a Hamiltonian $H \in \mathfrak{g}_0$ will coincide [BFFLS]. The typical example is the Moyal product on $W = \mathbb{R}^{2\ell}$.

2.2.3.1. *Spectrality.* Physicists want to get numbers matching experimental results, e.g. for energy levels of a system. That is usually achieved by describing the spectrum of a given Hamiltonian $\hat{H}$ supposed to be a self-adjoint operator so as to get a real spectrum and so that the evolution operator (the exponential of $it\hat{H}$) is unitary (thus preserves probability amplitude). A similar spectral theory can be done here, in an *autonomous manner*. The most efficient way to achieve it is to consider [BFFLS] the *star exponential* (corresponding to the evolution operator)

$$\mathrm{Exp}(Ht) \equiv \sum_{n=0}^{\infty} \frac{1}{n!} \left(\frac{t}{i\hbar} \right)^n (H\star)^n \tag{2.12}$$

where $(H\star)^n$ means the nth star power of the Hamiltonian $H \in N$ (or $N[[\lambda]]$). Then one writes its Fourier-Stieltjes transform $d\mu$ (in the distribution sense) as $\mathrm{Exp}(Ht) = \int e^{\lambda t/i\hbar} d\mu(\lambda)$ and defines *the spectrum of* $(H/\hbar)$ as the *support* S *of* $d\mu$ (incidentally this is the definition given by L. Schwartz for the spectrum of a distribution, out of motivations coming from Fourier analysis). In the particular case when H has discrete spectrum, the integral can be written as a sum (see the top equation in (2.14) below for a typical example): the distribution $d\mu$ is a sum of "delta functions" supported at the points of S multiplied by the symbols of the corresponding eigenprojectors.

In different orderings with various weight functions w in (2.8) one gets in general different operators for the same classical observable H, thus different spectra. For $X = \mathbb{R}^{2\ell}$ all orderings are mathematically equivalent (to Moyal under the Fourier transform T_w of the weight function w). This means that every observable H will have the same spectrum under Moyal ordering as $T_w H$ under the equivalent ordering. But this does not imply physical equivalence, i.e. the fact that H will have the same spectrum under both orderings. In fact the opposite is true: if two equivalent star-products are isospectral (give the same spectrum for a large family of observables and all $\hbar$), they are identical [CFGS].

It is worth mentioning that our definition of spectrum permits to define a spectrum even for symbols of non-spectrable operators, such as the derivative on a half-line which has different deficiency indices; this corresponds to an infinite potential barrier (see also [UU] for detailed studies of similar questions). That is one of the many advantages of our autonomous approach to quantization.

2.2.3.2. *Applications.* In quantum mechanics it is preferable to work (for $X = \mathbb{R}^{2\ell}$) with the star-product that has maximal symmetry, i.e. $\mathfrak{sp}(\mathbb{R}^{2\ell}) \cdot \mathfrak{h}_\ell$ as algebra of preferred observables: the Moyal product. One indeed finds [BFFLS] that the star exponential of these observables (polynomials of order ≤ 2) is proportional to the usual exponential. More precisely, if $H = \alpha p^2 + \beta pq + \gamma q^2 \in \mathfrak{sl}(2)$ with $p, q \in \mathbb{R}^\ell$, $\alpha, \beta, \gamma \in \mathbb{R}$, setting $d = \alpha\gamma - \beta^2$ and $\delta = |d|^{1/2}$ one gets, summing the star exponential series and taking its Fourier decompositions (the sums and integrals appearing in the various expressions being convergent as distributions, both in phase-space variables and in t or λ)

$$\mathrm{Exp}(Ht) = \begin{cases} (\cos\delta t)^{-l}\exp((H/i\hbar\delta)\tan(\delta t)) & \text{for} \quad d > 0 \\ \exp(Ht/i\hbar) & \text{for} \quad d = 0 \\ (\cosh\delta t)^{-l}\exp((H/i\hbar\delta)\tanh(\delta t)) & \text{for} \quad d < 0 \end{cases} \tag{2.13}$$

$$\mathrm{Exp}(Ht) = \begin{cases} \sum_{n=0}^{\infty} \Pi_n^{(\ell)} e^{(n+\frac{\ell}{2})t} & \text{for} \quad d > 0 \\ \int_{-\infty}^{\infty} e^{\lambda t/i\hbar}\Pi(\lambda, H)d\lambda & \text{for} \quad d < 0 \end{cases} \tag{2.14}$$

We get the discrete spectrum $(n + \frac{\ell}{2})\hbar$ of the *harmonic oscillator* and the continuous spectrum $\mathbb{R}$ for the dilation generator pq. The eigenprojectors $\Pi_n^{(\ell)}$ and $\Pi(\lambda, H)$ are given [BFFLS] by known special functions on phase-space (generalized Laguerre and hypergeometric, multiplied by some exponential). Formulas (2.13) and (2.14) can, by analytic continuation, be extended outside singularities and (as distributions) for singular values of t.

Other examples can be brought to this case by functional manipulations [BFFLS]. For instance the Casimir element C of $\mathfrak{so}(\ell)$ representing *angular momentum*, which can be written $C = p^2q^2 - (pq)^2 - \ell(\ell-1)\frac{\hbar^2}{4}$, has $n(n+(\ell-2))\hbar^2$ for spectrum. For the *hydrogen atom*, with Hamiltonian $H = \frac{1}{2}p^2 - |q|^{-1}$, the Moyal product on $\mathbb{R}^{2\ell+2}$ ($\ell = 3$ in the physical case) induces a star-product on $X = T^*S^\ell$; the energy levels, solutions of $(H - E) \star \phi = 0$, are found from (2.14) and the preceding calculations for angular momentum to be (as they should, with $\ell = 3$) $E = \frac{1}{2}(n+1)^{-2}\hbar^{-2}$ for the discrete spectrum, and $E \in \mathbb{R}^+$ for the continuous spectrum.

We thus have recovered, in a completely autonomous manner entirely within deformation quantization, the results of "conventional" quantum mechanics in those typical examples (many more can be treated similarly). It is worth noting that the term $\frac{\ell}{2}$ in the harmonic oscillator spectrum, obvious source of divergences in the infinite-dimensional case, disappears if the normal star-product is used instead of Moyal – which is one of the reasons it is preferred in field theory.

2.2.3.3. *Remark on convergence and the physical meaning of deformation quantization.* We have always considered star-products as formal series and looked for convergence only in specific examples, generally in the sense of distributions. The same applies to star exponentials, as long as each coefficient in the formal series is well defined. In the case of the harmonic oscillator or more generally the preferred observables H in Weyl ordering, the study is facilitated by the fact that the powers $(H\star)^n$ are polynomials in H.

Integral formulas for star-products and star exponentials can be used to give a meaning to these expressions beyond the domains where the sums are convergent or summable. *The sums must not be considered as perturbation series*, rather taken as a whole, whether expressed as integral formulas or formal series.

We stress that deformation quantization should not be seen as a mere reformulation of quantum mechanics or quantum theories in general. At the conceptual level, it is the true mathematical formulation of physical reality whenever quantum effects have to be taken into account. The above examples show that one can indeed perform important quantum mechanical calculations, in an autonomous manner, entirely within deformation quantization – and get the results obtained in conventional quantum mechanics. Whether one uses an operatorial formulation or some form of deformation quantization formulation is thus basically a practical question, which formulation is the most effective, at least in the cases where a Weyl or Wigner map exists. When such a map does not exist, a satisfactory operatorial formulation will be very hard to find, except locally on phase space, and deformation quantization is the solution.

One can of course (and should in practical examples, as we have done here, and also for algebraic varieties) look for small domains (in N) where one has convergence. We can then speak of "strict" deformation quantization. In particular we can look for domains where pointwise convergence can be proved; this was done e.g. for Hermitian symmetric spaces [CGR]. But it should be clearly understood that one can consider wider classes of observables – in fact, the latter tend to be physically more interesting – than those that fit in a strict C^*-algebraic approach.

2.2.3.4. *A 2-parameter star-product and statistical mechanics.* A relation similar to (2.7) had been considered for star-products in [BFLS], in connection with statistical mechanics. Indeed, in view of our philosophy on deformations, a natural question to ask is their *stability*: Can deformations be further deformed, or does "the buck stops there"? As we indicated at the beginning, the answer to that question may depend on the context. Quantum groups are an example, when dealing with Hopf algebras; properties of Harrison cohomology require going to noncontinuous cochains if one wants nontrivial abelian deformations (see (4.3.3.3) and [Fr]). Here is a simpler example.

If one looks for deformations of the Poisson bracket Lie algebra (N, P) one finds (under a mild assumption) that a further deformation of the Moyal bracket, with another deformation parameter ρ, is again a Moyal bracket for a ρ-deformed Poisson structure; in particular, for $X = \mathbb{R}^{2\ell}$, *quantum mechanics viewed as a deformation is unique and stable.*

Now, for the associative algebra N, the only *local* associative composition laws are of the form $(u, v) \mapsto ufv$ for some $f \in N$. If we take $f = f_\beta \in N[[\beta]]$ we get a 0-differentiable deformation (with parameter β) of the usual product. We were thus lead [BFLS] to look, starting from a product $\star_\lambda$, for a new composition law $(u, v) \mapsto u\tilde{\star}_{\lambda,\beta}v = u \star_\lambda f_{\lambda,\beta} \star_\lambda v$ with $f_{\lambda,\beta} = \sum_{r=0}^{\infty} \lambda^{2r} f_{2r,\beta} \in N[[N[[\lambda^2]], \beta]]$, where $f_{0,\beta} \equiv f_\beta \neq 0$ and $f_0 = 1$. The transformation $u \mapsto T_{\lambda,\beta}u = f_{\lambda,\beta} \star_\lambda u$ intertwines $\star_\lambda$ and $\tilde{\star}_{\lambda,\beta}$ but it is not an equivalence of star-products because $\tilde{\star}_{\lambda,\beta}$ is not a star-product: it is a (λ, β)-deformation of the usual product with at first order in λ the driver given by $P_\beta(u, v) = f_\beta P(u, v) + uP(f_\beta, v) - P(f_\beta, u)v$, a conformal Poisson bracket associated with a *conformal symplectic structure* given by the 2-tensor $\Lambda_\beta = f_\beta\Lambda$ and the vector $E_\beta = [\Lambda, f_\beta]_{SN}$.

We then start with a star-product, denoted $\star$, on some algebra $\mathcal{A}$ of observables, and take $f_{\lambda,\beta} = \exp_\star(c\beta H)$ with $c = -\frac{1}{2}$. The star exponential $\mathrm{Exp}(Ht)$ defines an automorphism $u \mapsto \alpha_t(u) = \mathrm{Exp}(-Ht) \star u \star \mathrm{Exp}(Ht)$. A *KMS state* σ on $\mathcal{A}$ is a state (linear functional) satisfying, $\forall a, b \in \mathcal{A}$, the Kubo–Martin–Schwinger condition $\sigma(\alpha_t(a) \star b) = \sigma(b \star \alpha_{t+i\hbar\beta}(a))$. Then the (quantum) KMS condition can be written [BFLS], with $[a, b]_\beta = (i/\hbar)(a\tilde{\star}_{\lambda,\beta}b - b\tilde{\star}_{\lambda,\beta}a)$, simply $\sigma(g_{-\beta} \star [a, b]_\beta) = 0$: up to a conformal factor, a KMS state is like a trace with respect to this new product. The (static) classical KMS condition is the limit for $\hbar = 0$ of the quantum one. So we can recover known features of statistical mechanics by introducing a new deformation parameter $\beta = (kT)^{-1}$ and the related conformal symplectic structure. This procedure commutes with usual deformation quantization. The question was recently treated from a more conventional point of view in deformation quantization (see e.g. [BRW, Wm]) using the notion of formal trace [NT].

2.2.4 Star representations, a short overview. Let G be a Lie group (connected and simply connected), acting by symplectomorphisms on a symplectic manifold X (e.g. coadjoint orbits in the dual of the Lie algebra $\mathfrak{g}$ of G). The elements $x, y \in \mathfrak{g}$ will be supposed to be realized by functions u_x, u_y in N so that their Lie bracket $[x, y]_\mathfrak{g}$ is realized by $P(u_x, u_y)$. Now take a G-covariant star-product $\star$, that is $P(u_x, u_y) = [u_x, u_y] \equiv (u \star v - v \star u)/2\lambda$, which shows that the map $\mathfrak{g} \ni x \mapsto (2\lambda)^{-1}u_x \in N$ is a Lie algebra morphism. The appearance of λ^{-1} here and in the trace (see 2.2.1) cannot be avoided and explains why we have often to take into account both λ and λ^{-1}. We can now define the *star exponential*

$$E(e^x) = \mathrm{Exp}(x) \equiv \sum_{n=0}^{\infty} (n!)^{-1}(u_x/2\lambda)^{\star n} \tag{2.15}$$

where $x \in \mathfrak{g}$, $e^x \in G$ and the power $\star n$ denotes the nth star-power of the corresponding function. By the Campbell–Hausdorff formula one can extend E to a *group homomorphism* $E : G \to (N[[\lambda, \lambda^{-1}]], \star)$ where, in the formal series, λ and λ^{-1} are treated as independent parameters for the time being. Alternatively, the values of E can be taken in the algebra $(\mathcal{P}[[\lambda^{-1}]], \star)$, where $\mathcal{P}$ is the algebra generated by $\mathfrak{g}$ with the $\star$-product (it is a representation of the enveloping algebra).

A *star representation* [BFFLS] of G is a distribution $\mathcal{E}$ (valued in ImE) on X defined by

$$D \ni f \mapsto \mathcal{E}(f) = \int_G f(g)E(g^{-1})dg$$

where D is some space of test-functions on G. The corresponding *character* χ is the (scalar-valued) distribution defined by $D \ni f \mapsto \chi(f) = \int_X \mathcal{E}(f)d\mu$, $d\mu$ being a quasi-invariant measure on X.

The character is one of the tools which permit a comparison with usual representation theory. For semi-simple groups it is singular at the origin in irreducible representations, which may require caution in computing the star exponential (2.15). In the case of the harmonic oscillator that difficulty was masked by the fact that the corresponding representation of $\mathfrak{sl}(2)$ generated by (p^2, q^2, pq) is integrable to a double covering of SL(2, $\mathbb{R}$) and decomposes into a sum $D(\frac{1}{4}) \oplus D(\frac{3}{4})$: the singularities at the origin cancel each other for the two components.

This theory is now very developed, and parallels in many ways the usual (operatorial) representation theory. A detailed account of all the results would take us too far, but among the most notable one may quote:

i) An exhaustive treatment of *nilpotent* or *solvable exponential* [AC] and even *general solvable* Lie groups [ACL]. The coadjoint orbits are there symplectomorphic to $\mathbb{R}^{2\ell}$ and one can lift the Moyal product to the orbits in a way that is adapted to the Plancherel formula. Polarizations are not required, and "star-polarizations" can always be introduced to compare with usual theory. Wavelets, important in signal analysis, are manifestations of star-products on the (2-dimensional solvable) affine group of $\mathbb{R}$ or on a similar 3-dimensional solvable group [BB].

ii) For *semi-simple* Lie groups an array of results is already available, including [ACG, Mr] a complete treatment of the *holomorphic discrete series* (this includes the case of compact Lie groups) using a kind of Berezin dequantization, and scattered results for specific examples. Similar techniques have also been used [CGR, Ka] to find invariant star-products on Kähler and Hermitian symmetric spaces (convergent for an appropriate dense subalgebra). Note however, as shown by recent developments of unitary representations theory (see e.g. [SW]), that for semi-simple groups the coadjoint orbits alone are no more sufficient for the unitary dual and one needs far more elaborate constructions.

iii) For semi-direct products, and in particular the Poincaré and Euclidean groups, an autonomous theory has also been developed (see e.g. [ACM]).

Comparison with the usual results of "operatorial" theory of Lie group representations can be performed in several ways, in particular by constructing an invariant Weyl transform generalizing (2.8), finding "star-polarizations" that always exist, in contradistinction with the geometric quantization approach (where at best one can find complex polarizations), study of spectra (of elements in the center of the enveloping algebra and of compact generators) in the sense of (2.2.3.1), comparison of characters, etc. Note also in this context that the pseudodifferential analysis and (non autonomous) connection with quantization developed extensively by Unterberger, first in the case of $\mathbb{R}^{2\ell}$, has been extended to the above invariant context [U2, UU]. But our main insistence is that the theory of star representations is an *autonomous* one that can be formulated completely within this framework, based on coadjoint orbits (and some additional ingredients when required).

3 Existence and classification: geometric approach

The first proof of the existence of star-products on any (real) symplectic manifold given by De Wilde and Lecomte [DL] is based on the idea of gluing local Moyal star-products defined on Darboux charts. Their proof is algebraic and cohomological in essence. A more geometrical proof was provided by the construction of Omori, Maeda, and Yoshioka [OMY1]. They have introduced the notion of Weyl bundle as a locally trivial fiber bundle whose fibers are Weyl algebras. A Weyl algebra is a unital $\mathbb{C}$-algebra generated by $1, \lambda, X_1, \ldots, X_n, Y_1, \ldots, Y_n$ satisfying the commutation relations $[X_i, Y_j] = -\delta_{ij}\lambda$ and all other commutators are trivial. The main step was to show the existence of a Weyl bundle W over any symplectic manifold M. This was done by defining a special class of sections (the Weyl functions), first locally and then globally by showing that the class of Weyl functions is stable under different choices of local trivialization. Then, the composition of Weyl sections induces a star-product on M. The construction of a Weyl bundle shares some features with the De Wilde-Lecomte approach: in both cases, derivations play an important role and the basic idea is still to glue together local objects defined on Darboux charts by using cohomological arguments.

In 1985, motivated by index theory and independently of the previous proofs, Fedosov [Fe1] announced a purely geometrical construction, also based on Weyl algebras, of star-products on a symplectic manifold. Unfortunately, the announcement appeared in a local volume with a summary in Doklady and has stayed almost unknown for several years until a complete version was published in an international journal [Fe]. The beautiful construction of Fedosov does provide a new proof of existence, but above all it gives a much better understanding of deformation quantization that allowed many major developments that will be discussed below. Let us first briefly review the Fedosov construction (see also [Wb]). For a comprehensive treatment and much more, we recommend Fedosov's book [Fe3].

3.1 The construction of Fedosov

Let (M, ω) be a $2m$-dimensional symplectic manifold. Let ∇ be a symplectic connection, i.e., a linear connection without torsion and preserving the symplectic form ω.

At each point $x \in M$, the symplectic form endows the tangent space T_xM with a (constant) symplectic structure and we can define the Moyal star-product $*_\lambda$ associated to ω_x on T_xM. The space of formal series in λ with polynomial coefficients on T_xM, endowed with $*_\lambda$, gives us the Weyl algebra W_x associated to T_xM. The algebras W_x, $x \in M$, can be smoothly patched and we get a fiber bundle $W = \bigcup_{x\in M} W_x$ on M, called the Weyl bundle over M. The fiberwise Moyal star-product endows naturally the space of sections $\Gamma(W)$ with a structure of unital associative algebra. The center of $\Gamma(W)$ can be identified as a vector space with $C^\infty(M)[[\lambda]]$. The basic idea of Fedosov's construction is to build from ∇ a flat connection D on W such that the algebra of horizontal sections for D induces a star-product on M. This "infinitesimal" approach bypasses the need of gluing together star-products defined on (large) Darboux charts.

In a local Darboux chart at $x \in M$, denote by $(y^1, \ldots, y^{2m})$ the coordinates on T_xM. A section of W can be locally written as

$$a(x, y) = \sum_{k\geq 0, |\alpha|\geq 0} \lambda^k a_{k,\alpha}(x) y^\alpha, \tag{3.1}$$

where $\alpha = (\alpha_1, \ldots, \alpha_{2m})$ is a multi-index. We shall need the bundle

$$\Gamma(W) \otimes \Lambda = \bigoplus_{0\leq q\leq 2m} \Gamma(W) \otimes \Lambda^q$$

of differential forms on M taking their values in W. A section of $\Gamma(W) \otimes \Lambda$ can be expressed locally as

$$a(x, y, dx) = \sum_{k\geq 0} \sum_{|\alpha|, |\beta|\geq 0} \lambda^k a_{k,\alpha,\beta}(x) y^\alpha (dx)^\beta, \tag{3.2}$$

where $(dx)^\beta \equiv dx^{\beta_1} \wedge \cdots \wedge dx^{\beta_q}$. The natural extension of the product in $\Gamma(W)$ obtained by taking the Moyal star-product $*_\lambda$ between the y^i's and the wedge product between the dx^i's, defines a product $\circ$ which makes $\Gamma(W) \otimes \Lambda$ an algebra. There is also a structure of graded Lie algebra on $\Gamma(W) \otimes \Lambda$ given by the bracket $[a, b] = a \circ b - (-1)^{pq} b \circ a$ for $a \in \Gamma(W) \otimes \Lambda^p$ and $b \in \Gamma(W) \otimes \Lambda^q$. It is just the natural extension of the bracket in $\Gamma(W)$ for the Moyal star-product $*_\lambda$ to the differential forms taking values in W.

The algebra $\Gamma(W) \otimes \Lambda$ is filtered in the following way. Define the degrees of λ and y^i by $\deg(\lambda) = 2$ and $\deg(y^i) = 1$. Then $\Gamma(W) \otimes \Lambda$ has a filtration with respect to the total degree $2k + |\alpha|$ of the terms appearing in (3.2):

$$\Gamma(W) \otimes \Lambda \supset \Gamma_1(W) \otimes \Lambda \supset \Gamma_2(W) \otimes \Lambda \supset \cdots.$$

Notice that the center of $\Gamma(W) \otimes \Lambda$ for the bracket $[\,,\,]$ is the set of sections independent of the y^i's. We shall denote by σ the projection of $\Gamma(W) \otimes \Lambda$ onto its center.

The exterior differential d on M extends to an antiderivation δ of the algebra $\Gamma(W) \otimes \Lambda$ by the formula $\delta(a) = \sum_j dx^j \wedge \partial_j a$, where $\partial_j = \partial/\partial_{y^j}$. The antiderivation δ reduces by one unit the degree of the filtration. Fedosov has also introduced another operation δ^* on $\Gamma(W)\otimes\Lambda$. It is defined by the formula $\delta^*(a) = \sum_j y^j \mathbf{i}(\partial_j)(a)$, where $\mathbf{i}$ denotes the interior product. One can check that $\delta^2 = \delta^{*\,2} = 0$ and $\delta\delta^* + \delta^*\delta = (p+q)\mathrm{Id}$ on homogeneous elements of degrees p in y^i and q in dx^i. There is a Hodge-like decomposition for the sections (3.2). Denote by a_{00} the term of the section a in (3.2) which is independent of the y^i's and the dx^i's. By defining δ^{-1} on (p,q)-homogeneous elements by $\delta^{-1} = 0$ if $p+q = 0$ and $\delta^{-1} = \delta^*/(p+q)$ if $p+q > 0$, every section a admits the decomposition $a = \delta\delta^{-1}(a) + \delta^{-1}\delta(a) + a_{00}$.

The next step is to introduce another connection on W. First, the symplectic connection ∇ defines on W a connection $\hat{\nabla}$: $\Gamma(W)\otimes\Lambda^q \to \Gamma(W)\otimes\Lambda^{q+1}$, by $\hat{\nabla}_i = \sum_i dx^i \wedge \nabla_i$. Fedosov introduces the new connection on W as follows: for an element $\gamma \in \Gamma(W)\otimes\Lambda^1$ such that $\sigma(\gamma) = 0$, define the connection $D = \hat{\nabla} - [\gamma, \cdot]/i\lambda$. The curvature of D is $\Omega = R + \hat{\nabla}\gamma - \gamma^2/i\lambda$, where R is the curvature of $\hat{\nabla}$. Ω is called the *Weyl curvature* of D. The usual relation $D^2 = -[\Omega, \cdot]/i\lambda$ holds.

Here comes the essential ingredient of Fedosov's construction. The idea is to make quantum corrections to D in order to make it flat, i.e., $D^2 = 0$. In other words, it consists to make the Weyl curvature Ω central (Abelian connection). Fedosov showed the existence of a flat Abelian connection on W, normalized by the condition $\sigma(r) = 0$, of the form $D = -\delta + \hat{\nabla} - [r, \cdot]/i\lambda$, with $r \in \Gamma_3(W)\otimes\Lambda^1$. The strategy used is to solve iteratively the equation $\delta(r) = R + \hat{\nabla}r - r^2/i\lambda$ with respect to the filtration of $\Gamma(W)\otimes\Lambda$. It can be shown that the preceding equation has a unique solution if the initial condition $\delta^{-1}(r) = 0$ is satisfied. It turns out that for such specific solution r the Weyl curvature is nothing else than $\Omega = -\omega$. Such a flat Abelian connection is called a *Fedosov connection* on W.

Given a Fedosov connection D on W, let us see how it induces a star-product on M. The set of horizontal sections, i.e., sections a such that $Da = 0$, is an algebra for the product $\circ$. We shall denote this algebra by $\Gamma_D(W)$. The restriction of the projection $\sigma\colon \Gamma(W) \to C^\infty(M)[[\lambda]]$ to $\Gamma_D(W)$ is a bijective map. To any $f \in C^\infty(M)[[\lambda]]$ we can then associate a horizontal section denoted by $\sigma^{-1}(f)$. Then $f \star g = \sigma(\sigma^{-1}(f)\circ\sigma^{-1}(g))$ defines a star-product on M.

The method of Fedosov extends to regular Poisson manifolds. An earlier proof along the lines of De Wilde and Lecomte and Omori, Maeda, and Yoshioka was published in 1992 by Masmoudi [Mas] where the existence of tangential star-products is established.

3.2 Classification, characteristic classes, and closed star-products

3.2.1 Classification. There are several occurrences in the literature of the second de Rham cohomology space $H^2_{dR}(M)$ in relation with the question of equivalence of star-products and deformed Poisson brackets.

For formal 1-differentiable deformations of the Poisson bracket (i.e., formal Poisson bracket) on a symplectic manifold, the first of these occurrences goes back to [BFFLS], where it is shown that the obstructions to equivalence (at each step in the power of the deformation parameter) are in $H^2_{dR}(M)$. By taking advantage of graded Lie structures related to deformations, Lecomte [Le] showed that equivalence classes of formal Poisson brackets are in one-to-one correspondence with $H^2_{dR}(M)[[\lambda]]$.

There are similar occurrences for star-products. For example, it is implicit in the work of Gutt [Gu1] and in [Gu2, Gth] where the second and third Chevalley–Eilenberg cohomology spaces have been computed. Also as a consequence of their proof of existence when $b_3(M) = 0$, Neroslavsky and Vlasov [NV] found that the classification for this class of star-products was given by sequences in $H^2_{dR}(M)$.

The Fedosov construction provides a more geometric point of view on the classification problem. Fedosov's method allows to canonically construct a star-product on (M, ω) whenever a symplectic connection ∇ is chosen. Any star-product obtained in such a way is called a *Fedosov star-product.* The iterative construction reviewed in Section 3.1 can be generalized to the case where one starts with a formal symplectic form on M, $\omega_\lambda = \omega + \sum_{k\geq 1} \lambda^k \omega_k$, the ω_k's being closed 2-forms on M. Hence we have a canonical method to construct a star-product $\star$ from the object $(M, \omega_\lambda, \nabla)$. Then the Weyl curvature is equal to $-\omega_\lambda$ and Fedosov has established that two Fedosov star-products $\star$ and $\star'$ are equivalent if and only if their Weyl curvatures belong to the same cohomology class in $H^2_{dR}(M)[[\lambda]]$.

Shortly after Fedosov's paper, Nest and Tsygan [NT] showed that any differentiable star-product on a symplectic manifold is equivalent to a Fedosov star-product. This fact allowed them to define the characteristic class of a star-product $\star$ to be the class of the Weyl curvature in $H^2_{dR}(M)[[\lambda]]$ of any Fedosov star-product equivalent to $\star$. As a consequence, one gets a parametrization of the equivalence classes of star-products on (M, ω) by elements in $H^2_{dR}(M)[[\lambda]]$.

The parametrization of equivalence classes of star-products has also been made explicit by Bertelson, Cahen and Gutt [BCG] (using older results in Gutt's thesis [Gth]) and by Xu [Xu]; the relation between classes of star-products and classes of 1-differentiable deformations of a Poisson bracket was described in [Bon]. Deligne [De] obtained the same classification result, without any reference to a particular method of construction of deformations, by using cohomological Čech techniques and gerbes. Deligne's approach has been studied further by Gutt and Rawnsley [Gu, GR].

3.2.2 Trace and closed star products. The prominent role played by traces of operators in index theory has its counterpart in deformation quantization. One can naturally define a trace on the deformed algebra of smooth functions having compact support $(C^\infty_c(M)[[\lambda]], \star)$. It is a $\mathbb{C}[[\lambda]]$-linear map $\mathrm{Tr}\colon (C^\infty_c(M)[[\lambda]] \to \mathbb{C}[\lambda^{-1}, \lambda]]$ such that $\mathrm{Tr}(f \star g) = \mathrm{Tr}(g \star f)$. The negative powers of λ appear already in the classical examples and Tr is allowed to take its values not only in formal series in λ but in Laurent series in λ^{-1} as well. Such a trace always exists and is essentially unique [NT, Fe3]. Around 1995, new algebraic versions of the (generalized) Atiyah–Singer theorem appeared: one is due to Nest and Tsygan [NT] where Gelfand–Fuks techniques play an important role, the other, by Fedosov, uses Thom isomorphism.

Recently, Fedosov [Fe4] published a new proof based on the topological invariants of symplectic connections of Tamarkin [Ta].

3.2.2.1. *Closed star-products.* The trace allows to distinguish a particular class of star-products on a symplectic manifold M. These are the so-called *closed star-products* introduced in [CFS]. A star-product $\star$ is closed if the map

$$T\colon f \mapsto \frac{1}{(2\pi\lambda)^m}\int f_m \frac{\omega^m}{m!}$$

is a trace ($2m$ is the dimension of M and f_m is the coefficient of λ^m of $f \in C_c^\infty(M)[[\lambda]]$). Closed star-products exist on any symplectic manifold [OMY2] and any star-product is equivalent to a closed star-product. Cyclic cohomology replaces the Hochschild one for closed star-products: obstructions to existence and equivalence live in the third and second cyclic cohomology spaces, correspondingly. Another important feature of closed star-products is the existence of an invariant ϕ defined in [CFS]. It is a cocycle for the differential of the cyclic bicomplex of M, i.e., Hochschild in one direction and cyclic in the other. Let $\theta(f, g) = f \star g - fg$; for an integer k define the map

$$\begin{aligned}\phi_{2k}\colon C_c^\infty(M)[[\lambda]]^{\otimes 2k+1} &\to \mathbb{R},\\ (f_0, \ldots, f_{2k}) &\mapsto \tau(f_0 \star \theta(f_1, f_2) \star \cdots \star \theta(f_{2k-1}, f_{2k}))\end{aligned}$$

where $\tau = (2\pi\lambda)^m m! T$ does not depend on λ. It turns out that $\phi_{2k} = 0$ for $k > m$ and the cocycle is defined by $\phi = \sum_k \phi_{2k}/k!$. The computation of ϕ, called the CFS-invariant, has been performed for the case where M is a cotangent bundle of a compact Riemannian manifold [CFS]. Notice that in that case, associative deformations correspond to the standard calculus of pseudo-differential operators, and the CFS-invariant is nothing else than the Todd class of M. This result has been generalized by Halbout [Ha] to the case of any symplectic manifold.

4 Metamorphoses

4.1 Poisson manifolds and Kontsevich's formality

The problem of deformation quantization of general Poisson differentiable manifolds has been left open for a long time. For the nonregular Poisson case, first examples of star-products appeared in [BFFLS] in relation with the quantization of angular momentum. They were defined on the dual of $\mathfrak{so}(n)$ endowed with its natural Kirillov–Poisson structure. The case for any Lie algebra follows from the construction given by Gutt [Gu3] of a star-product on the cotangent bundle of a Lie group G. This star-product restricts to a star-product on the dual of the Lie algebra of G. It translates the associative structure of the enveloping algebra in terms of functions on the dual of the Lie algebra of G. Omori, Maeda and Yoshika [OMY3] have constructed quantization for a class of quadratic Poisson structures.

There is no equivalent of Darboux theorem for Poisson structures and this fact is at the origin of the main difficulty to construct (even local) star-products in the Poisson case. The situation changed drastically in 1997, when Kontsevich [Ko2] proved his Formality Conjecture [Ko1]. The existence of star-products on any smooth Poisson manifold appears as a consequence of the Formality Theorem.

We shall review the construction of a star-product on $\mathbb{R}^d$ given in [Ko2] and then present the essential steps of the Formality Theorem.

4.1.1 Kontsevich star-product. Consider $\mathbb{R}^d$ endowed with a Poisson bracket α. We denote by $(x^1, \dots, x^d)$ the coordinate system on $\mathbb{R}^d$; the Poisson bracket of two smooth functions f, g is given by $\alpha(f, g) = \sum_{1 \le i,j \le n} \alpha^{ij} \partial_i f \partial_j g$, where ∂_k denotes the partial derivative with respect to x^k. Instead of the whole of $\mathbb{R}^d$, one may consider an open subset of it.

The formula for the Kontsevich star-product is conveniently defined by considering, for each $n \ge 0$, a family of oriented graphs G_n. To a graph $\Gamma \in G_n$ is associated a bidifferential operator B_Γ and a weight $w(\Gamma) \in \mathbb{R}$. The sum $\sum_{\Gamma \in G_n} w(\Gamma) B_\Gamma$ gives the coefficient of λ^n, i.e., the cochain C_n of the star-product.

For later use, we shall reproduce the general definitions of [Ko2]. They are more general than the ones needed in this section for the star-product.

An oriented graph Γ belongs to $G_{n,m}$, with $n, m \ge 0$, $2n - m + 2 \ge 0$, if:

i) The set of vertices V_Γ of Γ has $n + m$ elements labeled $\{1, \dots, n; \bar{1}, \dots, \bar{m}\}$. The vertices $\{1, \dots, n\}$ are said of the first type and $\{\bar{1}, \dots, \bar{m}\}$ are said of the second type.

ii) E_Γ is the set of oriented edges of Γ. E_Γ has $2n - m + 2$ edges. There is no edge starting at a vertex of the second type. The set of edges starting at a vertex of the first type k is denoted by $\mathrm{Star}(k)$ and its cardinality by $\sharp(k)$, hence $\sum_{1 \le k \le n} \sharp(k) = 2n + m - 2$. The edges of Γ starting at vertex k will be denoted by $\{e_k^1, \dots, e_k^{\sharp(k)}\}$. When it is needed to make explicit the starting and ending vertices of some edge v, we shall write $v = (s(v), e(v))$, where $s(v) \in \{1, \dots, n\}$ is its starting vertex and $e(v) \in \{1, \dots, n; \bar{1}, \dots, \bar{m}\}$ is its ending vertex.

iii) Γ has no loop (edge starting at some vertex and ending at that vertex) and no parallel multiple edges (edges sharing the same starting and ending vertices).

Let us specialize the preceding definition to the case of star-products. Here $m = 2$ and only the family of graphs $G_{n,2}$, $n \ge 0$, will be of interest. Moreover there are exactly 2 edges starting at each vertex of the first type, i.e., $\sharp k = 2$, $\forall k \in \{1, \dots, n\}$. Hence E_Γ has $2n$ edges for $\Gamma \in G_{n,2}$. For notational convenience we shall write the pair of edges $\{e_k^1, e_k^2\}$ as $\{i_k, j_k\}$. Also $\{i_k, j_k\}$ will stand for indices of the Poisson tensor $\alpha^{i_k j_k}$, with $1 \le i_k, j_k \le d$. The set of graphs in $G_{n,2}$ has $(n(n+1))^n$ elements. For $n = 0$, $G_{0,2}$ has only one element: the graph having as set of vertices $\{\bar{1}, \bar{2}\}$ and no edges.

A bidifferential operator $(f, g) \mapsto B_\Gamma(f, g)$, $f, g \in C^\infty(\mathbb{R}^d)$, is associated to each graph $\Gamma \in G_{n,2}$, $n \geq 1$, in the following way. To each vertex of the first type k from where the edges $\{i_k, j_k\}$ start, one associates the components $\alpha^{i_k j_k}$ of the Poisson tensor; f is associated to the vertex $\bar{1}$ and g to the vertex $\bar{2}$. Each edge, e.g. i_k, acts by partial differentiation with respect to x^{i_k} on its ending vertex.

For example, the graph in Figure 1 gives the bidifferential operator

$$B_\Gamma(f, g) = \sum_{0 \leq i_*, j_* \leq d} \alpha^{i_1 j_1}\, \partial_{j_1 j_3} \alpha^{i_2 j_2}\, \partial_{i_2} \alpha^{i_3 j_3}\, \partial_{i_1 j_2} f\, \partial_{i_3} g.$$

Notice that for $n = 0$, we simply have the usual product of f and g.

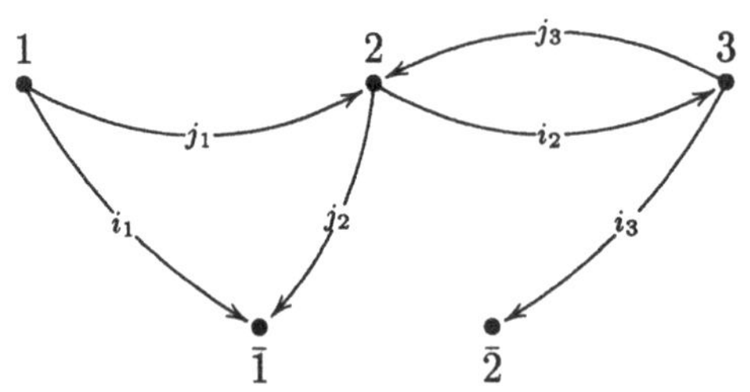

Figure 1. A graph in $G_{3,2}$.

The general formula for $B_\Gamma(f, g)$ for $\Gamma \in G_{n,2}$ is given by a sum over all maps $I\colon E_\Gamma \to \{1, \ldots, d\}$,

$$B_\Gamma(f, g) = \sum_I \left(\left(\prod_{k=1}^{n} \prod_{k'=1}^{n} \partial_{I(k',k)} \alpha^{I(e_k^1) I(e_k^2)} \right) \left(\prod_{k_1=1}^{n} \partial_{I(k_1,\bar{1})} f \right) \left(\prod_{k_2=1}^{n} \partial_{I(k_2,\bar{2})} g \right) \right), \tag{4.1}$$

where the usual convention for missing factors is used (i.e., whenever an edge is absent in E_Γ, the corresponding factor is 1).

Finally the weight $w(\Gamma)$ of a graph Γ is defined as follows (a general definition will be given in Section 4.1.2). Let $\mathcal{H} = \{z \in \mathbb{C} \mid \operatorname{Im}(z) > 0\}$ be the upper half-plane with its Lobachevsky hyperbolic metric. The configuration space of n distinct points in $\mathcal{H}$ is denoted by $\mathcal{H}_n$. It is an open submanifold of $\mathbb{C}^n$ with its natural orientation.

Let $\phi\colon \mathcal{H}_2 \to \mathbb{R}/2\pi\mathbb{Z}$ be the function (angle function):

$$\phi(z_1, z_2) = \frac{1}{2\sqrt{-1}} \operatorname{Log}\left(\frac{(z_2 - z_1)(\bar{z}_2 - z_1)}{(z_2 - \bar{z}_1)(\bar{z}_2 - \bar{z}_1)} \right). \tag{4.2}$$

$\phi(z_1, z_2)$ is extended by continuity for $z_1, z_2 \in \mathbb{R}$, $z_1 \neq z_2$. The function ϕ measures the angle at z_1 between the geodesic passing by z_1 and the point at infinity, and the geodesic passing by z_1 and z_2 for the hyperbolic metric.

For a graph $\Gamma \in G_{n,2}$, the vertex k, $1 \leq k \leq n$, is associated to a variable $z_k \in \mathcal{H}$, the vertex $\bar{1}$ to $0 \in \mathbb{R}$, and the vertex $\bar{2}$ to $1 \in \mathbb{R}$. Doing so, we can associate a function $\tilde{\phi}_v$ on $\mathcal{H}_n$ to an edge $v \in E_\Gamma$ by $\tilde{\phi}_v = \phi(s(v), e(v))$.

The weight $w(\Gamma)$ is defined by integrating a $2n$-form over $\mathcal{H}_n$ (in fact on some compactified space, cf. Section 4.1.2):

$$w(\Gamma) = \frac{1}{n!(2\pi)^{2n}} \int_{\mathcal{H}_n} \bigwedge_{1\le k\le n} (d\tilde{\phi}_{e_k^1} \wedge d\tilde{\phi}_{e_k^2}), \tag{4.3}$$

where $\{e_k^1, e_k^2\}$, $1 \le k \le n$, is the set of edges of Γ. The weights are universal in the sense that they do not depend on the Poisson structure or the dimension d.

The above construction goes over to any formal Poisson bracket $\alpha_\lambda = \sum_{k\ge 0} \lambda^k \alpha_k$.

Theorem 4.1 (Kontsevich [Ko2]). *The map* $\star\colon C^\infty(\mathbb{R}^d) \times C^\infty(\mathbb{R}^d) \to C^\infty(\mathbb{R}^d)[[\lambda]]$ *defined by:*

$$(f, g) \mapsto f \star g = \sum_{n\ge 0} \lambda^n \sum_{\Gamma\in G_{n,2}} w(\Gamma) B_\Gamma(f, g)$$

defines a differential star-product on $(\mathbb{R}^d, \alpha)$. *The classes of equivalence of star-products are in a one-to-one correspondence with the classes of equivalence of formal Poisson brackets.*

The first cochains of the Kontsevich star-product can be computed by elementary means. Of course we have $C_0(f, g) = fg$. At first order, $G_{1,2}$ has 2 graphs: one has weight $\frac{1}{2}$ and as bidifferential operator the Poisson bracket $\{f, g\}_\alpha = \alpha(df \wedge dg)$, the other graph has weight $-\frac{1}{2}$ and bidifferential operator $\{g, f\}_\alpha$. The sum gives $C_1(f, g) = \{f, g\}_\alpha$. All the graphs contributing to C_2, up to symmetries, are shown in Figure 2, and the second order term is given by:

$$\begin{aligned} C_2(f, g) = {} & \frac{1}{2} \alpha^{i_1 j_1} \alpha^{i_2 j_2} \partial_{i_1 i_2} f \, \partial_{j_1 j_2} g \\ & + \frac{1}{3} \alpha^{i_1 j_1} \partial_{i_1} \alpha^{i_2 j_2} (\partial_{j_1 j_2} f \, \partial_{i_2} g + \partial_{i_2} f \, \partial_{j_1 j_2} g) \\ & - \frac{1}{6} \partial_{j_2} \alpha^{i_1 j_1} \partial_{j_1} \alpha^{i_2 j_2} \partial_{i_1} f \, \partial_{i_2} g, \end{aligned} \tag{4.4}$$

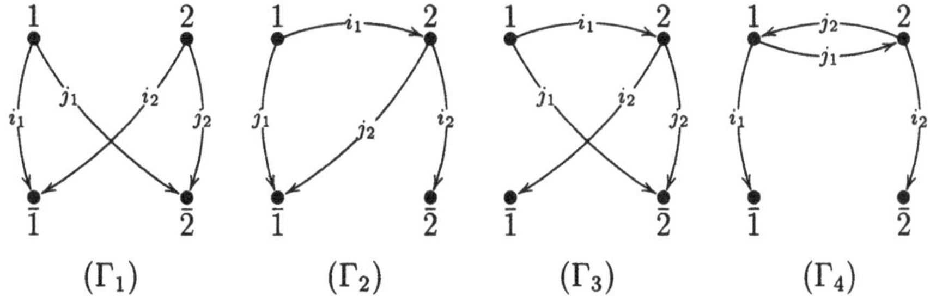

Figure 2. Graphs in $G_{2,2}$ contributing to C_2.

where summation over repeated indices is understood. For higher cochains, there is no systematic way to compute weights of graphs, especially in the presence of cycles (e.g. graph Γ_4 in Fig. 2), but we should mention that Polyak [Po] has provided an interpretation of weights in term of degree of maps which allowed him to "easily" compute a large class of graphs and all of them for the linear Poisson structure case.

The Kontsevich star-product has been extensively studied for linear Poisson structures, i.e., on the dual of a Lie algebra. An important result in that context is due to Shoikhet [Sho] who showed the vanishing of the weight of all even wheel graphs. As a consequence, on the dual of a Lie algebra, the Kontsevich star-product coincides with the one given by the Duflo isomorphism. Moreover, the Kontsevich star-product induces a convolution on the space of distributions supported at 0 in the Lie algebra $\mathfrak{g}$. This has allowed [ADS] to prove the conjecture on invariant distributions on a Lie group formulated by Kashiwara and Vergne in 1978.

The linear Poisson case allows to quantize the class of quadratic Poisson brackets that are in the image of a map introduced by Drinfeld which associates a quadratic Poisson to a linear Poisson bracket; not all quadratic brackets arise in this way [MMR].

4.1.2 Formality Theorem. Theorem 4.1 is a particular consequence of a much more general result called Formality Theorem. Given a smooth manifold M one may consider two algebraic structures constructed on the associative algebra $A \equiv C^\infty(M)$, namely, the Hochschild complex $C^\bullet(A, A)$ and its cohomology $H^\bullet(A, A)$. They both have a structure of differential graded Lie algebra. Roughly speaking, the Formality Theorem states that these two differential graded Lie algebras are "equivalent" in a sense we now make precise.

4.1.2.1. *Differential graded Lie algebras.* Let $\mathbb{K}$ be a field of characteristic zero. All vector spaces are over $\mathbb{K}$. Let V be a $\mathbb{Z}$-graded vector space: $V = \oplus_{k\in\mathbb{Z}} V^k$. An element $x \in V$ is called homogeneous of degree k if $x \in V^k$. The degree of a homogeneous element x is denoted $|x|$. From now on, "graded" will always mean "$\mathbb{Z}$-graded".

A *differential graded Lie algebra* over $\mathbb{K}$ (DGLA for short) is a graded vector space $\mathfrak{g} = \oplus_{k\in\mathbb{Z}}\mathfrak{g}^k$ endowed with Lie superbracket $[\cdot\,,\cdot]$ and a differential d of degree 1. It means that $(\mathfrak{g}, [\cdot\,,\cdot], d)$ fulfills (a, b, c are homogeneous elements):

1) $[\cdot\,,\cdot]$ is a bilinear map, $[\cdot\,,\cdot]\colon \mathfrak{g}^i \times \mathfrak{g}^j \to \mathfrak{g}^{i+j}$, satisfying:

   $[a, b] = -(-1)^{|a||b|}[b, a]$,

   $(-1)^{|a||c|}[a, [b, c]] + (-1)^{|b||a|}[b, [c, a]] + (-1)^{|c||b|}[c, [a, b]] = 0.$

2) $d\colon \mathfrak{g}^i \to \mathfrak{g}^{i+1}$ is a linear map satisfying $d^2 = 0$ and:

   $d[a, b] = [da, b] + (-1)^{|a|}[a, db].$

The Hochschild complex $C^\bullet(A, A) = \oplus_{k\geq 0} C^k(A, A)$ consists of polydifferential operators on M, $C^k(A, A) = \{D \mid D\colon A^{\otimes^k} \to A\}$. In the following, we will detail the ingredients that make the Hochschild complex a DGLA. We shall look at the Hochschild complex as a graded vector space $D^{\bullet}_{\text{poly}}(M) = \oplus_{k\in\mathbb{Z}} D^k_{\text{poly}}(M)$, where $D^k_{\text{poly}}(M)$ is equal to $C^{k+1}(A, A)$ for $k \geq -1$ and equal to $\{0\}$ otherwise. Notice the shift by one unit in the grading of $D^{\bullet}_{\text{poly}}(M)$ with respect to that of $C^{\bullet}(A, A)$: $|D| = k$ if $D \in C^{k+1}(A, A)$.

The *Gerstenhaber bracket* $[\cdot\,,\cdot]_G$ on $D^{\bullet}_{\text{poly}}(M)$ is defined for $D_i \in D^{k_i}_{\text{poly}}(M)$ by:

$$[D_1, D_2]_G = D_1 \circ D_2 - (-1)^{k_1 k_2} D_2 \circ D_1,$$

where $\circ\colon D^{k_1}_{\text{poly}}(M) \times D^{k_2}_{\text{poly}}(M) \to D^{k_1+k_2}_{\text{poly}}(M)$ is a nonassociative composition law for polydifferential operators:

$$\begin{aligned}(D_1 \circ D_2)&(f_0, \dots, f_{k_1+k_2})\\ &= \sum_{0\leq j\leq k_1} (-1)^{jk_2} D_1(f_0, \dots, f_{j-1}, D_2(f_j, \dots, f_{j+k_2}), f_{j+k_2+1}, \dots f_{k_1+k_2}).\end{aligned}$$

Finally, we have the usual Hochschild differential b defined by (2.1). It can be expressed in terms of the Gerstenhaber bracket. If $m\colon A \otimes A \to A$ denotes the product of A (i.e., usual product of functions on M), then one can verify that $b = -[\,\cdot\,, m]_G$. The usual Hochschild differential is not a differential of the Gerstenhaber bracket, but $\delta = [m, \cdot\,]_G$ is a differential of degree 1 (for $D \in D^k_{\text{poly}}(M)$, we have $bD = (-1)^k \delta D$). It is an important fact that $(D^{\bullet}_{\text{poly}}(M), [\cdot\,,\cdot]_G, \delta)$ is a DGLA.

The cohomology $H^{\bullet}(A, A)$ of the complex $C^{\bullet}(A, A)$ (with b or δ as differential) is the space of polyvectors on M: $\Gamma(\wedge^{\bullet} TM)$ [Ve, HKR]. We denote by $T^{\bullet}_{\text{poly}}(M)$ the graded vector space $T^{\bullet}_{\text{poly}}(M) = \oplus_{k\in\mathbb{Z}} T^k_{\text{poly}}(M)$, where $T^k_{\text{poly}}(M) = \Gamma(\wedge^{k+1} TM)$ for $k \geq -1$ and is equal to $\{0\}$ otherwise. On $T^{\bullet}_{\text{poly}}(M)$ one has the Schouten–Nijenhuis bracket $[\cdot\,,\cdot]_{SN}$ which is the natural extension of the bracket of vector fields. On decomposable tensors it is defined by:

$$\begin{aligned}&[\xi_0 \wedge \cdots \wedge \xi_k, \eta_0 \wedge \cdots \wedge \eta_l]_{SN}\\ &\quad = \sum_{0\leq i\leq k}\sum_{0\leq j\leq l} (-1)^{i+j} [\xi_i, \eta_j] \wedge \xi_0 \wedge \cdots \wedge \hat{\xi}_i \wedge \cdots \wedge \xi_k \wedge \eta_0 \wedge \cdots \wedge \hat{\eta}_j \wedge \cdots \wedge \eta_l.\end{aligned} \tag{4.5}$$

The Schouten–Nijenhuis bracket makes $(T^{\bullet}_{\text{poly}}(M), [\cdot\,,\cdot]_{SN})$ into a graded Lie algebra. On it, take as the differential 0, then $(T^{\bullet}_{\text{poly}}(M), [\cdot\,,\cdot]_{SN}, 0)$ is trivially a DGLA.

4.1.2.2. *L_∞-morphism.* We recall some definitions needed to state the Formality Theorem. A *graded coalgebra* $\mathfrak{c}$ is a graded vector space with a coproduct $\Delta\colon \mathfrak{c}_i \to \oplus_{m+n=i} \mathfrak{c}_m \otimes \mathfrak{c}_n$. To any graded vector space $\mathfrak{g}$ is associated a graded coalgebra $\mathcal{C}(\mathfrak{g}) = \otimes_{k\geq 1} (\wedge^k \mathfrak{g})[k]$, where $[k]$ indicates the shift to the left by k units in the grading. An *L_∞-algebra* $(\mathfrak{g}, Q)$ is a graded vector space $\mathfrak{g}$ with a codifferential Q satisfying $Q^2 = 0$ on $\mathcal{C}(\mathfrak{g})$. L_∞-algebras are also known as *strong(ly) homotopy Lie algebras* (see e.g. [Sta]). The differential Q defines a sequence $\{Q_k\}$ of morphisms $Q_k\colon \wedge^k(\mathfrak{g}) \to \mathfrak{g}[2-k]$, $k \geq 1$ and the relation $Q^2 = 0$ puts quadratic constraints on the Q_k's. These constraints imply that $Q_1 \colon \mathfrak{g} \to \mathfrak{g}[1]$ is a differential of degree 1, hence $(\mathfrak{g}, Q_1)$ is a complex, and $Q_2 \colon \wedge^2(\mathfrak{g}) \to \mathfrak{g}$ defines a superbacket on $\mathfrak{g}$ for which Q_1 is compatible. Hence any L_∞-algebra $(\mathfrak{g}, Q)$ such that $Q_k = 0$ for $k \geq 3$ is a DGLA.

An L_∞-morphism $\mathcal{F}$ between two L_∞-algebras $(\mathfrak{g}_1, Q^{(1)})$ and $(\mathfrak{g}_2, Q^{(2)})$ is a morphism between the associated coalgebras such that $\mathcal{F}\, Q^{(1)} = Q^{(2)} \mathcal{F}$. We have similar constraints on the sequence of morphisms $\{\mathcal{F}_k\}$: $\mathcal{F}_k \colon \wedge^k(\mathfrak{g}_1) \to \mathfrak{g}_2[1-k]$, $\quad k \geq 1$. At first order, they imply that the formality $\mathcal{F}$ is a morphism of complexes $(\mathfrak{g}_1, Q_1^{(1)}) \to (\mathfrak{g}_2, Q_1^{(2)})$, but the constraint at order 2 shows that $\mathcal{F}_2$ is not a homomorphism of DGLA's. Finally, an L_∞-morphism is a quasi-morphism if the morphism of complexes $\mathcal{F}_1 \colon (\mathfrak{g}_1, Q_1^{(1)}) \to (\mathfrak{g}_2, Q_1^{(2)})$ induces an isomorphism between the corresponding cohomology spaces (i.e., quasi-isomorphism of complexes in the usual sense).

The two DGLA's $(T^\bullet_{\text{poly}}(M), [\cdot\,,\cdot]_{SN}, 0)$ and $(D^\bullet_{\text{poly}}(M), [\cdot\,,\cdot]_G, \delta)$ are L_∞-algebras by definition. The natural injection $\mathcal{I} \colon T^\bullet_{\text{poly}}(M) \to D^\bullet_{\text{poly}}(M)$ defined by

$$\mathcal{I}(\xi_1 \wedge \cdots \wedge \xi_k).(f_1, \ldots, f_k) = \frac{1}{k!} \det(\xi_i(f_j)),$$

is a quasi-isomorphism of complexes, but is not a homomorphism of DGLA's since it is incompatible with the brackets. But Kontsevich has shown that:

Theorem 4.2 (Formality). *There is a quasi-isomorphism*

$$\mathcal{K} \colon (T^\bullet_{\text{poly}}(M), [\cdot\,,\cdot]_{SN}, 0) \to (D^\bullet_{\text{poly}}(M), [\cdot\,,\cdot]_G, \delta)$$

such that $\mathcal{K}_1$ is equal to $\mathcal{I}$.

The main part of the proof consists in writing down explicitly a quasi-isomorphism for the case $M = \mathbb{R}^d$. The quasi-isomorphism $\mathcal{K}$ is expressed in terms of graphs and weights and it generalizes the formula for the star-product given in Section 4.1.1.

4.1.2.3. *Construction of $\mathcal{K}$ for $M = \mathbb{R}^d$.* We have already defined the family of graphs $G_{n,m}$. Here we need a generalization of the bidifferential operators B_Γ of the star-product formula. For a graph Γ in $G_{n,m}$, take $(\alpha_1, \ldots, \alpha_n)$ n polyvectors such that $\alpha_k \in T^{\sharp(k)-1}_{\text{poly}}(M)$. By definition of $G_{n,m}$ we have $\sum_{1 \leq k \leq n} \sharp(k) = 2n + m - 2$, and consequently $\alpha_1 \wedge \cdots \wedge \alpha_n$ is in $T^{n+m-2}_{\text{poly}}(M)$. Define a polydifferential operator $B_\Gamma(\alpha_1 \wedge \cdots \wedge \alpha_n)$ belonging to $T^{m-1}_{\text{poly}}(M)$ by a formula similar to (4.1). For m smooth functions $(f_1, \ldots, f_m)$ on $\mathbb{R}^d$, we have:

$$\begin{aligned} &B_\Gamma(\alpha_1 \wedge \cdots \wedge \alpha_n).(f_1, \ldots, f_m) \\ &\quad = \sum_I \Big(\Big(\prod_{k=1}^{n} \prod_{k'=1}^{n} \partial_{I(k',k)} \alpha_k^{I(e_k^1) \cdots I(e_k^{\sharp(k)})}\Big)\Big(\prod_{i=1}^{m} \prod_{j=1}^{n} \partial_{I(j,\bar{i})} f_i\Big)\Big), \end{aligned} \tag{4.6}$$

where the sum is on all maps $I \colon E_\Gamma \to \{1, \ldots, d\}$.

The weight of a graph is defined by integration on compactified configuration spaces that we now make precise. For $2n + m - 2 \geq 0$, let

$$\mathsf{Conf}_{n,m} = \{(p_1, \dots, p_n; q_1, \dots, q_m) \in \mathcal{H}^n \times \mathbb{R}^m \mid p_i \neq p_j \text{ for } i \neq j; \\ q_i \neq q_j \text{ for } i \neq j\}.$$

$\mathsf{Conf}^+_{n,m}$ will denote the connected component of $\mathsf{Conf}_{n,m}$ for which $q_1 < \cdots < q_m$. $\mathsf{Conf}_{n,m}$ is acted upon by the group G of transformations $z \mapsto az + b$, $a > 0$, $b \in \mathbb{R}$. Define $\mathsf{C}_{n,m} = \mathsf{Conf}_{n,m}/G$: It inherits a natural orientation. $\mathsf{C}^+_{n,m}$ will denote the connected component corresponding to $\mathsf{Conf}^+_{n,m}$. The compactified spaces $\bar{\mathsf{C}}^+_{n,m}$ are manifolds with corners. The weight of a graph $\Gamma \in G_{n,m}$ is defined by:

$$w(\Gamma) = \prod_{1 \leq k \leq n} \frac{1}{(\sharp k)!} \frac{1}{(2\pi)^{2n+m-2}} \int_{\bar{\mathsf{C}}^+_{n,m}} \bigwedge_{e \in E_\Gamma} d\tilde{\phi}_e \tag{4.7}$$

Notice that the order in the set of edges E_Γ gives the order in the wedge product. We can now write down the sequence of morphisms $\{\mathcal{K}_n\}$ of the quasi-isomorphism $\mathcal{K}$:

$$\mathcal{K}_n = \sum_{m \geq 0} \sum_{\Gamma \in G_{n,m}} w(\Gamma) B_\Gamma, \quad n \geq 1.$$

The verification that $\mathcal{K}$ is a quasi-isomorphism amounts to check an infinite number of quadratic equations for the weights, which are essentially a consequence of Stokes theorem applied to the weights. One has to keep track of all the signs mixed by the gradings and the orientations of the configuration spaces. A detailed description of these computations can be found in [AMM].

4.2 Field theory and path integral formulas

4.2.1 Star-products in infinite dimension. The deformation quantization of a given classical field theory consists in giving a proper definition for a star-product on the infinite-dimensional manifold of initial data for the classical field equation and constructing with it, as rigorously as possible, whatever physical expressions are needed. As in other approaches to field theory, here also one faces serious divergence difficulties as soon as one is considering interacting fields theory, and even at the free field level if one wants a mathematically rigorous theory. But the philosophy in dealing with the divergences is significantly different and one is in position to take advantage of the cohomological features of deformation theory to perform what can be called *cohomological renormalization* [Dth].

Poisson structures are known on infinite-dimensional manifolds since a long time and there is an extensive literature on this subject. A typical structure, for our purpose, is a weak symplectic structure such as that defined by Segal [Se] (see also [Ko74]) on the space of solutions of a classical field equation like $\Box\Phi = F(\Phi)$, where $\Box$ is the d'Alembertian. Now if one considers scalar-valued functionals Ψ over such a space of solutions, i.e., over the phase space of initial conditions $\varphi(x) = \Phi(x, 0)$ and $\pi(x) = \frac{\partial}{\partial t}\Phi(x, 0)$, one can consider a Poisson bracket defined by

$$P(\Psi_1, \Psi_2) = \int \left(\frac{\delta\Psi_1}{\delta\varphi} \frac{\delta\Psi_2}{\delta\pi} - \frac{\delta\Psi_1}{\delta\pi} \frac{\delta\Psi_2}{\delta\varphi} \right) dx \tag{4.8}$$

where δ denotes the functional derivative. The problem is that while it is possible to give a precise mathematical meaning to (4.8) by specifying an appropriate algebra of functionals, the formal extension to powers of P, needed to define the Moyal bracket, is highly divergent, already for P^2. This should not be so surprising for physicists who know from experience that the correct approach to field theory starts with normal ordering, and that there are infinitely many inequivalent representations of the canonical commutation relations (as opposed to the von Neumann uniqueness in the finite-dimensional case, for projective representations), even if in recent physical literature some are working formally with Moyal product.

Starting with some star-product $\star$ (e.g., an infinite-dimensional version of a Moyal-type product or, better, a star-product similar to the normal star-product (4.9)) on the manifold of initial data, one would interpret various divergences appearing in the theory in terms of coboundaries (or cocycles) for the relevant Hochschild cohomology. Suppose that we are suspecting that a term in a cochain of the product $\star$ is responsible for the appearance of divergences. Applying an iterative procedure of equivalence, we can try to eliminate it, or at least get a lesser divergence, by subtracting at the relevant order a divergent coboundary; we would then get a better theory with a new star-product, "equivalent" to the original one. Furthermore, since in this case we can expect to have at each order an infinity of non equivalent star-products, we can try to subtract a cocycle and then pass to a nonequivalent star-product whose lower order cochains are identical to those of the original one. We would then make an analysis of the divergences up to order $\hbar^r$, identify a divergent cocycle, remove it, and continue the procedure (at the same or hopefully a higher order). Along the way one should preserve the usual properties of a quantum field theory (Poincaré covariance, locality, etc.) and the construction of adapted star-products should be done accordingly. The complete implementation of this program should lead to a cohomological approach to renormalization theory. It would be interesting to formulate the Connes–Kreimer [CK] rigorous renormalization procedure in a way that would fit in this pattern.

A very good test for this approach would be to start from classical electrodynamics, where (among others) the existence of global solutions and a study of infrared divergencies were recently rigorously performed [FST], and go towards mathematically rigorous QED. Physicists may think that spending so much effort in trying to give complete mathematical sense to recipes that work so well is a waste of time, but the fact it has for so long resisted a rigorous formulation is a good indication that the new mathematical tools needed will prove very efficient.

4.2.1.1. *Normal star-product.* In the case of free fields, one can write down an explicit expression for a star-product corresponding to normal ordering. Consider a (classical) free massive scalar field Φ with initial data (ϕ, π) in the Schwartz space $\mathcal{S}$. The initial data (ϕ, π) can advantageously be replaced by their Fourier modes $(\bar{a}, a)$, which for a massive theory are also in $\mathcal{S}$ seen as a real vector space. After quantization $(\bar{a}, a)$ become the usual creation and annihilation operators, respectively. The normal star-product $\star_N$ is formally equivalent to the Moyal product and an integral representation for $\star_N$ is given by [Dit]:

$$(F \star_N G)(\bar{a}, a) = \int_{\mathcal{S}' \oplus \mathcal{S}'} d\mu(\bar{\xi}, \xi) F(\bar{a}, a + \xi) G(\bar{a} + \bar{\xi}, a), \tag{4.9}$$

where, by Bochner–Minlos theorem, μ is the Gaussian measure on $\mathcal{S}' \oplus \mathcal{S}'$ defined by the characteristic function $\exp(-\frac{1}{\hbar} \int dk\, \bar{a}(k) a(k))$ and F, G are holomorphic functions with semi-regular kernels. The reader may object that creation and annihilation operators are operator-valued distributions and one should consider $(\bar{a}, a)$ in $\mathcal{S}' \oplus \mathcal{S}'$ instead. In fact the distribution aspect is present in the very definition of the cochains of the star-product through (4.9).

Likewise, Fermionic fields can be cast in that framework by considering functions valued in some infinite dimensional Grassmann algebra and super-Poisson brackets (for the deformation quantization of the latter see e.g. [Bm]).

For the normal product (4.9) one can formally consider interacting fields. It turns out that the star-exponential of the Hamiltonian is, up to a multiplicative well-defined function, equal to Feynman's path integral. For free fields, we have a mathematical meaningful equality between the star-exponential and the path integrals as both of them are defined by a Gaussian measure, and hence well-defined. In the interacting fields case, giving a rigorous meaning to either of them would give a meaning to the other.

The interested reader will find in [Dit] calculations performing some steps in the above direction, for free scalar fields and the Klein–Gordon equation, and an example of cancellation of some infinities in $\lambda\phi_2^4$-theory via a λ-dependent star-product formally equivalent to a normal star-product.

Recently many field theorists became interested in deformation quantization in relation with M-theory and string theory. The main hope seems to be that deformation quantization may give a good framework for noncommutative field theory. We refer the reader to the thorough review by Douglas and Nekrasov [DN] for more details.

4.2.2 Path integral formula of Cattaneo and Felder. Kontsevich [Ko2] suggested that his formula comes from a perturbation theory for a related bidimensional topological field theory. The physical origin of the weights and graphs has been elucidated by Cattaneo and Felder [CF]. They have constructed a topological field theory on a disc whose perturbation series makes Kontsevich graphs and weights appear explicitly after a finite renormalization. The field model in question is the so-called Poisson sigma model [Ik]; it is made up of the following ingredients:

a) the closed unit disc Σ in $\mathbb{R}^2$,

b) a d-dimensional Poisson real manifold (M, α),

c) a map C^1 $X\colon \Sigma \to M$,

d) a 1-form η on Σ taking values in the space of 1-forms on M, i.e,

$$\eta \in \Gamma(\Sigma, T^*\Sigma \otimes T^*M).$$

In physics, Σ is called the worldsheet, and M the target space.

The action functional for the Poisson sigma model is given by the integration of the sum of two 2-forms on Σ. Let V_1 and V_2 be two vector fields on Σ, then $\langle \eta, V_i \rangle$, $i = 1, 2$, is a 1-form on M. We can define a 2-form β on Σ as follows:

$$\langle \beta, V_1 \wedge V_2 \rangle = \frac{1}{2}\big(\langle\langle \eta, V_1\rangle, DX.V_2\rangle - \langle\langle \eta, V_2\rangle, DX.V_1\rangle\big),$$

where $DX\colon T\Sigma \to TM$ is the tangent map of X. The 2-form β is also denoted $\langle \eta, DX\rangle$. A second 2-form γ on Σ involves the Poisson tensor α, and is given by:

$$\langle \gamma, V_1 \wedge V_2 \rangle = \frac{1}{2}\langle \alpha \circ X, \langle \eta, V_1\rangle \wedge \langle \eta, V_2\rangle\rangle,$$

and we shall write $\gamma = \frac{1}{2}\langle \alpha \circ X, \eta \wedge \eta\rangle$. If (u^1, u^2) is a chart at $s \in \Sigma$, and $(x^1, \ldots, x^d)$ is a chart at $x = X(s)$, the local expressions for β and γ read:

$$\beta(s) = \sum_{i,\mu,\nu} \eta_{\mu i}(s) \frac{\partial X^i}{\partial u^\nu}(s)\, du^\mu \wedge du^\nu$$

$$\gamma(s) = \frac{1}{2} \sum_{i,j,\mu,\nu} \alpha^{ij}(x)\eta_{\mu i}(s)\eta_{\nu j}(s)\, du^\mu \wedge du^\nu$$

Finally, the action integral is defined by:

$$S[X, \eta] = \int_\Sigma \Big(\langle \eta, DX\rangle + \frac{1}{2}\langle \alpha \circ X, \eta \wedge \eta\rangle\Big). \tag{4.10}$$

The 1-form η is required to vanish on $T(\partial\Sigma)$ (tangent vectors of Σ tangential to the boundary $\partial\Sigma = S^1$).

The dynamics of $S[X, \eta]$ is one of a constrained Hamiltonian system governing the propagation of an open string. Let $\alpha^\sharp\colon T^*M \to TM$ be the canonical homomorphism of vector bundles induced by α. The variation of (4.10) under the constraint $\eta|_{\partial\Sigma} = 0$ gives us the critical point (X, η) which is a solution of (here $x = X(s)$):

$$\begin{aligned} &DX(s) + \langle \alpha^\sharp(x), \eta(s)\rangle = 0, \\ &d_s\eta(s) + \frac{1}{2} d_x\langle \alpha(x), \eta(s) \wedge \eta(s)\rangle = 0, \end{aligned} \tag{4.11}$$

where d_s (resp. d_x) is the de Rham differential on Σ (resp. M). Equations (4.11) admit the trivial solution: X is a constant map equal to x, and $\eta = 0$.

The integral formula of Cattaneo and Felder is defined through path integration over (X, η) subject to the following conditions: Take three cyclicly counterclockwise ordered points a, b, c on the unit circle, the boundary of Σ. $X(c) = x$ a fixed point in M, $\eta|_{\partial\Sigma} = 0$; then the functional integral

$$(f \star_\hbar g)(x) = \int \mathcal{D}(X)\mathcal{D}(\eta) f(X(a))g(X(b))e^{\frac{i}{\hbar}S[X,\eta]}.$$

where f, g are smooth functions on M, "defines" a differential star-product on (M, α). Of course, the meaning of the integral has to be taken with a grain of salt, but its heuristic power cannot be denied (here the star-product itself is expressed as a path integral, for a zero Hamiltonian).

The proof of this formula is given by the quantization of this Poisson sigma model, which is a delicate question. The nature of the constraints do not allow to use BRST-methods and quantization is performed in the Batalin–Vilkovisky framework. We shall only say here that the perturbative expansion of this theory provides (essentially) the graphs and weights of Kontsevich (the angle function appears as a propagator). Loop graphs (or tadpole), which are excluded in Kontsevich formula, do appear in this perturbative expansion, but it is shown that they disappear by renormalization.

We conclude this section by remarking that the methods used by Cattaneo and Felder proved to be very efficient. For example, they have provided a simple way to understand the globalization of Kontsevich star-products in more geometrical terms [CFT]. The proof goes much along the lines of Fedosov's construction.

4.3 Recent metamorphoses: operadic approach, algebraic varieties

This section will deal mostly with the metamorphoses that occurred in deformation quantization during the past three years. About one month after the ICM98 Congress in Berlin, Tamarkin [Tam] found a new short derivation (for $X = \mathbb{R}^n$) of the Formality Theorem, based on a very general result concerning all associative algebras: For any algebra A (over a field of characteristic 0), its cohomological Hochschild complex $C^\bullet(A, A)$ and its Hochschild cohomology $B \equiv H^\bullet(A, A)$ (this graded space being considered as a complex with zero differential) are algebras over the same operad (up to homotopy). However abstruse this statement may seem to a physicist, when Kontsevich saw Tamarkin's result he thought to himself, like Commissaire Maigret in Simenon's detective stories, "Bon sang, mais c'est bien sûr ...". In fact, he had been close to such a result in some works of his in the early nineties but somehow missed the point at that time. He then devoted a lot of his time and efforts to understand fully Tamarkin's result and generalize it, which he did to a considerable extent. That is probably the main reason why he did not finalize in print his fundamental 1997 e-print [Ko2], nor write his ICM98 lecture on the relations between deformations, motives (in particular, the motivic Galois group) and the Grothendieck–Teichmüller group: all of this became much more clear when expressed in the new language. Instead he pushed a lot further the various ramifications of the new metamorphosis in two fundamental papers [Ko3, KoS], the latter with Soibelman with whom he is working on what will certainly be a very seminal book on these subjects (and many more).

Recently he accomplished still another leap forward [Ko4] by going from the C^∞ situation to the algebro-geometric setting, performing what can be called 'noncommutative algebraic geometry'. In the previous case, operads and deformations of algebras over operads were essential notions. In the new setting he had to use sheaves of algebroids, a slightly more complicated structure. We shall not attempt here to summarize in a couple of pages the very dense content of several long and complicated papers, rather to try and convey a few touches of (what we understand of) the flavor that emerges from them, giving only very basic definitions. Both developments have a wide array of ramifications and implications in a variety of frontier mathematics.

4.3.1 Operadic approach. To fix ideas we shall give a short definition of an operad and of related notions as in [Ko3]. More abstract (and general) definitions can be found in [KoS]. The language of operads, convenient for descriptions and constructions of various algebraic structures, became recently quite popular in theoretical physics because of the emergence of many new types of algebras related with quantum field theories.

Definition 4.3. An *operad* (of vector spaces) consists of the following:

O_1) a collection of vector spaces $P(n)$, $n \geq 0$,

O_2) an action of the symmetric group S_n on $P(n)$ for every n,

O_3) an identity element $\mathrm{id}_P \in P(1)$,

O_4) compositions $m_{(n_1,\dots,n_k)}$:

$$P(k) \otimes \big(P(n_1) \otimes P(n_2) \otimes \cdots \otimes P(n_k)\big) \longrightarrow P(n_1 + \cdots + n_k) \tag{4.12}$$

for every $k \geq 0$ and $n_1, \dots, n_k \geq 0$ satisfying a natural list of axioms.

If we replace vector spaces by topological spaces, $\otimes$ by $\times$, and require continuity of all maps, we get a *topological operad*.

An *algebra over an operad* P consist of a vector space A and a collection of polylinear maps $f_n : P(n) \otimes A^{\otimes n} \longrightarrow A$ for all $n \geq 0$ satisfying the following list of axioms:

A_1) for any $n \geq 0$ the map f_n is S_n-equivariant,

A_2) for any $a \in A$ we have $f_1(\mathrm{id}_P \otimes a) = a$,

A_3) all compositions in P map to compositions of polylinear operations on A.

The *little discs operad* C_d is a topological operad such that $C_d(0) = \emptyset$, $C_d(1) =$ point $= \{\mathrm{id}_{C_d}\}$, and for $n \geq 2$ the space $C_d(n)$ is the space of configurations of n disjoint discs $(D_i)_{1\leq i\leq n}$ inside the standard disc D_0. The composition $C_d(k) \times C_d(n_1) \times \cdots \times C_d(n_k) \longrightarrow C_d(n_1 + \cdots + n_k)$ is obtained by applying the affine transformations in $AF(\mathbb{R}^d)$ associated with discs $(D_i)_{1\leq i\leq k}$ in the configuration in $C_d(k)$ (the translations and dilations transforming D_0 to D_i) to configurations in all $C_d(n_i)$, $i = 1, \dots, k$ and putting the resulting configurations together. The action of the symmetric group S_n on $C_d(n)$ is given by renumerations of indices of discs $(D_i)_{1\leq i\leq n}$.

An associative algebra is an algebra over some operad (denoted Assoc in [Ko3]). This explains the universality of associative algebras and suggests how to develop the deformation theory of algebras over operads, which is the right context for the deformation theory of many algebraic structures.

Now, for a topological space X, denote by $\mathsf{Chains}(X)$ the complex concentrated in $\mathbb{Z}_{\leq 0}$ whose $(-k)$-th component for $k = 0, 1, \ldots$ consists of the formal finite additive combinations $\sum_{i=1}^{N} n_i \cdot f_i, n_i \in \mathbb{Z}, N \in \mathbb{Z}_{\geq 0}$ of continuous maps $f_i : [0, 1]^k \longrightarrow X$ (singular cubes in X), modulo the following relations: $f \circ \sigma = \operatorname{sign}(\sigma) f$ for any $\sigma \in S_k$ acting on the standard cube $[0, 1]^k$ by permutations of coordinates, and $f' \circ \mathrm{pr}_{k \to (k-1)} = 0$ where $\mathrm{pr}_{k \to (k-1)} : [0, 1]^k \longrightarrow [0, 1]^{k-1}$ is the projection onto first $(k-1)$ coordinates, and $f' : [0, 1]^{k-1} \longrightarrow X$ is a continuous map. The boundary operator on cubical chains is defined in the usual way. A d-algebra is an algebra over the operad $\mathsf{Chains}(C_d)$.

If P is a topological operad then the collection of complexes $\big(\mathsf{Chains}(P(n))\big)_{n \geq 0}$ has a natural operad structure in the category of complexes of Abelian groups. The compositions in $\mathsf{Chains}(P)$ are defined using the external tensor product of cubical chains. Passing from complexes to their cohomology we obtain an operad $H_\bullet(P)$ of $\mathbb{Z}$-graded Abelian groups (complexes with zero differential), the homology operad of P.

A remarkable fact is the relationship of the deformation theory of associative algebras to the geometry of configuration spaces of points on surfaces. One of its incarnations is the so-called Deligne conjecture. A formulation of the conjecture is that there is a natural action of $\mathsf{Chains}(C_2)$, the chain operad of the little discs operad C_2, on the Hochschild complex $C^\bullet(A, A)$ for an arbitrary associative algebra A (we refer to [Ko3, KoS] for a more precise formulation). The conjecture was formulated in 1993, deemed proved in 1994 until in the Spring of 1998 Tamarkin found a very subtle flaw in these proofs (he had used the result in an earlier version of his new proof of the Formality Theorem). New proofs can be found in [KoS] and other papers published around that time. The power of those notions can be grasped if we realize that the Formality Theorem in (4.1.2) becomes a relatively easy consequence of the formality of $\mathsf{Chains}(C_2)$, which shows that the Hochschild complex of an associative algebra is a homotopic Gerstenhaber algebra. A general result is [Ko3]:

Theorem 4.4. *The operad* $\mathsf{Chains}(C_d) \otimes \mathbb{R}$ *of complexes of real vector spaces is quasi-isomorphic to its cohomology operad endowed with the zero differential.*

Then Tamarkin's result (which implies the Formality Theorem; see [TaT] for further developments) can be formulated as

Theorem 4.5. *Let* $A := \mathbb{R}[x_1, \ldots, x_n]$ *be the algebra of polynomials considered merely as an associative algebra. Then the Hochschild complex* $\mathsf{Hoch}(A)$ *is quasi-isomorphic as* 2*-algebra to its cohomology* $B := H^\bullet(\mathsf{Hoch}(A))$, *the space of polynomial polyvector fields on* $\mathbb{R}^n$, *considered as a Gerstenhaber algebra, hence a* 2*-algebra.*

Pursuing further in the directions hinted at in [Ko3], he could show in [KoS], among other, that the Grothendieck–Teichmüller group acts homotopically on the moduli space of structures on 2-algebras on the Hochschild complex. The weights $w(\Gamma)$ (4.7) that appear in Kontsevich's quasi-isomorphism are examples of some very special numbers, the periods [Kper] of rational algebraic varieties, a countable set of numbers lying between rational algebraic numbers and all complex numbers. At the end of [KoS] there is a theory of singular chains, suitable for working with manifolds with corners, where one sees appearing a notion of piecewise algebraic spaces.

4.3.2 Algebraic varieties. There have been a number of studies of star-products in the complex domain, starting with e.g. [BG] (even if the authors had not realized at that time that this was in fact what they were studying). A more recent development, with a number of interesting examples, can be found in [Bo]. However the most systematic study appeared very recently with Kontsevich's paper [Ko4] devoted to peculiarities of the deformation quantization in the algebro-geometric context.

A direct application of the Formality Theorem to an algebraic Poisson manifold gives a canonical sheaf of categories deforming coherent sheaves. The global category is very degenerate in general. For a general algebraic Poisson manifold one gets a canonical presheaf of *algebroids* and eventually an Abelian category, that can be quite degenerate. Thus, he introduces a new notion of a *semi-formal deformation*, a replacement in algebraic geometry of a weakened notion of strict deformation quantization (versus a formal one), which is quite natural here because of the context. To give a flavor of the kind of metamorphosis undergone here by deformation quantization, we reproduce the precise definition of [Ko4]:

Definition 4.6. Let R be a complete pro-Artin local ring with residue field $\mathbf{k}$. A *deformation* over $\mathrm{Spec}(R)$ of a $\mathbf{k}$-algebra A is an algebra $\widehat{A}$ over R, topologically free as R-module, considered together with an identification of $\mathbf{k}$-algebras $A \simeq \widehat{A} \otimes_R \mathbf{k}$. A *semi-formal deformation* over R of a finitely generated associative algebra $A/\mathbf{k}$ is an algebra $\widehat{A}_{\text{finite}}$ over R endowed with an identification of $\mathbf{k}$-algebras $A \simeq \widehat{A}_{\text{finite}} \otimes_R \mathbf{k}$ such that there exists an exhaustive increasing filtration of $\widehat{A}_{\text{finite}}$ by finitely generated free R-modules $\widehat{A}_{\leq n}$, compatible with the product, admitting a splitting as a filtration of R-modules, and such that the Rees ring of the induced filtration of A is finitely generated over $\mathbf{k}$.

Deformed algebras obtained by semi-formal deformations are Noetherian and have polynomial growth. Kontsevich gives constructions of semi-formal quantizations (noncommutative deformations over the ring of formal power series) of projective and affine algebraic Poisson manifolds satisfying certain natural geometric conditions. Projective symplectic manifolds (e.g. K3 surfaces and abelian varieties) do not satisfy those conditions (in contradistinction with the C^∞ situation where quantum tori are important examples), but projective spaces with quadratic Poisson brackets and Poisson–Lie groups can be semi-formally quantized. In other words, in that framework, semi-formal deformation quantization is either canonical or impossible.

4.3.3 Generalized deformations, singularities. We shall end this (long) review by various attempts, motivated by physical problems, to go beyond DrG (continuous) deformations. The first two involve deformations that are formally different from the latter, while the last one (mentioned only for completeness) involves going to noncontinuous cochains in a context which has a flavor of the preceding Section 4.3.2.

4.3.3.1. *Nambu mechanics and its "Fock–Zariski" quantization.* We mention this aspect here mainly as an example of generalized deformation. Details can be found in [DFST], so we shall just briefly indicate a few highlights.

In 1973 Nambu published some calculations which he had made a dozen years before: with quarks in the back of his mind he started with a kind of "Hamilton equations" on $\mathbb{R}^3$ with two "Hamiltonians" g, h functions of r. In this new mechanics the evolution of a function f on $\mathbb{R}^3$ is $\frac{df}{dt} = \frac{\partial(f,g,h)}{\partial(x,y,z)}$, a 3-bracket, where the right-hand side is the Jacobian of the mapping $\mathbb{R}^3 \to \mathbb{R}^3$ given by $(x, y, z) \mapsto (f, g, h)$. That expression was easily generalized to n functions f_i, $i = 1, \dots, n$. One introduces an n-tuple of functions on $\mathbb{R}^n$ with composition law given by their Jacobian, linear canonical transformations $\mathrm{SL}(n, \mathbb{R})$ and a corresponding $(n-1)$-form which is the analogue of the Poincaré–Cartan integral invariant. The Jacobian has to be interpreted as a generalized Poisson bracket: It is skew-symmetric with respect to the f_i's, satisfies an identity (called the fundamental identity) which is an analogue of the Jacobi identity and is a derivation of the algebra of smooth functions on $\mathbb{R}^n$ (i.e., the Leibniz rule is verified in each argument). There is a complete analogy with the Poisson bracket formulation of Hamilton equations, including the important fact that the components of the $(n-1)$-tuple of "Hamiltonians" $(f_2, \dots, f_n)$ are constants of motion. Shortly afterwards it was shown that Nambu mechanics could be seen as a coming from constrained Hamiltonian mechanics; e.g. for $\mathbb{R}^3$ one starts with $\mathbb{R}^6$ and an identically vanishing Hamiltonian, takes a pair of second class constraints to reduce it to some $\mathbb{R}^4$ and one more first-class Dirac constraint, together with time rescaling, will give the reduction. This "chilled" the domain for almost 20 years – and gives a physical explanation to the fact that Nambu could not go beyond Heisenberg quantization.

In order to quantize the Nambu bracket, a natural idea is to replace, in the definition of the Jacobian, the pointwise product of functions by a deformed product. For this to make sense, the deformed product should be Abelian, so we are lead to consider commutative DrG-deformations of an associative and commutative product. Looking first at polynomials we are lead to the commutative part of Hochschild cohomology called Harrison cohomology, which is trivial. Dealing with polynomials, a natural idea is to factorize them and take symmetrized star-products of the irreducible factors. More precisely we introduce an operation α which maps a product of factors into a symmetrized tensor product (in a kind of Fock space) and an evaluation map T which replaces tensor product by star-product. Associativity will be satisfied if α annihilates the deformation parameter λ (there are still λ-dependent terms in a product due to the last action of T); intuitively one can think of a deformation parameter which is λ times a Dirac γ matrix. This fact brought us to generalized deformations, but even that was not enough. Dealing with distributivity of the product with respect to addition, and with derivatives, posed difficult problems. In the end we took for observables Taylor developments of elements of the algebra of the semi-group generated by irreducible polynomials ("polynomials over polynomials", inspired by second quantization techniques) and were then able to perform a meaningful quantization of these Nambu–Poisson brackets (cf. [DFS] for more details and subsequent developments). We call this approach "Fock–Zariski" to underline the decomposition of polynomials into irreducible factors and the role of these factors, similar to one-particle states in Fock quantization. Related cohomologies were studied [Ga].

4.3.3.2. *Generalized deformations.* The fact that in the Fock–Zariski quantization, the deformation parameter behaves almost as if it was nilpotent, has very recently induced Pinczon [Pi] and Nadaud [NaD] to generalize the Gerstenhaber theory to the case of a deformation parameter which *does not commute with the algebra.* For instance one can have [Pi], for $\tilde{a} = \sum_n a_n \lambda^n$, $a_n \in A$, a left multiplication by λ of the form $\lambda \cdot \tilde{a} = \sum_n \sigma(a_n)\lambda^{n+1}$ where σ is an endomorphism of A. A similar theory can be done in this case, with appropriate cohomologies. While that theory does not yet reproduce the above mentioned Nambu quantization, it gives new and interesting results. In particular [Pi], while the Weyl algebra W_1 (generated by the Heisenberg Lie algebra $\mathfrak{h}_1$) is known to be DrG-rigid, it can be nontrivially deformed in such a *supersymmetric deformation theory* to the supersymmetry enveloping algebra $\mathcal{U}(\mathfrak{osp}(1,2))$. More recently [NaD], on the polynomial algebra $\mathbb{C}[x, y]$ in 2 variables, Moyal-like products of a new type were discovered; a more general situation was studied, where the relevant Hochschild cohomology is still valued in the algebra but with "twists" on both sides for the action of the deformation parameter on the algebra. Though this more balanced generalization of deformations also does not (yet) give Nambu mechanics quantization, it opens a whole new direction of research for deformation theory. This is another example of a physically motivated study which goes beyond a generally accepted framework and opens new perspectives.

4.3.3.3. *Singularities.* We have already mentioned the case of manifolds with boundaries or with corners, where deformation quantization can be extended [NT], as was the pseudodifferential calculus by the Melrose b-calculus [Me]. More general situations are being studied. An interesting new idea, where a nontrivial Harrison cohomology can be expected, is to try and use noncontinuous cochains. When the manifold has singularities (like a cone, a most elementary example), this is a reasonable thing to do [Fr].

Acknowledgement. The authors wish to thank Gilles Halbout for his persistent and very persuasive efforts that brought them to eventually write down this panorama.

References

[AW] Agarwal, G. S., and Wolf, E., Calculus for functions of noncommuting operators and general phase-space methods in quantum mechanics I, II, III, *Phys. Rev.* D2 (1970), 2161–2186, 2187–2205, 2206–2225.

[ADS] Andler, M., Dvorsky, A., and Sahi, S., Kontsevich quantization and invariant distributions on Lie groups, `math.QA/9910104` (1999).

[ACG] Arnal, A., Cahen, M., and Gutt, S., Representations of compact Lie groups and quantization by deformation, *Bull. Acad. Royale Belg.* 74 (1988), 123–141; Star exponential and holomorphic discrete series, *Bull. Soc. Math. Belg.* 41 (1989), 207–227.

[AC] Arnal, D., and Cortet, J.-C., Nilpotent Fourier transform and applications, *Lett. Math. Phys.* 9 (1985), 25–34; Star-products in the method of orbits for nilpotent Lie groups, *J. Geom. Phys.* 2 (1985), 83–116; Représentations star des groupes exponentiels, *J. Funct. Anal.* 92 (1990), 103–135.

[ACL] Arnal, D., Cortet, J.-C., and Ludwig, J., Moyal product and representations of solvable Lie groups, *J. Funct. Anal.* 133 (1995), 402–424.

[ACM] Arnal, D., Cortet J.-C., and Molin, P., Star-produit et représentation de masse nulle du groupe de Poincaré, *C.R. Acad. Sci. Paris Sér. A* 293, 309–312 (1981).

[ACMP] Arnal, D., Cortet, J.-C., Molin, P., and Pinczon, G., Covariance and geometrical invariance in star-quantization, *J. Math. Phys.* 24 (1983), 276–283.

[AMM] Arnal, D., Manchon, D., and Masmoudi, M., Choix des signes pour la formalité de M. Kontsevich, `math/0003003` (2000).

[AS] Atiyah, M. F., and Singer, I. M., The index of elliptic operators on compact manifolds, *Bull. Amer. Math. Soc.* 69 (1963) 422–433.

[BFLS] Basart, H., Flato, M., Lichnerowicz, A., and Sternheimer, D., Deformation theory applied to quantization and statistical mechanics, *Lett. Math. Phys.* 8 (1984), 483–494.

[BFFLS] Bayen, F., Flato, M., Frønsdal, C., Lichnerowicz, A., and Sternheimer, D., Quantum mechanics as a deformation of classical mechanics, *Lett. Math. Phys.* 1 (1977), 521–530, Deformation theory and quantization I, II, *Ann. Phys.* (NY) 111 (1978), 61–110, 111–151.

[Be] Berezin, F. A., General concept of quantization, *Comm. Math. Phys.* 40 (1975), 153–174; Quantization, *Izv. Akad. Nauk SSSR Ser. Mat.* 38 (1974), 1116–1175; Quantization in complex symmetric spaces, *Izv. Akad. Nauk SSSR Ser. Mat.* 39 (1975), 363–402, 472. [English translations: *Math. USSR-Izv.* 38 (1975) 1109–1165 and 39 (1976) 341–379].

[BS] Berezin, F. A., and Šubin, M. A., Symbols of operators and quantization, in *Hilbert space operators and operator algebras* (Tihany 1970), Colloq. Math. Soc. János Bolyai 5, North-Holland, Amsterdam 1972, 21–52 .

[BCG] Bertelson, M., Cahen, M., and Gutt, S., Equivalence of star products, *Class. Quantum Grav.* 14 (1997), A93–A107.

[BB] Bertrand, J., and Bertrand, P., Symbolic calculus on the time-frequency half-plane, *J. Math. Phys.* 39 (1998), 4071–4090.

[Bon] Bonneau, P., Fedosov star products and one-differentiable deformations, *Lett. Math. Phys.* **45** (1998), 363–376.

[BFGP] Bonneau, P., Flato, M., Gerstenhaber, M., and Pinczon, G., The hidden group structure of quantum groups: strong duality, rigidity and preferred deformations, *Comm. Math. Phys.* 161 (1994), 125–156.

[Bm] Bordemann, M., The deformation quantization of certain super-Poisson brackets and BRST cohomology, in *Quantization, Deformations, and Symmetries* (Dito, G., Sternheimer, D., eds.), Conférence Moshé Flato 1999, Math. Phys. Stud. 22, Kluwer Academic Publishers, Dordrecht, Boston, London 2000, 45–68.

[BRW] Bordemann, M., Römer, H., and Waldmann, S., A Remark on Formal KMS States in Deformation Quantization, *Lett. Math. Phys.* 45 (1998), 49–61, math/9801139.

[Bo] Boutet de Monvel, L., Complex star algebras, *Math. Phys. Anal. Geom.* 2 (1999), 113–139.

[BG] Boutet de Monvel, L., and Guillemin, V., *The spectral theory of Toeplitz operators*, Ann. of Math. Stud. 99, Princeton University Press, Princeton 1981.

[CFGS] Cahen M., Flato M., Gutt, S., and Sternheimer, D., Do different deformations lead to the same spectrum?, *J. Geom. Phys.* 2 (1985), 35–48.

[CGR] Cahen M., Gutt S., and Rawnsley J., Quantization of Kähler manifolds IV, *Lett. Math. Phys.* 34 (1995), 159–168.

[CF] Cattaneo, A. S., and Felder, G., A path integral approach to the Kontsevich quantization formula, *Comm. Math. Phys.* 212 (2000), 591–611; On the globalization of Kontsevich's star product and the perturbative Poisson sigma model, hep-th/0111028 (2001).

[CFT] Cattaneo, A. S., Felder, G., and Tomassini, L., From local to global deformation quantization of Poisson manifolds, math.QA/0012228 (2000); Fedosov connections on jet bundles and deformation quantization, in *Deformation Quantization* (G. Halbout, ed.), IRMA Lectures in Math. Theoret. Phys. 1, Walter de Gruyter, Berlin 2002, 191–202, math.QA/0111290 (2001).

[CE] Chevalley, C., and Eilenberg, E., Cohomology theory of Lie groups and algebras, *Trans. Amer. Math. Soc.* 63 (1948), 85–124.

[Co] Connes, A. *Noncommutative Geometry*, Academic Press, San Diego 1994; Noncommutative differential geometry, *Inst. Hautes Études Sci. Publ. Math.* 62 (1985), 257–360.

[CDV] Connes, A., and Dubois-Violette, M., Noncommutative finite-dimensional manifolds. I. Spherical manifolds and related examples, math.QA/0107070 (2001).

[CFS] Connes, A., Flato, M., and Sternheimer, D., Closed star-products and cyclic cohomology, *Lett. Math. Phys.* 24 (1992), 1–12.

[CK] Connes, A., and Kreimer, D., Renormalization in quantum field theory and the Riemann-Hilbert problem. I: The Hopf algebra structure of graphs and the main theorem; II: The β-function, diffemorphisms and the renormalization group, *Comm. Math. Phys.* 210 (200), 249–273; 216 (2001), 215–241.

[De] Deligne, P., Déformations de l'Algèbre des Fonctions d'une Variété Symplectique: Comparaison entre Fedosov et De Wilde, Lecomte, *Selecta Math.* (N.S.) 1 (1995), 667–697.

[DL] De Wilde, M., and Lecomte, P. B. A., Existence of star-products and of formal deformations of the Poisson Lie algebra of arbitrary symplectic manifolds, *Lett. Math. Phys.* 7 (1983), 487–496.

[DLG] De Wilde, M., Lecomte, P. B. A., and Gutt, S., A propos des deuxième et troisième espaces de cohomologie de l'algèbre de Lie de Poisson d'une variété symplectique, *Ann. Inst. H. Poincaré Sect. A* (N.S.) 40 (1984), 77–83.

[Di] Dirac, P.,A.,M. *Lectures on Quantum Mechanics*, Belfer Graduate School of Sciences Monogr. Ser. 2, Yeshiva University, New York 1964; Generalized Hamiltonian dynamics, *Canad. J. Math.* 2 (1950), 129–148.

[Dit] Dito, J., Star-product approach to quantum field theory: the free scalar field, *Lett. Math. Phys.* 20 (1990), 125–134; Star-products and nonstandard quantization for K-G equation, *J. Math. Phys.* 33 (1992), 791-801; An example of cancellation of infinities in star-quantization of fields, *Lett. Math. Phys.* 27 (1993), 73–80.

[Dth] Dito, J. *Star-produits en dimension infinie: le cas de la théorie quantique des champs*, Thesis, Université de Bourgogne, Dijon 1993; Deformation quantization of covariant fields, in *Deformation Quantization* (G. Halbout, ed.), IRMA Lectures in Math. Theoret. Phys. 1, Walter de Gruyter, Berlin 2002, 55–66.

[DFS] Dito, G., Flato, M., and Sternheimer, D., Nambu mechanics, n-ary operations and their quantization, in *Deformation Theory and Symplectic Geometry* (D. Sternheimer, J. Rawnsley and S. Gutt, eds.), Math. Physics Stud. 20, Kluwer Academic Publishers, Dordrecht 1997, 43–66.

[DFST] Dito, G., Flato, M., Sternheimer, D., and Takhtajan, L., Deformation Quantization and Nambu Mechanics, *Comm. Math. Phys.* 183 (1997) 1–22.

[DN] Douglas, M. R., and Nekrasov, N. A., Noncommutative field theory, `hep-th/0106048` (2001), to be published in *Rev. Mod. Phys.*

[Dr] Drinfeld, V., Quantum Groups, in *Proceedings ICM86, Berkeley*, Vol. 1, Amer. Math. Soc., Providence 1987, 101–110.

[ES] Etingof, P., and Schiffmann, O., *Lectures on quantum groups*, Lectures in Mathematical Physics, International Press, Boston 1998.

[FRT] Faddeev, L. D., Reshetikhin, N. Yu., and Takhtajan, L. A., Quantization of Lie groups and Lie algebras, in *Algebraic analysis*, Vol. I, Academic Press, Boston 1988, 129–139.

[Fe1] Fedosov, B.,V., Formal quantization, in *Some Topics of Modern Mathematics, and Their Application to Problems of Mathematical Physics*, Moscow 1985, 129–136; Quantization and index, *Dokl. Akad. Nauk SSSR* 291 (1986), 82–86.

[Fe] Fedosov, B.,V., A simple geometrical construction of deformation quantization, *J. Differential Geom.* 40 (1994), 213–238.

[Fe3] Fedosov, B. V. *Deformation quantization and index theory*. Math. Top. 9., Akademie Verlag, Berlin 1996.

[Fe4] Fedosov, B. V., The Atiyah-Bott-Patodi method in deformation quantization, *Comm. Math. Phys.* 209 (2000), 691–728.

[Fl] Flato, M., Deformation view of physical theories, *Czechoslovak J. Phys.* B32 (1982), 472–475; Two disjoint aspects of the deformation programme: quantizing Nambu mechanics; singleton physics, in *Particles, Fields, and Gravitation, Łodz 1998* (J. Rembieliński, ed.), AIP Press, New York 1998, 49–52, hep-th/9809073.

[FLS1] Flato, M., Lichnerowicz, A., and Sternheimer, D., Déformations 1-différentiables d'algèbres de Lie attachées à une variété symplectique ou de contact, *C.R. Acad. Sci. Paris Sér. A* 279 (1974), 877–881; *Compositio Math.* 31 (1975), 47–82.

[FLSq] Flato, M., Lichnerowicz, A., and Sternheimer, D., Crochets de Moyal-Vey et quantification, *C.R. Acad. Sci. Paris Sér. A* 283 (1976), 19–24.

[FST] Flato, M., Simon, J. C. H., and Taflin, E., The Maxwell-Dirac equations: the Cauchy problem, a symptotic completeness and the infrared problem, *Mem. Amer. Math. Soc.* 127 (no. 606), 1997.

[Fr] Frønsdal, C., Harrison cohomology and abelian deformation quantization on algebraic varieties, in *Deformation Quantization* (G. Halbout, ed.), IRMA Lectures in Math. Theoret. Phys. 1, Walter de Gruyter, Berlin 2002, 151–163.

[Ga] Gautheron, P., Some remarks concerning Nambu mechanics, *Lett. Math. Phys.* 37 (1996), 103–116; Simple facts concerning Nambu algebras, *Comm. Math. Phys.* 195 (1998), 417–434.

[Ge] Gerstenhaber, M., The cohomology structure of an associative ring, *Ann. Math.* 78 (1963), 267–288; On the deformation of rings and algebras, *ibid.* 79 (1964), 59–103; and (IV), *ibid.* 99 (1974), 257–276.

[GG] Gerstenhaber, M., and Giaquinto, A., Compatible deformations, in *Trends in the Representation Theory of Finite Dimensional Algebras* (E. L. Green and B. Huisgen-Zimmermann, eds.), Contemp. Math. 229, American Mathematical Society, Providence, RI, 1998, 159–168.

[GGS] Gerstenhaber, M., Giaquinto, A., and Schaps, M. E., The Donald-Flanigan Problem for Finite Reflection Groups, *Lett. Math. Phys.* 56 (2001), 41–72.

[GS] Gerstenhaber, M., and Schack, S. D., Algebraic cohomology and deformation theory, in *Deformation Theory of Algebras and Structures and Applications* (M. Hazewinkel and M. Gerstenhaber, eds.), NATO ASI Ser. C 247, Kluwer Academic Publishers, Dordrecht 1988, 11–264.

[GZ] Giaquinto, A., and Zhang, J. J., Quantum Weyl Algebras, *J. Algebra* 176 (1995), 861–881.

[Gu1] Gutt, S., Equivalence of deformations and associated $\star$-products, *Lett. Math. Phys.* 3 (1979), 297–309.

[Gth] Gutt, S., *Déformations formelles de l'algèbre des fonctions différentiables sur une variété symplectique*, Thesis, Université Libre de Bruxelles 1980.

[Gu2] Gutt, S., Second et troisième espaces de cohomologie différentiable de l'algèbre de Lie de Poisson d'une variété symplectique, *Ann. Inst. H. Poincaré Sect. A* (N.S.) 33 (1980), 1–31.

[Gu3] Gutt, S., An explicit $\star$-product on the cotangent bundle of a Lie group, *Lett. Math. Phys.* 7 (1983), 249–258.

[Gu] Gutt, S., Variations on deformation quantization, in *Quantization, Deformations, and Symmetries* (Dito, G., Sternheimer, D., eds.), Conférence Moshé Flato 1999, Math. Phys. Stud. 21, Kluwer Academic Publishers, Dordrecht, Boston, London 2000, 217–254,

[GR] Gutt, S., and Rawnsley, J., Equivalence of star-products on a symplectic manifold: an introduction to Deligne's Čech cohomology class, *J. Geom. Phys.* 29 (1999), 347–392.

[Gr] Groenewold, A., On the principles of elementary quantum mechanics, *Physica* 12 (1946), 405–460.

[Ha] Halbout, G., Calcul d'un invariant de star produit fermé sur une variété symplectique, *Comm. Math. Phys.* 205 (1999), 53–67.

[HKR] Hochschild, G., Kostant, B., and Rosenberg, A., Differential forms on regular affine algebras, *Trans. Amer. Math. Soc.* 102 (1962), 383–406.

[Ik] Ikeda, N., Two-dimensional gravity and nonlinear gauge theory, *Ann. Phys.* 235 (1994), 435–464.

[WI] Inönü, E., and Wigner, E. P., On the contraction of groups and their representations, *Proc. Nat. Acad. Sci. U. S. A.* 39 (1953), 510–524.

[Ji] Jimbo, M., A q-difference algebra of $\mathcal{U}(\mathfrak{g})$ and the Yang-Baxter equation, *Lett. Math. Phys.* 10 (1985), 63–69.

[Ka] Karabegov, A. V., Cohomological classification of deformation quantizations with separation of variables, *Lett. Math. Phys.* 43 (1998), 347–357; Berezin's quantization on flag manifolds and spherical modules, *Trans. Amer. Math. Soc.* 350 (1998), 1467–1479.

[KS] Kodaira, K., and Spencer, D. C., On deformations of complex analytic structures, *Ann. Math.* 67 (1958), 328–466.

[Ko1] Kontsevich, M., Formality conjecture, in *Deformation theory and symplectic geometry, Ascona 1996* (D. Sternheimer, J. Rawnsley and S. Gutt, eds.), Math. Phys. Stud. 20, Kluwer Academic Publishers, Dordrecht 1997, 139–156.

[Ko2] Kontsevich, M., Deformation quantization of Poisson manifolds I, `q-alg/9709040` (1997).

[Ko3] Kontsevich, M., Operads and motives in deformation quantization, *Lett. Math. Phys.* 48 (1999), 35–72.

[Kper] Kontsevich, M., Periods, Mathématique et physique 28–39, SMF Journ. Annu. 1999, Soc. Math. France, Paris 1999.

[KoS] Kontsevich, M., and Soibelman, Y., Deformations of algebras over operads and the Deligne conjecture, in *Quantization, Deformations, and Symmetries* (Dito, G., Sternheimer, D., eds.), Conférence Moshé Flato 1999, Math. Phys. Stud. 22, Kluwer Academic Publishers, Dordrecht, Boston, London 2000, 255–307.

[Ko4] Kontsevich, M., Deformation quantization of algebraic varieties, *Lett. Math. Phys.* 56 (2001), 271–294.

[Ko74] Kostant, B., Symplectic spinors, in *Symposia Mathematica* 14, Academic Press London, 1974, 139–152.

[Ko] Kostant, B., Quantization and unitary representations. I: Prequantization, in: Lecture Notes in Math. 170, Springer-Verlag, Berlin 1970, 87–208; On the definition of quantization, in *Géométrie symplectique et physique mathématique* (Colloq. Int. CNRS, N^{0} 237), Éds. CNRS, Paris 1975, 187–210; Graded manifolds, graded Lie theory, and prequantization, in *Differential geometrical methods in mathematical physics* (K. Beuler and A. Reetz, eds.), Lecture Notes in Math. 570, Springer-Verlag, Berlin 1977, 177–306.

[KR] Kulish, P. P., and Reshetikhin, N. Yu., Quantum linear problem for the sine-Gordon equation and higher representations, *Zap. Nauch. Sem. LOMI* 101 (1981), 101–110 (English translation in *J. Sov. Math.* 23 (1983), 24–35).

[Le] Lecomte, P. B. A., Application of the cohomology of graded Lie algebras to formal deformations of Lie algebras, *Lett. Math. Phys.* 13 (1987), 157–166.

[Li1] Lichnerowicz, A., Cohomologie 1-différentiable des algèbres de Lie attachées à une variété symplectique ou de contact, *J. Math. Pures Appl.* 53 (1974), 459–483.

[LiP] Lichnerowicz, A., Les variétés de Poisson et leurs algèbres de Lie associées, *J. Diff. Geom.* 12 (1977), 253–300; Variétés de Poisson et feuilletages, *Ann. Fac. Sci. Toulouse Math.* (5) 4 (1983), 195–262.

[Maj] Majid, S., *Foundations of quantum group theory*, Cambridge University Press, Cambridge 1995.

[MMR] Manchon, D., Masmoudi, M., and Roux, A., On quantization of quadratic Poisson structures, *Comm. Math. Phys.* 225 (2002), 121–130, `math.QA/0105068` (2001).

[Ma] Manin, Yu., Theta functions, quantum tori and Heisenberg groups, *Lett. Math. Phys.* 56 (2001), 295–320.

[Mas] Masmoudi, M., Tangential formal deformations of the Poisson bracket and tangential star products on a regular Poisson manifold, *J. Geom. Phys.* 9 (1992), 155–171.

[Me] Melrose, R. *The Atiyah-Patodi-Singer index theorem.*) Res. Notes . Math. 4, A K Peters Ltd., Wellesley, MA, 1993.

[Mr] Moreno, C., Invariant star products and representations of compact semi-simple Lie groups, *Lett. Math. Phys.* 12 (1986), 217–229.

[Mo] Moyal, J. E., Quantum mechanics as a statistical theory, *Proc. Cambridge Philos. Soc.* 45 (1949), 99–124.

[NaD] Nadaud F., Generalized deformations, Koszul resolutions, Moyal products, *Rev. Math. Phys.* 10 (1998), 685–704; *Thèse*, Dijon (janvier 2000); Generalized Deformations and Hochschild Cohomology, *Lett. Math. Phys.* 58 (2001), 41–55.

[Na] Nadaud, F., On continuous and differentiable Hochschild cohomology, *Lett. Math. Phys.* 47 (1999), 85–95.

[NT] Nest, R., and Tsygan, R., Algebraic index theorem, *Comm. Math. Phys.* 172 (1995), 223–262; Algebraic index theorem for families, *Adv. Math.* 113 (1995), 151–205; Formal deformations of symplectic manifolds with boundary, *J. Reine Angew. Math.* 481 (1996), 27–54.

[NV] Neroslavsky, O. M., and Vlasov, A. T., Sur les déformations de l'algèbre des fonctions d'une variété symplectique, *C. R. Acad. Sc. Paris Sér. I* 292 (1981), 71–76.

[OMY1] Omori, H., Maeda, Y., and Yoshioka, A., Weyl manifolds and deformation quantization, *Adv. Math.* 85 (1991), 225–255.

[OMY2] Omori, H., Maeda, Y., and Yoshioka, A., Existence of a closed star product, *Lett. Math. Phys.* 26 (1992), 285–294.

[OMY3] Omori, H., Maeda, Y., and Yoshioka, A., Deformation quantizations of Poisson algebras, *Contemp. Math.* 179 (1994), 213–240.

[Pi] Pinczon, G., Non commutative deformation theory, *Lett. Math. Phys.* 41 (1997), 101–117.

[Po] Polyak, M., Quantization of linear Poisson structures and degrees of maps, to appear in *Lett. Math. Phys.* (2001).

[SW] Schmid, W., Character formulas and localization of integrals, in *Deformation theory and symplectic geometry, Ascona 1996* (D. Sternheimer, J. Rawnsley and S. Gutt, eds.), Math. Phys. Stud. 20, Kluwer Academic Publishers, Dordrecht 1997, 259–270.

[Se] Segal, I. E., Symplectic structures and the quantization problem for wave equations, in *Symposia Math.* 14, Academic Press, London 1974, 79–117.

[SSt] Shnider, S., and Sternberg, S., *Quantum Groups*, Vol. II, Grad. Texts in Math. Phys., International Press, Boston & Hong-Kong 1993.

[Sho] Shoikhet, B., Vanishing of the Kontsevich Integrals of the Wheels, *Lett. Math. Phys.* 56 (2001), 141–149.

[Sta] Stasheff, J., Deformation theory and the Batalin-Vilkovisky master equation, in *Deformation theory and symplectic geometry, Ascona 1996* (D. Sternheimer, J. Rawnsley and S. Gutt, eds.), Math. Phys. Stud. 20, Kluwer Academic Publishers, Dordrecht 1997, 271–284.

[SD] Sternheimer, D., Deformation quantization: Twenty years after, in *Particles, Fields, and Gravitation, Łodz 1998* (J. Rembieliński, ed.), AIP Press, New York 1998, 107–145, `math.QA/9809056`; In retrospect: A personal view on Moshé Flato's scientific legacy, in *Quantization, Deformations, and Symmetries* (Dito, G., Sternheimer, D., eds.), Conférence Moshé Flato 1999, Math. Phys. Stud. 22, Kluwer Academic Publishers, Dordrecht, Boston, London 2000, 31–41.

[Ta] Tamarkin, D. E., Topological invariants of connections on symplectic manifolds, *Funct. Anal. Appl.* 29 (1996), 258–267.

[Tam] Tamarkin, D. E., Another proof of M. Kontsevich formality theorem, `math.QA/9803025` (1998).

[TaT] Tamarkin, D. E., and Tsygan, B., Cyclic Formality and Index Theorems, *Lett. Math. Phys.* 56 (2001), 85–97.

[U2] Unterberger, A. & J., Quantification et analyse pseudo-différentielle, *Ann. Sci. École Norm. Sup.* (4) 21 (1988), 133–158; La série discrète de SL(2, $\mathbb{R}$) et les opérateurs pseudo-différentiels sur une demi-droite, *Ann. Sci. Éc. Norm. Sup.* (4) 17 (1984), 83–116.

[UU] Unterberger, A., and Upmeier, H., *Pseudodifferential analysis on symmetric cones*, Stud. Adv. Math., CRC Press, Boca Raton, FL, 1996; The Berezin transform and invariant differential operators, *Comm. Math. Phys.* 164 (1994), 563–597.

[Ve] Vey, J., Déformation du crochet de Poisson sur une variété symplectique, *Comment. Math. Helv.* 50 (1975), 421–454.

[Wb] Weinstein, A., Deformation quantization, Séminaire Bourbaki, exposé 789 (juin 1994), *Astérisque* 227, 389–409.

[Wm] Weinstein, A., The modular automorphism group of a Poisson manifold, *J. Geom. Phys.* 23 (1997), 379–394.

[We] Weyl, H., *The theory of groups and quantum mechanics*, Dover, New-York 1931, translated from *Gruppentheorie und Quantenmechanik*, Hirzel Verlag, Leipzig 1928; Quantenmechanik und Gruppentheorie, *Z. Physik* 46 (1927), 1–46.

[Wi] Wigner, E. P., Quantum corrections for thermodynamic equilibrium, *Phys. Rev.* 40 (1932), 749–759.

[Wo] Woodhouse, N. M. J., *Geometric quantization*, 2nd edition, Oxford Science Publications, Oxford Math. Monogr.. Oxford University Press, New York 1992.

[Xu] Xu, P., Fedosov $\star$-Products and Quantum Momentum Maps, *Comm. Math. Phys.* 197 (1998) 167–197.

Deformation quantization of covariant fields

Giuseppe Dito

Laboratoire Gevrey de Mathématique Physique, CNRS UMR 5029
Département de Mathématiques, Université de Bourgogne
BP 47870, F-21078 Dijon Cedex, France
e-mail: `Giuseppe.Dito@u-bourgogne.fr`

Abstract. After sketching recent advances and subtleties in classical relativistically covariant field theories, we give in this short Note some indications as to how the deformation quantization approach can be used to solve or at least give a better understanding of their quantization.

2000 Mathematics Subject Classification: 53D55, 81T70, 81R20, 35G25

1 Introduction

We would like to discuss here a possible application of deformation quantization, viz. a nonperturbative construction of star-products (along the lines indicated in [Di93]), to the quantization of interacting field theory. It is based on recent advances in the linearization programme for covariant wave equations proposed in [FS80a, FS80b] that uses nonlinear representation theory [FPS77]. We shall present this approach on an example. It can now be formulated in a more general setting thanks to results obtained by Flato, Simon, and Taflin concerning the existence of global solutions, scattering theory, and linearization of several covariant wave equations relevant to physics [ST85, FST87, ST93, FST97] (see [ST00] for a recent nontechnical survey of these results).

2 A sketch of the linearization programme

We shall give only the most basic facts and describe a few results, relevant for our discussion of the linearization programme [FS80a, FS80b] for covariant nonlinear equations. We skip the technical details: the reader can find them in the references listed at the end of the paper. At this stage we shall only briefly present the context.

For the reader's, convenience, we have collected in an appendix the basic notions on nonlinear representation theory of Lie groups and algebras introduced and developed in [FPS77] and some later works.

Let G be a Lie group and E be some vector space of functions (endowed with a Banach or Fréchet topology). We assume that there is an analytic action of G on a neighborhood $\mathcal{O}$ of the origin in E and that $0 \in E$ is a fixed point for the action. Let $H(\mathcal{O}, \mathcal{O})$ be the space of analytic maps from $\mathcal{O}$ to $\mathcal{O}$ [1]. Hence there is a map $g \in G \mapsto S_g \in H(\mathcal{O}, \mathcal{O})$ such that $S_g(\mathcal{O}) \subset \mathcal{O}$, $S_{g_1 g_2} = S_{g_1} S_{g_2}$, $S_{g^{-1}} = S_g^{-1}$, and $S_g(0) = 0$. The pair $(S, \mathcal{O})$ is called here an *analytic nonlinear representation* of G in $\mathcal{O}$.

At the infinitesimal level, by differentiating the action of $S_{\exp(tX)}$, where X is an element of the Lie algebra $\mathfrak{g}$ of G, one would expect to find "vector fields" T_X on $\mathcal{O}$ (or on a open subset of it) satisfying $T_{[X,Y]} = T_X T_Y - T_Y T_X$. The pair $(T, \mathcal{O})$ is called an *analytic nonlinear Lie algebra representation* of $\mathfrak{g}$ on $\mathcal{O}$.

One may think of the maps S and T be given by their Taylor coefficients: $S = \sum_{n\geq 1} S^n$ and $T = \sum_{n\geq 1} T^n$. Then the first order term S^1 (resp. T^1) is a linear representation of G (resp. $\mathfrak{g}$). Nonlinear representations are built by successive extensions of a linear representation by its successive tensor powers: they can be viewed as a kind of deformation of their linear part. Interactions described by covariant nonlinear equations (see below) can then be seen, as they should, as deformations of the free (noninteracting) situation, in line with Flato's deformation philosophy [Fl82].

2.1 Covariant nonlinear equations

We shall be concerned with relativistic equations on Minkowski spacetime with direct physical relevance and "covariant" will always mean covariant under the action of the Poincaré group $\mathcal{P} = \mathrm{SL}(2, \mathbb{C}) \ltimes \mathbb{R}^4$. The Lie algebra of $\mathcal{P}$ will be denoted by $\mathfrak{p} = \mathfrak{sl}(2, \mathbb{C}) \dot{+} \mathbb{R}^4$. We use standard conventions: Einstein summation convention is used throughout for upper and lower indices (Latin indices run from 1 to 3 and Greek ones from 0 to 3). The metric tensor $g_{\mu\nu}$ has signature $(+---)$. For convenience of notation we shall write $x \in \mathbb{R}^4$ as $x = (t, \vec{x})$, where $t = x^0$, $\vec{x} = (x^1, x^2, x^3)$. For a function $\phi\colon \mathbb{R}^4 \to V$ where V is a vector space, we set $\phi(t)\colon \mathbb{R}^3 \to V$ by $\phi(t)(\vec{x}) \equiv \phi(t, \vec{x}) \equiv \phi(x)$. An element $g \in \mathcal{P}$ is written as $g = (\Lambda, a)$ where $\Lambda \in \mathrm{SL}(2, \mathbb{C})$ and $a \in \mathbb{R}^4$. Let $\Lambda \mapsto \tilde{\Lambda}$ be the canonical homomorphism of $\mathrm{SL}(2, \mathbb{C})$ onto the group $\mathrm{SO}_0(1, 3)$. Then the group law in $\mathcal{P}$ is $(\Lambda_1, a_1).(\Lambda_2, a_2) = (\Lambda_1\Lambda_2, a_1 + \tilde{\Lambda}_1 a_2)$. The natural action of the Poincaré group on spacetime will be denoted by $g\cdot x = \tilde{\Lambda}x + a$. Finally, the standard basis of $\mathfrak{p}$ is $\{P_\mu, M_{\alpha\beta}\}$ with $\alpha < \beta$.

Most of the classical wave equations met in physics can be cast into evolution equations governed by some Hamiltonian operator (see e.g. [RS79]). Suppose that one starts with a vector space E of initial data (usually a Banach or Fréchet space).

[1]This kind of holomorphic mappings space between topological vector spaces can be given several topologies and the theory is well developed, see [Din99].

For simplicity, elements of E are functions defined on the whole of $\mathbb{R}^3$ and take their values in a finite-dimensional vector space V. Let H be an operator on E given by a (non)linear differential operator such that $H(0) = 0$. We are interested in the solutions of the following evolution equation:

$$\frac{d\psi(t)}{dt} = H(\psi(t)), \quad \psi(0) = \psi_0 \in E. \tag{2.1}$$

Now suppose that the previous evolution equation is associated to some covariant nonlinear (wave) equation. One has a finite-dimensional representation of SL(2, $\mathbb{C}$) in V, $\Lambda \mapsto A(\Lambda)$, and given a solution ψ of Eq. (2.1), then $\psi^g(x) \equiv A(\Lambda)\psi(\tilde{\Lambda}^{-1}(x-a))$ is also a solution for any $g = (\Lambda, a) \in \mathcal{P}$. There is a local action S of $\mathcal{P}$ on E having the origin in E as fixed point and given by $S_g(\psi_0)(\vec{x}) = \psi^g(0, \vec{x})$, where $\psi_0(\vec{x}) = \psi(0, \vec{x})$, i.e., $S_{g_1 g_2} = S_{g_1} S_{g_2}$. In particular, if $t \mapsto U_t$ is the evolution operator of Eq. (2.1), $U_{t+t'} = U_t U_{t'}$, $\psi(t) = U_t(\psi_0)$ and we have $U(t) = S_{\exp(tP_0)}$, where $P_0 \in \mathfrak{p}$ is the generator of time translations. Hence we can associate to the covariant equation (2.1) what looks very much like a nonlinear representation of $\mathcal{P}$ in E. Conversely, if one has a nonlinear representation (S, E) of $\mathcal{P}$, by formally differentiating $\psi(t) = S_{\exp(tP_0)}(\psi(0))$ with respect to t, one would expect to recover a covariant evolution equation as in Eq. (2.1).

The linearization programme [FS80a, FS80b] is based on the idea to take into account the full covariance of a nonlinear equation by associating to it a nonlinear representation, and then study its linearizability. In the case where the nonlinear representation is linearizable, i.e., equivalent to its linear part which corresponds to the associated linear (free) equation, then one can directly deduce global existence in time of the solutions of the nonlinear equation.

As one would expect from deformation theory, the linearization involves the computation of some cohomology spaces. It is remarkable that formal nonlinear massive representations of the Poicaré group are always formally linearizable [Ta84]. The covariance provides the right spaces where to perform the study (spaces of differentiable or analytic vectors for the linear part) and then to look for convergence of the expressions involved when going to the analytic setting.

Notice that the Maxwell–Dirac system in $(1 + 3)$ dimensions (i.e., classical electrodynamics) presents many subtleties due to the zero mass of the photon. Even at the formal level, one cannot get a (local) nonlinear representation of Lie algebra by using the techniques of [FPS77]. Further technical refinements are needed and it is a real *tour de force* that was achieved in [FST97] by solving the long standing problem of global solutions for classical electrodynamics and asymptotic completeness. It turns out that Maxwell–Dirac system cannot be linearized completely because of the long-range nature of the vector potential. The asymptotic fields do not transform according to a linear representation of the Poincaré group and consequently they are not asymptocally free. Nevertheless, one gets tractable expressions which provide new insights on the infrared problem in quantum electrodynamics.

2.2 Nonlinear Klein–Gordon equation

We present an example that we shall need in Section 3 for our discussion on deformation quantization. It is based on results obtained in [ST93]. Consider the massive nonlinear Klein–Gordon equation:

$$\Box\Phi + m^2\Phi = P(\Phi, \partial_t\Phi, \nabla\Phi), \quad m > 0, \tag{2.2}$$

where $\Box = \partial^\mu\partial_\mu = \partial_t^2 - \Delta$, and P is a C^∞ function covariant under $\mathcal{P}$ and vanishes at 0 along with its first derivatives.

Simon and Taflin [ST85, ST93] solved the Cauchy problem for Eq. (2.2), in $(1+n)$-dimensional spacetime with $n \geq 2$, by using linearization techniques. They have shown global existence and asymptotic completeness for Cauchy data in some neighborhood of the origin of an Hilbert space.

Eq. (2.2) can be put in the form of an evolution equation (2.1) by taking

$$\psi = \begin{pmatrix}\Phi\\ \Pi\end{pmatrix}, \qquad H(\psi) = \begin{pmatrix}0 & I\\ \Delta - m^2 & 0\end{pmatrix}\begin{pmatrix}\Phi\\ \Pi\end{pmatrix} + \begin{pmatrix}0\\ P(\Phi, \Pi, \nabla\Phi)\end{pmatrix}. \tag{2.3}$$

We shall use the following decomposition into Fourier modes:

$$\begin{aligned}\Phi(t)(\vec{x}) &= \int \frac{d^3\vec{k}}{2(2\pi)^{3/2}\omega(\vec{k})}\left(\overline{a}(t)(\vec{k})e^{-i\vec{k}\cdot\vec{x}} + a(t)(\vec{k})e^{i\vec{k}\cdot\vec{x}}\right),\\ \Pi(t)(\vec{x}) &= i\int \frac{d^3\vec{k}}{2(2\pi)^{3/2}}\left(\overline{a}(t)(\vec{k})e^{-i\vec{k}\cdot\vec{x}} - a(t)(\vec{k})e^{i\vec{k}\cdot\vec{x}}\right),\end{aligned} \tag{2.4}$$

where $\omega(\vec{k}) = \sqrt{|\vec{k}|^2 + m^2}$, along with

$$a_+(t)(\vec{x}) = i\int \frac{d^3\vec{k}}{(2\pi)^{3/2}}\overline{a}(t)(\vec{k})e^{-i\vec{k}\cdot\vec{x}}, \quad a_-(t)(\vec{x}) = -i\int \frac{d^3\vec{k}}{(2\pi)^{3/2}}a(t)(\vec{k})e^{i\vec{k}\cdot\vec{x}}. \tag{2.5}$$

In terms of $\mathbf{a} = (a_+, a_-)$, the evolution equation for the nonlinear Klein–Gordon equation reads

$$\frac{d}{dt}\begin{pmatrix}a_+(t)\\ a_-(t)\end{pmatrix} = i\omega(-i\nabla)\begin{pmatrix}a_+(t)\\ -a_-(t)\end{pmatrix} + \begin{pmatrix}F(\mathbf{a}(t))\\ F(\mathbf{a}(t))\end{pmatrix}, \tag{2.6}$$

where $F(\mathbf{a}(t)) = P(\Phi(t), \Pi(t), \nabla\Phi(t))$ through the substitutions (2.4) and (2.5) and $\omega(-i\nabla)$ is the operator acting as multiplication by $\omega(\vec{k})$ after Fourier transformation.

The free Klein–Gordon equation (when $F = 0$) induces a Lie algebra linear representation T^1 of $\mathfrak{p}$ in $E_\infty = \mathcal{S}(\mathbb{R}^3, \mathbb{C}) \oplus \mathcal{S}(\mathbb{R}^3, \mathbb{C})$. For $(f_+, f_-) \in E_\infty$ it is given by

$$\begin{aligned}T^1_{P_0}(f_+, f_-) &= i\omega(-i\nabla)(f_+, -f_-),\\ T^1_{P_j}(f_+, f_-) &= \partial_j(f_+, f_-),\\ T^1_{M_{ij}}(f_+, f_-) &= (x_i\partial_j - x_j\partial_i)(f_+, f_-),\\ T^1_{M_{0j}}(f_+, f_-) &= i\omega(-i\nabla)(x_j f_+, -x_j f_-).\end{aligned}$$

The representation T^1 exponentiates to a continuous linear representation of $\mathcal{P}$ in the Hilbert space $E = L^2(\mathbb{R}^3, \mathbb{C}) \oplus L^2(\mathbb{R}^3, \mathbb{C})$, for which E_∞ is the set of C^∞-vectors.

The nonlinear representation $T = \sum_{n\geq 1} T^n$ of $\mathfrak{p}$ in E_∞ associated to Eq. (2.6) is determined by the interaction term F. By writing $T = T^1 + \tilde{T}$, on $\mathbf{f} = (f_+, f_-) \in E_\infty$, it is defined by

$$\begin{aligned}
\tilde{T}_{P_0}(f_+, f_-) &= (F(\mathbf{f}), F(\mathbf{f})),\\
\tilde{T}_{P_j}(f_+, f_-) &= 0,\\
\tilde{T}_{M_{ij}}(f_+, f_-) &= 0,\\
\tilde{T}_{M_{0j}}(f_+, f_-) &= (x_j F(\mathbf{f}), x_j F(\mathbf{f})).
\end{aligned}$$

The coefficients T^n, $n \geq 2$, are the homogeneous terms of degree n in $\tilde{T}$. With these notations, Eq. (2.6) now reads

$$\frac{d\mathbf{f}(t)}{dt} = (T^1_{P_0} + \tilde{T}_{P_0})(\mathbf{f}(t)).$$

It is shown in [ST93] that this nonlinear representation of $\mathfrak{p}$ can be exponentiated to a nonlinear representation $U = \sum_{n\geq 1} U^n$ of $\mathcal{P}$ on some neighborhood of $0 \in E$ and that U is linearizable. We shall be interested in the particular case where the interaction term P in (2.2) is a covariant polynomial of the field Φ only, without constant term or terms of degree 1. Also we require that the classical Hamiltonian associated to Eq. (2.2) is a positive functional. In that particular case, it follows directly from more general results proved in [ST93] that:

(1) U is an analytic map $\mathcal{P} \times E_\infty \to E_\infty$ and $g \mapsto U^1_{g^{-1}} U_g$ is a continuous map from $\mathcal{P}$ to the space of Banach analytic maps $\mathcal{H}(E_\infty, E_\infty)$.

(2) There exists a unique invertible analytic $\Omega_\epsilon \colon E_\infty \to E_\infty$, $\varepsilon = \pm$, such that $U_g \Omega_\varepsilon = \Omega_\varepsilon U^1_g$ for $g \in \mathcal{P}$.

(3) $\lim_{t\to\pm\infty} \|U_{\exp(tP_0)}(\Omega_\pm(\psi)) - U^1_{\exp(tP_0)}(\psi)\|_E = 0$, $\psi \in E_\infty$.

(4) If the Cauchy data $(\Phi(0), \partial_t \Phi(0))$ are in E_∞, then $\Phi(t)$ is in $\mathcal{S}(\mathbb{R}^3, \mathbb{C})$ for all $t \in \mathbb{R}$ and $\Phi \in C^\infty(\mathbb{R}^4)$.

3 Application of deformation quantization

Consider a finite-dimensional manifold M with a Poison bracket $\{\cdot, \cdot\}$ defined on it. Recall that a deformation quantization of M is given by a formal associative deformation $\star$ (star-product) of the usual multiplication of smooth functions on M:

$$f \star g = fg + \sum_{n\geq 1} \lambda^n C_n(f, g), \tag{3.1}$$

such that $[f, g]_\star = \frac{1}{2\lambda}(f \star g - g \star f)$ is a formal Lie algebra deformation of the Poisson bracket on M. All the series involved are formal series in the deformation parameter λ with coefficients in the space of smooth functions on M (see e.g. [DS] for details). The cochains C_n are usually required to be bidifferential operators vanishing on constants. In quantum-mechanical models, one takes $\lambda = \frac{i\hbar}{2}$ and the formal series often converge in the distribution sense.

In order to treat field theoretic problems, one would naturally go over to infinite-dimensional manifolds of fields. Then one would consider a family of functionals of the fields and try to deform it via a star-product. Very quickly, one will be faced with the problem of giving a meaning to (3.1) as a formal series, i.e., to have well-defined cochains C_n on the family of functionals considered. Of course it is always possible to take a small enough family of functionals such that the star-product is well defined. For example, if the manifold $M = H \oplus H$ where H is a Hilbert space, take as algebra of functionals $\mathcal{A}$ the one generated by (finite) tensor products of continuous linear forms on H, then it is easy to write down a well-defined star-product such that $\mathcal{A} \star \mathcal{A} \subset \mathcal{A}[\lambda]$. But this is of no interest if one has in mind applications to field quantization: even free Hamiltonians do not belong to this family of functionals. So our very first concern is to construct star-products defined on a sufficiently rich family of functionals which should at least includes the Poincaré generators.

For more than a half-century, physicists have known how to handle divergences appearing in physical interacting quantum models, especially in QED, and developed perturbative renormalization methods. The idea underlying most of the renormalization methods is to first regularize singular expressions (e.g. with a cut-off), then identify and extract terms becoming singular when the regularization is removed. This is done perturbatively with respect to various relevant parameters of the model considered.

In the deformation quantization framework, one will obviously meet divergences in the cochains of the star-product and try to remove them in a more controlled way. The idea here is to identify the diverging terms as singular cocycles or coboundaries of the Hochschild cohomology and get rid of them by passing to an equivalent star-product. This approach is perturbative in $\hbar$ and if one can manage to construct a "renormalized" star-product in such a way up to some order in $\hbar$, then it will necessarily be a multi-parameter deformation (the deformation parameters being the physical parameters of the model under study, e.g. coupling constant). The systematization of this idea would lead to a cohomological approach to renormalization (see [DS], Section 4.2.1).

3.1 Normal star-product

Consider the free massive real scalar field Φ. It satisfies the linear Klein–Gordon equation $(\Box + m^2)\Phi(t,\vec{x}) = 0, m > 0$. Let $E_\infty = \mathcal{S}(\mathbb{R}^3,\mathbb{R}) \oplus \mathcal{S}(\mathbb{R}^3,\mathbb{R})$. The initial data $(\phi,\pi) \in E_\infty$ are decomposed into (definite energy) Fourier modes by

$$\begin{aligned}\phi(\vec{x}) &= \int \frac{d^3\vec{k}}{2(2\pi)^{3/2}\omega(\vec{k})}\left(\overline{a}(\vec{k})e^{-i\vec{k}\cdot\vec{x}} + a(\vec{k})e^{i\vec{k}\cdot\vec{x}}\right),\\ \pi(\vec{x}) &= i\int \frac{d^3\vec{k}}{2(2\pi)^{3/2}}\left(\overline{a}(\vec{k})e^{-i\vec{k}\cdot\vec{x}} - a(\vec{k})e^{i\vec{k}\cdot\vec{x}}\right),\end{aligned} \tag{3.2}$$

where as usual $\omega(\vec{k}) = (|\vec{k}|^2 + m^2)^{1/2}$. Since $m > 0$ and $(\phi,\pi) \in E_\infty$ it follows that $(\overline{a}, a)$ is in $W_\infty \equiv \mathcal{S}(\mathbb{R}^3,\mathbb{C}) \oplus \mathcal{S}(\mathbb{R}^3,\mathbb{C})$. We look at W_∞ as a real vector space.

Denote by $\mathcal{A} = \mathcal{H}_s(W_\infty)$ the space of analytic $\mathbb{C}$-valued mappings on W_∞ having semiregular kernels as tempered distributions. An element $F \in \mathcal{H}_s(W_\infty)$ can be written as

$$F(\overline{a}, a) = \sum_{m,n\geq 0} \langle F^{mn}, \overline{a}^{\otimes^m} \otimes a^{\otimes^n}\rangle, \tag{3.3}$$

where the kernel $F^{mn} \in \mathcal{S}'(\mathbb{R}^{3m},\mathbb{C}) \oplus \mathcal{S}'(\mathbb{R}^{3n},\mathbb{C})$ is regular with respect to either its first $3m$ variables or its last $3n$ variables. Semiregularity implies that usual functional derivatives of F are in $\mathcal{S}$ as long as one does not consider mixed derivatives. Notice that the free Hamiltonian $H_0(\overline{a}, a) = \int \frac{d^3\vec{k}}{2}\overline{a}(\vec{k})a(\vec{k})$ belongs to $\mathcal{A}$ along with all of the free Poincaré generators.

We define normalized functional derivatives as

$$D_{a(\vec{k})} = (2\omega(\vec{k}))^{1/2}\frac{\delta}{\delta a(\vec{k})}, \qquad D_{\overline{a}(\vec{k})} = (2\omega(\vec{k}))^{1/2}\frac{\delta}{\delta \overline{a}(\vec{k})},$$

and the standard Poisson bracket reads

$$\{F, G\} = \frac{2}{i}\int d^3\vec{k}\left(D_{a(\vec{k})}(F)\, D_{\overline{a}(\vec{k})}(G) - D_{\overline{a}(\vec{k})}(F)\, D_{a(\vec{k})}(G)\right). \tag{3.4}$$

Clearly, $\mathcal{A}$ endowed with this bracket is a Poisson algebra (i.e., it is closed for the bracket operation), and $(W_\infty, \mathcal{A}, \{\cdot\,,\cdot\})$ is a Poisson space.

The normal star-product $\star_N$ on $(W_\infty, \mathcal{A})$ is defined by

$$F \star_N G = FG + \sum_{n\geq 1} \hbar^n C_n^N(F, G),$$

where

$$C_n^N(F, G) = \frac{1}{n!}\int d^3\vec{k}_1 \cdots d^3\vec{k}_n \left(D_{a(\vec{k}_1)} \cdots D_{a(\vec{k}_n)}(F)\, D_{\overline{a}(\vec{k}_1)} \cdots D_{\overline{a}(\vec{k}_n)}(G)\right). \tag{3.5}$$

Notice that the deformation parameter is $\frac{i\hbar}{2}$, the factor $\frac{i}{2}$ has been absorbed in the definition of the cochains, so the star-bracket is $[f, g]_{\star_N} = \frac{2}{i\hbar}(F \star_N G - G \star_N F)$.

The normal star-product realizes a deformation quantization of $(W_\infty, \mathcal{A}, \{\cdot, \cdot\})$ and it will be used in the next section as starting point for the construction of other star-products.

3.2 Deformation quantization for interacting fields

For our discussion of the applications of the linearization programme to deformation quantization of interacting fields, we shall consider a massive real scalar field theory in $(1+3)$-dimensions with polynomial interaction such that the Hamiltonian is positive:

$$(\Box + m^2)\Phi(t, \vec{x}) + V'(\Phi(t, \vec{x})) = 0 \quad m > 0, \tag{3.6}$$

where V' is the derivative of the potential in the Hamiltonian. V is assumed here to fulfill $V(0) = V'(0) = V''(0) = 0$. We denote by $\{P_\mu, M_{\alpha\beta}\}$ the Poincaré generators for Eq. (3.6) seen as functionals of the Cauchy data (ϕ, π) or their Fourier modes $(\overline{a}, a)$ defined above. Also $\{P^f_\mu, M^f_{\alpha\beta}\}$ will be the corresponding free generators.

For initial data $\Phi(0) = \phi$ and $\partial_t\Phi(0) = \pi$ in $E_\infty = \mathcal{S}(\mathbb{R}^3, \mathbb{R}) \oplus \mathcal{S}(\mathbb{R}^3, \mathbb{R})$, it follows from the results of [ST93], reviewed in Section 2.2, that there exists a unique global solution Φ with free asymptotic fields $\Phi^\pm$:

$$\lim_{t\to\pm\infty} \|\Phi(t) - \Phi^\pm(t)\|_E = 0,$$

where $\|\cdot\|_E$ denotes the Hilbertian norm of $E = L^2(\mathbb{R}^3, \mathbb{R}) \oplus L^2(\mathbb{R}^3, \mathbb{R})$.

Denote by $(\phi^\pm, \pi^\pm)$ the initial data of $\Phi^\pm$. The wave operators $\Omega^\pm : (\phi^\pm, \pi^\pm) \mapsto (\phi, \pi)$ are invertible analytic mappings from E_∞ onto E_∞, they preserve the Poisson bracket (Poisson maps), and the classical scattering operator $S = (\Omega^+)^{-1}\Omega^-$ is also analytic on E_∞. Moreover, and more importantly for us, the wave operators linearize the Poincaré generators:

$$P_\mu \circ \Omega^\pm = P^f_\mu, \quad M_{\alpha\beta} \circ \Omega^\pm = M^f_{\alpha\beta}. \tag{3.7}$$

Notice that in general the generators $\{P_\mu, M_{\alpha\beta}\}$ do not belong to the space $\mathcal{A}$ (e.g. for Φ^4-theory). However for the kind of potential we are considering, the first cochain of the normal star-product C^N_1 makes still sense on the Poincaré generators, the remark applies to the Poisson bracket as well. However, the second cochain C^N_2 is not defined on the Poincaré generators $\{P_\mu, M_{\alpha\beta}\}$ in general.

We now introduce two star-products from the normal one:

$$(F \star^\pm G) \circ \Omega^\pm = (F \circ \Omega^\pm) \star_N (G \circ \Omega^\pm). \tag{3.8}$$

They are defined on $\mathcal{A}^\pm$ the push-forward of $\mathcal{A}$ by $\Omega^\pm$. Since $\Omega^\pm$ are Poisson maps the star-brackets of $\star^\pm$ are also deformations of the standard Poisson bracket (3.4). Since $\Omega^\pm$ depends (in a non trivial way!) on the parameters (coupling constants) that may appear in the potential V in (3.6), so will the cochains $C^\pm_n$ of $\star^\pm$. One should also remark that (3.8) does not necessarily imply the equivalence of the star-products.

From Eq. (3.7), we deduce in particular for the k-th $\star^{\pm}$-power of the interacting Hamiltonian $H = P_0$:

$$(\star^{\pm} H)^k = (\star^N H_0)^k \circ (\Omega^{\pm})^{-1}, \tag{3.9}$$

where $H_0 = P_0^f$ is the free Hamiltonian. It is easy to check that $(\star^N H_0)^k$ belongs to $\mathcal{A}$ for all $k \geq 1$, and, since $(\Omega^{\pm})^{-1}$ sends W_∞ onto W_∞, the right-hand side of (3.9) is well defined for all $n \geq 0$. Consequently all the powers $(\star^{\pm} H)^n$ are defined on W_∞. Obviously, the same holds for any finite $\star^{\pm}$-products of the $\{P_\mu, M_{\alpha\beta}\}$. Therefore $\mathcal{A}^{\pm}$ contains a $\star^{\pm}$-subalgebra isomorphic to the universal enveloping algebra of the Poincaré-Lie algebra.

A fundamental tool in deformation quantization is the star-exponential:

$$\mathrm{Exp}_\star\left(\frac{tF}{i\hbar}\right) = \sum_{n\geq 0} \frac{1}{n!}\left(\frac{t}{i\hbar}\right)^n (\star F)^n. \tag{3.10}$$

In many quantum-mechanical models, this series is convergent as a distribution, and allows to determine the spectrum and eigenstates of some Hamiltonian or any physically relevant observable. This is accomplished by performing a Fourier analysis of $\mathrm{Exp}_\star(\frac{tH}{i\hbar})$ for some Hamiltonian. In field theory, for the free fields, one can easily sum up the series and get expressions for relevant observables in a closed form. For the example we are treating, each term in (3.10) is well defined, e.g. for the Hamiltonian. However due to the presence of the factor $\frac{1}{\hbar}$ one cannot conclude that the series is a formal series in both $\hbar$ and $\hbar^{-1}$, because their coefficients involve infinite sums. The linearization maps would be useful for these considerations

The star-products $\star^{\pm}$ are best suited than the normal product as they are free of a large class of divergences by construction. It is not clear how the fields behave in the framework we have presented. A deeper study of these constructions should open more avenues to treat divergences problem in field theory and make more explicit the interplay between classical and quantum field theories.

We end this Note by making a remark about Electrodynamics. Maxwell–Dirac equations cannot be linearized and the asymptotic fields are not free (the vector potential is free, but the Dirac spinor is not). The existence of modified wave operators has been established in [FST97] on a space of small initial data in neighborhood of the origin in some Hilbert space. These modified operators do linearize the Poincaré generators (seen as functional of the fields) and the electric current, but before applying a procedure similar to the one we have sketched here, one will first have to solve the deformation quantization of the asymptotic fields themselves.

Acknowledgements. The author thanks D. Sternheimer and E. Taflin for very useful discussions and remarks.

Appendix: Notions on nonlinear representations

Let E be a Fréchet space. Denote by $\mathcal{L}^n(E, E)$ the space of continuous symmetric n-linear maps on E taking their values in E. The natural identification $\mathcal{L}^n(E, E) \sim \mathcal{L}(\hat{\otimes}^n E, E)$, where $\hat{\otimes}$ is the completed symmetric projective tensor product, will be implicitly assumed. For $f \in \mathcal{L}^n(E, E)$, we shall denote by $\hat{f}$ the associated homogeneous polynomial on E, i.e, $\hat{f}(x) = f(x, \ldots, x)$. Let $\mathcal{F}(E)$ be the vector space of formal series $F = \sum_{n\geq 1} f^n$, where $f^n \in \mathcal{L}^n(E, E)$. There is an associative product on $\mathcal{F}(E)$, denoted $\circ$, given by the composition of formal series. For $F = \sum_{n\geq 1} f^n$ and $H = \sum_{n\geq 1} h^n$, the associated homogeneous polynomial of the term of degree n in $F \circ H$ reads

$$(\widehat{F \circ H})^n(x) = \sum_{1\leq p\leq n} f^p\Big(\sum_{n_1+\cdots+n_p=n} \hat{h}^{n_1}(x) \otimes \cdots \otimes \hat{h}^{n_p}(x)\Big), \quad x \in E. \tag{A.1}$$

Let $\mathcal{F}^{\mathrm{inv}}(E)$ be the group of invertible elements in $(\mathcal{F}(E), \circ)$.

Definition A.1. A formal nonlinear representation (S, E) of a Lie group G in E is a homomorphism $S\colon G \to \mathcal{F}^{\mathrm{inv}}(E)$ such that $(g; x_1, \ldots, x_n) \mapsto S_g^n(x_1, \ldots, x_n)$ are continuous maps from $G \times E^n$ to E, $\forall n \geq 1$.

As a consequence of Eq. (A.1), we have $S^1_{gg'}(x) = S^1_g(S^1_{g'}(x))$ for $g, g' \in G$, $x \in E$. Thus the first order term of $S_g = \sum_{n\geq 1} S_g^n$ is a linear representation of G in E, denoted (S^1, E) and called the free part of (S, E).

Definition A.2. Two formal nonlinear representations (S, E) and (S', E) of a Lie group G in E are said to be formally equivalent, if there exists an element $A \in \mathcal{F}^{\mathrm{inv}}(E)$ such that $S_g = A \circ S'_g \circ A^{-1}, \forall g \in G$. A formal nonlinear representation (S, E) is called formally linearizable, if it is formally equivalent to the linear representation (S^1, E).

Roughly speaking, a formal nonlinear representation can be seen as a formal action of G in E having $0 \in E$ as a fixed point. By looking at the infinitesimal action, one would expect to encounter formal vector fields and their Lie bracket. It is indeed the case. For $F, H \in \mathcal{F}(E)$, define $F \bullet H \in \mathcal{F}(E)$ by

$$(\widehat{F \bullet H})^n(x) = \sum_{1\leq p\leq n} f^p\Big(\sum_{1\leq i\leq p} \big(x^{\otimes^{i-1}}\big) \otimes \hat{h}^{n-p+1}(x) \otimes \big(x^{\otimes^{p-i}}\big)\Big), \quad x \in E. \tag{A.2}$$

The bracket $[F, H] \equiv F \bullet H - H \bullet F$ endows $\mathcal{F}(E)$ with a Lie algebra structure, and we have the corresponding of Def. A.1 for formal nonlinear representations of a Lie algebra.

Definition A.3. A formal nonlinear representation (T, E) of a Lie algebra $\mathfrak{g}$ in E is a Lie algebra homomorphism $T\colon \mathfrak{g} \to \mathcal{F}(E)$.

By writing $T = \sum_{n\geq 1} T^n$, one can check that $T^1\colon \mathfrak{g} \to \mathcal{L}(E, E)$ is a linear Lie algebra representation of $\mathfrak{g}$ in E, and (T^1, E) is called the free part of (T, E).

There are analytic counterparts to the previous definitions. One would speak of analytic nonlinear representations of Lie groups and algebras, analytic equivalence of analytic nonlinear representations, etc. if the formal series involved are analytic in some neighborhood of the origin in E.

If one has a continuous linear representation π of a real Lie group G in a Banach space E, then one has a linear representation $d\pi$ of the Lie algebra $\mathfrak{g}$ of G in E_∞, the Fréchet space of C^∞-vectors of π. It is a remarkable fact that a similar statement holds for nonlinear representations. More precisely [FPS77, ST95]: any (formal or analytic) nonlinear representation (S, E) of a real Lie group G on a Banach space E can be differentiated and gives a nonlinear representation (T, E_∞) of the Lie algebra $\mathfrak{g}$ on the space of C^∞-vectors for the free part (S^1, E) of (S, E). Then T^1 is equal to dS^1. Conversely, if G is moreover connected and simply connected, any nonlinear representation (T, E_∞) of $\mathfrak{g}$ such that its free part (T^1, E_∞) is the differential of a linear representation π of G in a Banach space E, can be exponentiated to a unique nonlinear representation (S, E_∞) of G such that $S^1 = \pi$ (under some regularity assumptions on (T, E_∞), (S, E_∞) can be extended to a nonlinear representation (S, E)).

References

[Din99] Dineen, S. *Complex analysis on infinite-dimensional spaces*. Springer-Verlag, London 1999.

[Di90] Dito, J., Star-product approach to quantum field theory: the free scalar field, *Lett. Math. Phys.* 20 (1990), 125–134; Star-products and nonstandard quantization for K-G equation, *J. Math. Phys.* 33 (1992), 791–801; An example of cancellation of infinities in star-quantization of fields, *Lett. Math. Phys.* 27 (1993), 73–80.

[Di93] Dito, J., *Star-produits en dimension infinie: le cas de la théorie quantique des champs*, Thesis, Université de Bourgogne, Dijon 1993.

[DS] Dito, G., and Sternheimer, D., Deformation quantization: genesis, developments and metamorphoses , in *Deformation Quantization* (G. Halbout, ed.), IRMA Lectures in Math. Theoret. Phys. 1, Walter de Gruyter, Berlin 2002, 9–54.

[Fl82] Flato, M., Deformation view of physical theories, *Czechoslovak J. Phys.* B32 (1982), 472–475.

[FPS77] Flato, M., Pinczon, G., and Simon, J., Non linear representations of Lie groups, *Ann. Sci. École Norm. Sup.* 10 (1977), 405–418.

[FS80a] Flato, M., and Simon, J., On a linearization program of nonlinear field equations, *Phys. Lett. B* 94 (1980), 518–522.

[FS80b] Flato, M., and Simon, J., Linearization of relativistic nonlinear wave equations, *J. Math. Phys.* 21 (1980), 913–917.

[FST87] Flato, M., Simon, J., and Taflin, E., On global solutions of the Maxwell-Dirac equations, *Comm. Math. Phys.* 112 (1987) 21–49.

[FST97] Flato, M., Simon, J. C. H., and Taflin, E., The Maxwell-Dirac equations: the Cauchy problem, asymptotic completeness and the infrared problem, *Mem. Amer. Math. Soc.* 127 (no. 606), 1997.

[RS79] Reed, M., and Simon, B., *Methods of Modern Mathematical Physics III: Scattering Theory*. Academic Press, Orlando 1979.

[ST85] Simon, J. C. H., and Taflin, E., Wave operators and analytic solutions for systems of nonlinear Klein-Gordon equations and of nonlinear Schrödinger equations, *Comm. Math. Phys.* 99 (1985), 541–562.

[ST93] Simon, J. C. H., and Taflin, E., The Cauchy problem for nonlinear Klein-Gordon equations, *Comm. Math. Phys.* 152 (1993), 433–478.

[ST95] Simon, J. C. H., and Taflin, E., Initial data for non-linear evolution equations and differentiable vectors of group representations, in *Modern Group Theoretical Methods in Physics* (J. Bertrand et al., eds.) , Math. Phys. Stud. 18, Kluwer Academic Publishers, Dordrecht 1995, 243–253.

[ST00] Simon, J. C. H., and Taflin, E., Nonlinear relativistic evolution equations: Survey of a new approach, in *Quantization, Deformations, and Symmetries* (Dito, G., Sternheimer, D., eds.), Conférence Moshé Flato 1999, Math. Phys. Stud. 22, Kluwer Academic Publishers, Dordrecht, Boston, London 2000, 389–402.

[Ta84] Taflin, E., Formal linearization of nonlinear massive representations of the connected Poincaré group, *J. Math. Phys.* 25 (1984), 765–771.

On the trace density in deformation quantization

*Boris Fedosov**

Institut für Mathematik
Universität Potsdam
14415 Potsdam, Postfach 601553, Germany
e-mail: `fedosov@math.uni-potsdam.de`

Abstract. We derive a variation formula for the trace functional resulting in a differential equation for the trace density. It enables one in principle to calculate completely the trace density. We illustrate this process computing several few terms.

2000 Mathematics Subject Classification: 53D55

1 Introduction

In this paper we consider a standard deformation quantization on a symplectic manifold M, ω and a canonical trace on the algebra of quantum observables, see e.g. [1, 2]. Our result consists in a variation formula for a trace density c which enables one to compute it in principle. We illustrate this process by computing several few terms.

In the simplest scalar-valued case we obtain

$$\begin{aligned} c(x,h) = 1 &+ \frac{h^2}{48}\partial_{i_1}\partial_{i_2}R_{j_1 j_2 i_3 j_3}\omega^{i_1 j_1}\omega^{i_2 j_2}\omega^{i_3 j_3} \\ &- \frac{h^2}{64}R_{i_1 i_2 i_3 j_3}R_{j_1 j_2 i_4 j_4}\omega^{i_1 j_1}\omega^{i_2 j_2}\omega^{[i_3 j_3}\omega^{i_4 j_4]} + O(h^3). \end{aligned} \tag{1.1}$$

Here R_{ijkl} is a curvature tensor of a symplectic connection $\partial = \partial^M$ on M, ω^{ij} is inverse to ω_{ij}, square brackets mean alternation (alternative averaging over the symmetric group S_4). Of course, we use the Einstein summation convention, moreover, in the sequel we will omit the tensor factors $\omega^{i_k j_k}$ meaning a contraction by $\omega^{i_k j_k}$ with respect to a pair of indices i_k, j_k with the same subscript k. Thus, (1.1) may be rewritten in a simplified form

$$c(x,h) = 1 + \frac{h^2}{48}\partial_{i_1}\partial_{i_2}R_{j_1 j_2 i_3 j_3} - \frac{h^2}{64}R_{i_1 i_2 [i_3 j_3}R_{|j_1 j_2| i_4 j_4]} + \cdots,$$

*Supported by the French Ministry of National Education

the vertical dashes mean that the indices $j_1 j_2$ between them are excluded from alternation.

Our quantization gives a star-product for which the identity function 1 is a unit with respect to $*$, that is

$$1 * a = a * 1 = a. \tag{1.2}$$

It is always possible to pass to an equivalent star-product $*'$ (the so-called closed star-product introduced in [3]) such that the integral

$$\mathrm{Tr}' a = \frac{(2\pi h)^{-n}}{n!} \int_M a(x, h) \omega^n$$

possesses a trace property

$$\mathrm{Tr}' a *' b = \mathrm{Tr}' b *' a.$$

Of course, in general the normalization (1.2) is violated for the closed product $*'$, but may be easily reconstructed. Indeed, let

$$c(x, h) = 1 + hc_1 + h^2 c_2 + \cdots \tag{1.3}$$

be a unit with respect to $*'$, then

$$a * b = ((ca) *' (cb)) \frac{1}{c}$$

defines a star-product $*$ satisfying (1.2), but the trace functional for $*$ looks more complicated:

$$\mathrm{Tr}\, a = \mathrm{Tr}' ca = \frac{(2\pi h)^{-n}}{n!} \int_M ca \frac{\omega^n}{n!}. \tag{1.4}$$

Thus, the unit in the $*'$-algebra gives a trace density in the $*$- algebra and vice versa. We will use the canonical deformation quantization for which (1.2) is fulfilled automatically and the trace functional is given by (1.4).

If M is compact, we obtain by the index theorem for deformation quantization

$$\mathrm{Tr}\, 1 = \frac{(2\pi h)^{-n}}{n!} \int_M c\omega^n = \int_M \exp \frac{\omega}{2\pi h} \det{}^{-1/2} \left(\frac{\sinh \kappa/2}{\kappa/2} \right)$$

where

$$\kappa = \kappa_M = \frac{1}{4\pi i} \omega^{im} R_{mjkl} dx^k \wedge dx^l$$

is the curvature 2-form of the connection ∂^M. In the last term of (1.1) one can easily recognize the Pontrjagin polynomial, being multiplied by $\omega \wedge \omega$ it is proportional to the first Pontrjagin form. The second term in (1.1) is a complete covariant derivative, so it vanishes under integration.

The computation of the higher-oder terms in the trace density is a challenging problem. We mean, of course, precise computation not only up to complete covariant derivatives leading to the index formula. Our direct computation of the lower-order terms may be of interest since it illustrates in detail the process of "miraculous cancelation" resulting in the elegant formula (1.1).

2 Preliminaries

To fix notation, let us recall necessary notions and constructions from [1, 2]. We will consider a canonical quantization with nontrivial coefficients, so we need a vector bundle E with a connection $\partial = \partial^E$. It defines a connection on the homomorphism bundle $K = \mathrm{Hom}(E, E)$ called a coefficient bundle. Further, W means a bundle over M whose fiber W_x is the Weyl algebra defined on the tangent space T_xM with respect to the constant symplectic form $1/2\omega_{ij}(x)dy^i \wedge dy^j$ defined on T_xM by the form ω on M. The symplectic connection $\partial = \partial^M$ on M gives a connection on the bundle W which in local Darboux coordinates has the form

$$\partial a = d_x a(x, y, h) + \left[\frac{i}{2h}\Gamma_{ijk} y^i y^j dx^k, a\right].$$

Here $a(x, y, h) \in C^\infty(M, W)$ is a section of W, Γ_{ijk} the connection coefficients of ∂^M. Next we consider the bundle $W \otimes K$ of Weyl algebras with coefficients in K with a connection $\partial = \partial^M \otimes 1 + 1 \otimes \partial^E$ on it. Finally, construct a connection

$$D = \delta + \partial + [s, \cdot\,]$$

where

$$\delta a = dx^i \frac{\partial a}{\partial y^i} = \frac{i}{h}[\omega_{ij} y^i dx^j, a],$$

and s subject to the following conditions:

(1) $s = ir/h$ where $r \in C^\infty(M, W \otimes K \otimes \Lambda^1)$ with $\deg r \geq 3$ (recall that $\deg y^i = 1$, $\deg h = 2$),

(2) $\delta^{-1} s = 0$, where δ^{-1} is defined as follows

$$\delta^{-1} a_{pq} = \frac{1}{p+q} y^k i\left(\frac{\partial}{\partial x^k}\right) a_{pq}$$

for a homogeneous form a_{pq} of degree p in y^i and q in dx^j if $p + q > 0$ and $\delta^{-1} a_{00} = 0$,

(3) $D^2 \equiv 0$ (the so-called Abelian property). More precisely, we impose a sufficient condition

$$\delta s = R + \partial s + s^2 \tag{2.1}$$

where

$$R = \frac{1}{2} R^E_{i_1 i_2} dx^{i_1} \wedge dx^{i_2} + \frac{i}{4h} R_{i_1 i_2 i_3 i_4} y^{i_1} y^{i_2} dx^{i_3} \wedge dx^{i_4}$$

is the curvature form of the connection ∂ on the bundle $W \otimes K$.

These conditions define D uniquely. It may be shown that they are equivalent to the following equation:

$$s = s_1 + \delta^{-1}(\partial s + s^2), \tag{2.2}$$

where

$$s_1 = \frac{1}{2} R^E_{i_1 i_2} y^{i_1} dx^{i_2} + \frac{i}{8h} R_{i_1 i_2 i_3 i_4} y^{i_1} y^{i_2} y^{i_3} dx^{i_4} \tag{2.3}$$

which may be solved by iterations.

We define W_D to be the algebra of flat sections of $W \otimes K$ with respect to D, that is

$$W_D = \{a \in C^\infty(M, W \otimes K) : Da \equiv 0\}.$$

We also consider a complex of differential forms with values in $W \otimes K$

$$0 \longrightarrow W \otimes K \xrightarrow{D} W \otimes K \otimes \Lambda^1 \xrightarrow{D} \cdots .$$

This complex is exact except the zero-order term where $\operatorname{Ker} D = W_D$ (see [2]). In particular, $Da = 0$ has a unique solution such that $a|_{y=0} = a_0 \in C^\infty(M, K)$. It may be found by iterations of the equation

$$a = a_0 + \delta^{-1}(\partial a + [s, a]). \tag{2.4}$$

The map

$$Q : a_0 \mapsto a \in W_D$$

will be called a canonical quantization map or simply quantization. From (2.4) we can obtain an "explicit" form of the operator Q:

$$Qa = \sum_{k=0}^{\infty} (\delta^{-1}(\partial + \operatorname{ad} s))^k a_0 = (1 - \delta^{-1}(\partial + \operatorname{ad} s))^{-1} a_0. \tag{2.5}$$

In such a form Q may be applied to any section of the bundle $W \otimes K$, the result will be not necessarily flat in general.

Further, the equation

$$Da = b$$

with a one-form b satisfying $Db = 0$ has a unique solution a such that $a|_{y=0} = 0$. We denote it $D^{-1}b$. It can be found from the equation similar to (2.4)

$$a = -\delta^{-1} b + \delta^{-1}(\partial a + [s, a]) \tag{2.6}$$

which is equivalent to a system

$$Da = b, \quad a|_{y=0} = 0$$

with a solvability condition $Db = 0$. Using the operator Q the result may be written in the form

$$D^{-1}b = -Q(\delta^{-1}b). \tag{2.7}$$

Next we recall the definition of the canonical trace on W_D. Locally, using the Darboux charts we can trivialize the manifold M and the bundle E and construct a connection D_0 on the restriction of the bundle $W \otimes K$ to the coordinate neighborhood $O \subset M$ starting with the de Rham differential d_x as an initial connection ∂.

It turns out that the two algebras $W_D|_O$ and $(W|_O)|_{D_0}$ defined over O are isomorphic, moreover, this isomorphism may be taken in the form

$$\begin{aligned} A : W_D|_O &\longrightarrow (W|_O)_{D_0} \\ Aa &= U \circ a \circ U^{-1}. \end{aligned} \tag{2.8}$$

Here $a \in W_D$, U is a section of $W|_O$ extended by negative powers of h so that the total degree of each term with respect to y^i and h is positive (see [1, 2]), the product $\circ$ is a fiberwise Weyl product in the bundle W. Applying D_0 to (2.8), we come to the following equation for U:

$$\left[U^{-1} \circ D_0 U - \Delta\gamma, a\right] = 0$$

with

$$\Delta\gamma = \frac{i}{2h}\Gamma_{ijk} y^i y^j dx^k + \Gamma_i^E dx^i + s.$$

We set

$$U^{-1} \circ D_0 U = \Delta\gamma. \tag{2.9}$$

This equation has a unique solution under the normalization condition $U|_{y=0} = 1$. In fact the equation (2.9) together with $U|_{y=0} = 1$ is equivalent to the following equation:

$$U = 1 + \delta^{-1}(dU + U \circ \Delta\gamma) \tag{2.10}$$

which may be solved by iterations.

Now, for a section $a \in W_D$ with support in a coordinate chart O we define

$$\operatorname{Tr} a = \frac{(2\pi h)^{-n}}{n!} \int_{\mathbb{R}^{2n}} \operatorname{tr}(Aa)\big|_{y=0} \omega^n, \tag{2.11}$$

where tr means coefficient trace in $K = \operatorname{Hom}(E, E)$. In general, we define $\operatorname{Tr} a$ by linearity using coordinate Darboux covering $\{O_i\}$, a partition of unity $\{\rho_i\}$ and local trivializations A_i defined by (2.8), (2.9). It turns out that this is a correct definition, in particular, (2.11) does not depend on the choice of local coordinates and trivialization isomorphisms. The trace (2.11) is unique up to a normalization factor, the choice of this factor as in (2.11) gives us the so-called canonical trace.

One can observe, carefully following the formulas (2.2), (2.4), (2.10), that the integrand in (2.11) is a linear combination of covariant derivatives of the original section $a_0 \in C^\infty(W, K)$, that is

$$\operatorname{tr}(Aa)\big|_{y=0} = \sum_I \operatorname{tr} c^I \partial_I a_0,$$

where $I = (i_1, i_2, \ldots, i_k)$ means a set of indices. Integration by parts in (2.11) gives

$$\operatorname{Tr} a = \frac{(2\pi h)^{-n}}{n!} \int_{\mathbb{R}^{2n}} \operatorname{tr} \sum_I (-1)^{|I|} \partial_{I'} c^I a_0, \tag{2.12}$$

where $I' = (i_k, i_{k-1}, \ldots, i_1)$. The function

$$c(x, h) = \sum_I (-1)^{|I|} \partial_{I'} c^I \in C^\infty(M, K) \tag{2.13}$$

is a correctly defined section of $C^\infty(M, K)$, in particular, it does not depend upon the choice of local coordinates and trivializations of the bundle E. Thus, we have come to the notion of the trace density which may be expressed by the formula

$$\operatorname{Tr} a = \frac{(2\pi h)^{-n}}{n!} \int_M \operatorname{tr} c a_0 \omega^n.$$

In principle, we are even able to compute it using (2.2)–(2.13), but very soon it becomes evident that the direct approach is rather hopeless. Nevertheless, these formulas lead to the following important conclusions:

1. the trace density depends only on connection coefficients Γ_{ijk} and Γ_i^E in local coordinates and local frames and their derivatives,

2. to compute $c(x, h)$ at a point $x \in M$, we can use arbitrary local coordinates and frames in a small neighborhood of x, the result does not depend on the choices,

3. using normal Darboux coordinates (see [2]) centered at $x \in M$, we see that $c(x, h)$ depends on jets of connection coefficients at the origin. They, in turn, may be expressed through curvature tensors and their covariant derivatives.

3 The variation of the trace

Let us fix a classical observable $a_0(x) \in C^\infty(M, K)$, quantize it obtaining $a = Q(a_0) \in W_D$ and take its canonical trace $\operatorname{Tr} a$. We are interested how this quantity depends on the connection ∂.

Proposition 3.1. *The following variation formula holds*

$$\frac{d}{dt}\operatorname{Tr} a = \operatorname{Tr}([D^{-1}\dot{\gamma}, a] - D^{-1}[\dot{\gamma}, a]), \tag{3.1}$$

where

$$\dot{\gamma} = \Delta\dot{\gamma} = \frac{i}{2h}\dot{\Gamma}_{ijk}y^i y^j dx^k + \dot{\Gamma}_i^E dx^i + \dot{s}.$$

Proof. Considering variations of the connection ∂ in a coordinate neighborhood O, we can use the local formula (2.11) implying

$$\frac{d}{dt}\operatorname{Tr} a = \frac{(2\pi h)^{-n}}{n!}\int_M \operatorname{tr}\frac{d}{dt}(Aa)\big|_{y=0}\omega^n.$$

Further, (2.8) yields

$$\frac{d}{dt}(Aa) = U \circ \{[U^{-1} \circ \dot{U}, a] + \dot{a}\} \circ U^{-1}. \tag{3.2}$$

Differentiating (2.9) and using that D_0 is a constant connection and thus independent of t, we obtain

$$U^{-1} \circ D_0\dot{U} - U^{-1} \circ \dot{U} \circ U^{-1} \circ D_0 U = \Delta\dot{\gamma}.$$

The first term may be transformed further

$$U^{-1} \circ D_0\dot{U} = D_0(U^{-1} \circ \dot{U}) + U^{-1} \circ D_0 U \circ U^{-1} \circ \dot{U},$$

so the left-hand side of the previous equation may be written as

$$\begin{aligned} &D_0(U^{-1} \circ \dot{U}) + [U^{-1} \circ D_0 U, U^{-1} \circ \dot{U}] \\ &\quad = D_0(U^{-1} \circ \dot{U}) + [\Delta\gamma, U^{-1} \circ \dot{U}] = D(U^{-1} \circ \dot{U}). \end{aligned}$$

Here we have made use that $U^{-1} \circ D_0 U$ is equal to $\Delta\gamma$ in virtue of (2.9) and further that

$$D_0 + [\Delta\gamma, \cdot\,] \equiv D.$$

Finally, we arrive at the equation for $U^{-1} \circ \dot{U}$

$$D(U^{-1} \circ \dot{U}) = \Delta\dot{\gamma}. \tag{3.3}$$

The solvability condition $D(\Delta\dot{\gamma}) = 0$ follows from the construction of D. In terms of the connection D_0 and the form $\Delta\gamma$ we can rewrite the equation (2.1) in the form

$$D_0(\Delta\gamma) + (\Delta\gamma)^2 = 0.$$

The differentiation yields

$$D_0(\Delta\dot{\gamma}) + [\Delta\gamma, \Delta\dot{\gamma}] = D(\Delta\dot{\gamma}) = 0.$$

Thus, from (3.3) we obtain

$$U^{-1} \circ \dot{U} = (D^{-1}(\Delta\dot{\gamma}) + b),$$

where b is an arbitrary flat section, that is $b \in W_D|_O$.

Now differentiating $Da = 0$, we come to the following equation for $\dot{a}$

$$D\dot{a} + \left[\Delta\dot{\gamma}, a\right] = 0.$$

The solvability condition is fulfilled since

$$D[\Delta\dot{\gamma}, a] = [D(\Delta\dot{\gamma}), a] + [\Delta\dot{\gamma}, Da] = 0.$$

We also have

$$\dot{a}|_{y=0} = \dot{a}_0 = 0,$$

implying

$$\dot{a} = -D^{-1}[\Delta\dot{\gamma}, a].$$

Thus, the right-hand side of (3.2) becomes

$$U \circ \left\{\left(\left[D^{-1}\dot{\gamma} + b, a\right] - D^{-1}\left[\dot{\gamma}, a\right]\right)\right\} \circ U^{-1},$$

where we have replaced $\Delta\dot{\gamma}$ by a shorter notation $\dot{\gamma}$.

Now we can return to global expressions. To this end observe that

$$[D^{-1}\dot{\gamma}, a] - D^{-1}[\dot{\gamma}, a]$$

is a flat section. Indeed, applying D, we get

$$D[D^{-1}\dot{\gamma}, a] = [\dot{\gamma}, a]$$

since $DD^{-1}\dot{\gamma} = \dot{\gamma}$ and $Da = 0$. On the other hand,

$$D(D^{-1}[\dot{\gamma}, a]) = [\dot{\gamma}, a],$$

implying the assertion.

Now for a flat section $s \in W_D$ with support in a neighborhood O we have by definition

$$\frac{(2\pi h)^{-n}}{n!} \int_{\mathbb{R}^{2n}} \operatorname{tr} U \circ s \circ U^{-1} \omega^n = \operatorname{Tr} s.$$

It means that the integral of the right-hand side of (3.2) is nothing but

$$\operatorname{Tr}([D^{-1}\dot{\gamma}, a] - D^{-1}[\dot{\gamma}, a] + [b, a]).$$

The last summand may be dropped since the trace of the commutator vanishes. □

To calculate the section $D^{-1}\dot{\gamma} \in C^\infty(M, W \otimes K)$ (which is not flat), let us write equation (2.6) for it. Denoting $u = D^{-1}\dot{\gamma}$, we have

$$u = -\delta^{-1}\dot{\gamma} + \delta^{-1}\left(\partial u + [s, u]\right).$$

Taking into account that $\delta^{-1}\dot{s} \equiv 0$, so that

$$\delta^{-1}\dot{\gamma} = \delta\dot{\Gamma} = \frac{i}{6h}\dot{\Gamma}_{ijk}y^i y^j y^k + \dot{\Gamma}_i^E y^i,$$

we obtain an equation

$$u = -\frac{i}{6h}\dot{\Gamma}_{ijk}y^i y^j y^k + \dot{\Gamma}_i^E y^i + \delta^{-1}(\partial u + [s, u]) \tag{3.4}$$

which gives a computational algorithm by the iteration method. The result may be represented as

$$D^{-1}\dot{\gamma} = -Q\dot{\Gamma} = -Q(\dot{\Gamma}_i^E y^i + \frac{i}{6h}\dot{\Gamma}_{ijk}y^i y^j y^k) \tag{3.5}$$

with Q given by (2.5). The formula (3.1) can be immediately translated into the language of trace density. Indeed, by the definition of the trace density the left-hand side of (3.1) is equal to (up to a normalizing factor)

$$\int_M \operatorname{tr} \dot{c} a_0 \omega^n.$$

Similarly, the right-hand side is

$$\int_M \operatorname{tr} c\left([D^{-1}\dot{\gamma}, a] - D^{-1}[\dot{\gamma}, a]\right)\big|_{y=0}\omega^n = \int_M \operatorname{tr} c\,[D^{-1}\dot{\gamma}, a]\big|_{y=0}\omega^n,$$

since $D^{-1}\ [\dot{\gamma}, a]|_{y=0} = 0$. Thus, the variation formula may be rewritten as

$$\int_M \operatorname{tr} \dot{c} a_0 \omega^n = \int_M \operatorname{tr} c\,[D^{-1}\dot{\gamma}, Q(a_0)\big|_{y=0}\omega^n. \tag{3.6}$$

Finally, we can get rid of a_0 since it is an arbitrary function and obtain an equation for c. To this end introduce a pairing

$$\langle a, b\rangle = \int_M \operatorname{tr} a \circ b|_{y=0}\,\omega^n$$

on the space of all sections $a, b \in C^\infty(M, W \otimes K)$

Lemma 3.2. *The pairing $\langle\cdot\,,\cdot\rangle$ has the following properties:*

(1) *it is non-degenerate, that is $\langle a, b\rangle = 0$ for any $b \in C^\infty(M, W \otimes K)$ if and only if $a = 0$,*

(2) *for homogeneous sections*

$$a = a_p = \frac{1}{p!}a_{i_1 i_2 \dots i_p}y^{i_1}y^{i_2}\dots y^{i_p}, \quad b = b_q = \frac{1}{q!}b_{j_1 j_2 \dots j_q}y^{j_1}y^{j_2}\dots y^{j_q}$$

$\langle a, b\rangle = 0$ if $p \neq q$ and

$$\langle a_p, b_p\rangle = (-1)^p\langle b_p, a_p\rangle, \tag{3.7}$$

(3)

$$\langle a, \delta^{-1}\partial b\rangle = -\frac{ih}{2}\langle i(\omega^{-1})\partial\delta e a, b\rangle, \tag{3.8}$$

where $ea_p = a_p/p$ *for* $p > 0$, $ea_0 = 0$ *and* $i(\omega^{-1})$ *means a contraction with the tensor* ω^{ij}.

Proof. The first two properties are obvious, so we prove only (3.8). Let a_p, b_{p-1} be two homogeneous sections of degree p and $p-1$ respectively. Then

$$\delta^{-1}\partial b = \frac{1}{p!}\partial_j b_{j_1\dots j_{p-1}} y^{j_1}\dots y^{j_{q-1}} y^j,$$

so, the left-hand side is equal to

$$\int_M \operatorname{tr}\left(-\frac{ih}{2}\right)^p \frac{1}{p!} a_{i_1\dots i_p}\partial_j b_{j_1\dots j_{p-1}}\omega^{i_1 j_1}\dots\omega^{i_{p-1}j_{p-1}}\omega^{i_p j}\omega^n.$$

On the right-hand side we have

$$\delta e a = \frac{1}{p!} a_{i_1\dots i_{p-1}i_p} y^{i_1}\dots y^{p-1} dx^{i_p},$$

$$\begin{aligned} i(\omega^{-1})\partial\delta e a &= i(\omega^{-1})\frac{1}{p!}\partial_j a_{i_1\dots i_{p-1}i_p} y^{i_1}\dots y^{i_{p-1}} dx^j \wedge dx^{i_p} \\ &= \frac{1}{p!}\partial_j a_{i_1\dots i_{p-1}i_p} y^{i_1}\dots y^{i_{p-1}}\omega^{j i_p}, \end{aligned}$$

so that

$$-\frac{ih}{2}\operatorname{tr}(i(\omega^{-1})\partial\delta e a)\circ b|_{y=0} = \left(-\frac{ih}{2}\right)^p \frac{1}{p!}\operatorname{tr}\partial_j a_{i_1\dots i_p} b_{j_1\dots j_{p-1}}\omega^{j i_p}\omega^{i_1 j_1}\dots\omega^{i_{p-1}j_{p-1}}.$$

After integration by parts we obtain the same expression as on the left-hand side. □

Non-degeneracy allows one to define a transposition for operators acting on sections, so that

$$\langle A'a, b\rangle = \langle a, Ab\rangle.$$

For example, if $A = \operatorname{ad} c_p$ where c_p is homogeneous in y of degree p then, using (3.7), we obtain

$$\langle a, (\operatorname{ad} c)b\rangle = \langle a, c\circ b - b\circ c\rangle = \langle a\circ c, b\rangle - (-1)^p\langle c\circ a, b\rangle,$$

so that

$$(\operatorname{ad} c_p)'a = a\circ c - (-1)^p c\circ a.$$

Now, we can remove the operator Q from a_0 using this transposition. We also use the fact that $a_0 \in C^\infty(M, K)$, that is does not depend on y. The latter observation implies

$$\delta^{-1}(\operatorname{ad} s)a_0 = (\operatorname{ad} \delta^{-1}s)a_0 = 0$$

because of the condition $\delta^{-1}s = 0$. Thus,

$$Qa_0 = a_0 + Q\delta^{-1}(\partial + \operatorname{ad} s)a_0 = a_0 + Q\delta^{-1}\partial a_0, \tag{3.9}$$

and we come to the following theorem.

Theorem 3.3. *The trace density satisfies the following differential equation*

$$\dot{c} = \frac{ih}{2}\omega^{ij}\partial_i \frac{\partial}{\partial y^j} Q'(\operatorname{ad} Q(\delta^{-1}\dot{\Gamma}))'c|_{y=0} \tag{3.10}$$

where

$$Q' = \left(1 + \frac{ih}{2} i(\omega^{-1})(\partial + (\operatorname{ad} s)'\delta e)\right)^{-1}.$$

Proof. The equation easily follows from (3.6), (3.5) and (3.9) by Lemma 3.2. □

This theorem gives an algorithm to compute the trace density. In a local chart introduce a family of connections with

$$\Gamma(t) = t\left(\Gamma_i^E dx^i + \frac{i}{2h}\Gamma_{ijk} y^i y^j dx^k\right), \quad 0 \le t \le 1. \tag{3.11}$$

It is clear that the coefficients of the trace density are polynomials in t, so the Taylor series gives a complete answer:

$$c = c(1) = \sum_{k=0}^{\infty} \frac{1}{k!} c^{(k)}(0).$$

On the other hand, the derivatives $c^{(k)}(0)$ may be computed successively from (3.10). This yields an expression for c in terms of connection coefficients and their derivatives. But we know that this expression is independent of the choice of local Darboux coordinates and local frames. So, using normal Darboux coordinates, we can rewrite this expression in terms of curvature tensors and their covariant derivatives.

4 Calculations

In this section we calculate the coefficients c_1 and c_2 of the expansion (1.3) using the differential equation (3.10) for the trace density. So, we introduce a family of connections in local coordinates and frames

$$\Gamma(t) = \Gamma_i^E(t)dx^i + \frac{i}{2h}\Gamma_{i_1 i_2 i_3}(t) y^{i_1} y^{i_2} dx^{i_3},$$

so that $\Gamma(0) = 0$ and $\Gamma(1)$ coincides with the given connection form Γ. For example, we may take a linear family (3.11).

Let

$$s = s_1 + s_2 + \cdots$$

be the solution of (2.2), (2.3) where s_k denotes the term of degree k in y, h. We calculate the right-hand side of (3.10) up to degree 4. So,

$$Q(\delta^{-1}\dot{\Gamma}(t)) = u_1 + u_2 + u_3 + \cdots$$

with

$$\begin{aligned} u_1 &= \delta\dot{\Gamma}(t) = \dot{\Gamma}_i^E(t)y^i + \frac{i}{6h}\dot{\Gamma}_{i_1i_2i_3}(t)y^{i_1}y^{i_2}y^{i_3} \\ u_2 &= \delta^{-1}\partial u_1 \\ u_3 &= \delta^{-1}[s_1, u_1] + \delta^{-1}\partial\delta^{-1}\partial u_1. \end{aligned} \tag{4.1}$$

Further, define $v = v_1 + v_2 + v_3 + \cdots$ to be

$$\begin{aligned} v &= (\operatorname{ad} Q(\delta^{-1}\dot{\Gamma}))'c \\ &= (1 + hc_1 + \cdots)(u_1 + u_2 + u_3 + \cdots) + (u_1 - u_2 + u_3 \cdots)(1 + hc_1 + \cdots) \\ &= 2u_1 + 2u_3 + h(u_1c_1 + c_1u_1) + \cdots, \end{aligned}$$

so that $v_2 \equiv 0$. Thus, up to degree 4 the right-hand side of (3.10) may be written in the form

$$\frac{ih}{2}\omega^{ij}\partial_i\frac{\partial}{\partial y^j}Q'v\big|_{y=0} = \frac{ih}{2}\omega^{ij}\partial_i\frac{\partial}{\partial y^j}\Big\{v_1 + v_3 - \frac{ih}{2}i(\omega^{-1})(\partial + (\operatorname{ad} s)')\delta e v_1 + \Big(\frac{ih}{2}i(\omega^{-1})\partial\delta e\Big)^2 v_1\Big\}\Big|_{y=0}. \tag{4.2}$$

The first observation is that only linear terms with respect to y in curly brackets do contribute to (4.2). Hence, the second term in (4.1) may be dropped since it is at least cubic in y. Retaining only linear terms in the first summand of (4.1), we obtain the following expression for v_3 in (4.2)

$$v_3 = 2\delta^{-1}\left([s_1, u_1]|_{y=0}\right) + h(c_1u_1 + u_1c_1). \tag{4.3}$$

Next, taking into account that v_1 (which is equal to $2u_1$) contains linear and cubic terms in y, we see that $\partial\delta e v_1$ contains only even-degree terms in y, so, this term does not contribute to (4.2).

Thus, retaining only essential terms, we rewrite (3.10) in the form

$$\begin{aligned} h\dot{c}_1 + h^2\dot{c}_2 + \cdots = {} & ih\omega^{ij}\partial_i \frac{\partial}{\partial y^j} u_1\big|_{y=0} \\ & + \frac{ih^2}{2}\omega^{ij}\partial_i \frac{\partial}{\partial y^j}(c_1 u_1 + u_1 c_1)\big|_{y=0} \\ & + ih\omega^{ij}\partial_i \frac{\partial}{\partial y_j}\delta^{-1}\big([s_1, u_1]\big|_{y=0}\big) \\ & + \frac{h^2}{2}\omega^{ij}\partial_i \frac{\partial}{\partial y^j} i(\omega^{-1})(\operatorname{ad} s_1)'\delta e u_1\big|_{y=0} \\ & - \frac{ih^3}{4}\omega^{ij}\partial_i \frac{\partial}{\partial y^j}(i(\omega^{-1})\partial\delta e)^2 u_1\big|_{y=0} + \cdots . \end{aligned} \tag{4.4}$$

We have replaced s by its leading term s_1, this does not affect the terms of degree ≤ 4.

Comparing linear terms in h, we get

$$\dot{c}_1 = i\omega^{ij}\partial_i \frac{\partial}{\partial y^j} u_1\big|_{y=0} = i\omega^{ij}\partial_i \dot{\Gamma}^E_j = \frac{i}{2}\omega^{ij}\big(\partial_i\dot{\Gamma}^E_j - \partial_j\dot{\Gamma}^E_i\big) = \frac{i}{2}\dot{R}^E_{ij}\omega^{ij},$$

so that

$$c_1 = \frac{i}{2}R^E_{ij}$$

(we again drop ω^{ij} in the final expression). Substitute this result into (4.4) and compare the coefficients at h^2. The second term in (4.4) yields

$$-\frac{1}{4}\partial_i\big(R^E_{i_1 j_1}\dot{\Gamma}^E_j + \dot{\Gamma}^E_j R^E_{i_1 j_1}\big) \tag{4.5}$$

(contraction in i, j and i_1, j_1 is understood). Passing to the third term, we first compute

$$\begin{aligned} [s_1, u_1]\big|_{y=0} = {} & \Big[\frac{1}{2}R^E_{i_1 i_2} y^{i_1} dx^{i_2}, \dot{\Gamma}^E_i y^i\Big]\Big|_{y=0} \\ & + \Big[\frac{i}{8h}R_{i_1 i_2 i_3 i_4} y^{i_1} y^{i_2} y^{i_3} dx^{i_4}, \frac{i}{6h}\dot{\Gamma}_{i_5 i_6 i_7} y^{i_5} y^{i_6} y^{i_7}\Big]\Big|_{y=0} \\ = {} & -\frac{ih}{4}\big(R^E_{i_1 i_2}\dot{\Gamma}^E_{j_1} + \dot{\Gamma}^E_{j_1} R^E_{i_1 i_2}\big)dx^{i_2} \\ & - \frac{ih}{32}R_{i_1 i_2 i_3 i_4}\dot{\Gamma}_{j_1 j_2 j_3} dx^{i_4}. \end{aligned}$$

Thus, the third term yields

$$\frac{1}{4}\partial_{i_2}\big(R^E_{i_1 j_2}\dot{\Gamma}^E_{j_1} + \dot{\Gamma}^E_{j_1} R^E_{i_1 j_2}\big) + \frac{1}{32}\partial_{i_4}\big(R_{i_1 i_2 i_3 j_4}\dot{\Gamma}_{j_1 j_2 j_3}\big). \tag{4.6}$$

The fourth term requires more work. First of all we have

$$(\operatorname{ad} s_1)'\delta e u_1 = \delta e u_1 \circ s_1 - s_1 \circ \delta e u_1,$$

the sign $-$ before the second summand appears since we have to change the sign twice: once because s_1 contains the terms of odd degrees in y, and once more because s_1 and $\delta e u_1$ are one-forms. The terms linear in y are

$$\frac{1}{2}(\dot\Gamma^E_{i_1} R^E_{i_2 i_3} + R^E_{i_2 i_3}\dot\Gamma^E_{i_1}) y^{i_2} dx^{i_1} \wedge dx^{i_3} + \frac{1}{16}\dot\Gamma_{i_1 i_2 i_3} R_{(j_1 j_2 i_4) i_5} y^{i_4} dx^{i_3} \wedge dx^{i_5},$$

where parentheses mean symmetrization. Thus, the fourth summand becomes

$$\frac{1}{4}\partial_{i_1}\left(\dot\Gamma^E_{i_2} R^E_{j_1 j_2} + R^E_{j_1 j_2}\dot\Gamma^E_{i_2}\right) + \frac{1}{32}\partial_{i_4}\left(\dot\Gamma_{i_1 i_2 i_3} R_{(j_1 j_2 j_4) j_3}\right).$$

The completely symmetric component $R_{((j_1 j_2 j_4) j_3)}$ of the tensor $R_{(j_1 j_2 j_4) j_3}$ vanishes, hence

$$R_{(j_1 j_2 j_4) j_3} + R_{(j_1 j_2 j_3) j_4} + R_{(j_1 j_3 j_4) j_2} + R_{(j_3 j_2 j_4) j_1} = 0.$$

After contraction with the completely symmetric tensor $\dot\Gamma_{i_1 i_2 i_3}$ this identity yields

$$3\dot\Gamma_{i_1 i_2 i_3} R_{(j_1 j_2 j_4) j_3} = -\dot\Gamma_{i_1 i_2 i_3} R_{j_1 j_2 j_3 j_4},$$

so that the fourth term in (4.4) becomes

$$\frac{1}{4}\partial_{i_1}\left(\dot\Gamma^E_{i_2} R^E_{j_1 j_2} + R^E_{j_1 j_2}\dot\Gamma^E_{i_2}\right) - \frac{1}{96}\partial_{i_4}\left(\dot\Gamma_{i_1 i_2 i_3} R_{j_1 j_2 j_3 j_4}\right). \tag{4.7}$$

Finally, the last term in (4.4) gives

$$\frac{1}{24}\partial_{i_1}\partial_{i_2}\partial_{i_3}\dot\Gamma_{j_1 j_2 j_3}. \tag{4.8}$$

Summing up (4.5)–(4.8), we obtain

$$\begin{aligned} \dot c_2 = &-\frac{1}{4}\partial_{i_1}\left(\dot\Gamma^E_{j_1} R^E_{i_2 j_2} + R^E_{i_2 j_2}\dot\Gamma^E_{j_1}\right) \\ &+ \frac{1}{2}\partial_{i_1}\left(\dot\Gamma^E_{i_2} R^E_{j_1 j_2} + R^E_{j_1 j_2}\dot\Gamma^E_{i_2}\right) \\ &+ \frac{1}{24}\left(\partial_{i_1}\partial_{i_2}\partial_{i_3}\dot\Gamma_{j_1 j_2 j_3} - \partial_{i_4}(\dot\Gamma_{i_1 i_2 i_3} R_{j_1 j_2 j_3 j_4})\right). \end{aligned} \tag{4.9}$$

Now we can watch an amazing interplay of the symmetry properties of connection coefficients and curvature tensors resulting in the fact that the right-hand side is a complete derivative in t. Clearly, the terms corresponding to ∂^E and ∂ may be treated separately. First consider the terms corresponding to ∂^E dropping temporarily the superscript E:

$$\begin{aligned} &-\frac{1}{4}(R_{i_1 j_1}\partial_{i_2}\dot\Gamma_{j_2} + \partial_{i_2}\dot\Gamma_{j_2} R_{i_1 j_1}) - \frac{1}{4}(\partial_{i_2} R_{i_1 j_1}\dot\Gamma_{j_2} + \dot\Gamma_{j_2}\partial_{i_2} R_{i_1 j_1}) \\ &\quad + \frac{1}{2}(\partial_{i_1}\dot\Gamma_{i_2} R_{j_1 j_2} + R_{j_1 j_2}\partial_{i_1}\dot\Gamma_{i_2}) + \frac{1}{2}(\dot\Gamma_{i_2}\partial_{i_1} R_{j_1 j_2} + \partial_{i_1} R_{j_1 j_2}\dot\Gamma_{i_2}). \end{aligned}$$

Note, that the second and the fourth terms cancel out. Indeed, changing summation indices, rewrite them as

$$-\frac{1}{4}(\partial_{i_2} R_{i_1 j_1} + 2\partial_{i_1} R_{j_1 i_2})\dot{\Gamma}_{j_2} - \frac{1}{4}\dot{\Gamma}_{j_2}(\partial_{i_2} R_{i_1 j_1} + 2\partial_{i_1} R_{j_1 i_2}).$$

Expressions in parentheses vanish by Bianchi's identity:

$$\begin{aligned}\partial_{i_2} R_{i_1 j_1} + 2\partial_{i_1} R_{j_1 j_2} &= \partial_{i_2} R_{i_1 j_1} + \partial_{i_1} R_{j_1 j_2} - \partial_{j_1} R_{i_1 j_2} \\ &= \partial_{i_2} R_{i_1 j_1} + \partial_{i_1} R_{j_1 j_2} + \partial_{j_1} R_{j_2 i_1} = 0.\end{aligned}$$

In the remaining summands everything may be expressed in terms of the curvature tensor. Since $R_{j_1 j_2}$ is antisymmetric and because j_1, j_2 are contracted with i_1, i_2, we may antisymmetrize with respect to i_1, i_2, that is

$$\begin{aligned}\partial_{i_1}\dot{\Gamma}_{i_2} R_{j_1 j_2} &= \tfrac{1}{2}(\partial_{i_1}\dot{\Gamma}_{i_2} - \partial_{i_2}\dot{\Gamma}_{i_1}) R_{j_1 j_2} \\ &= \tfrac{1}{2}\dot{R}_{i_1 i_2} R_{j_1 j_2}.\end{aligned}$$

Thus, the remaining terms give

$$\begin{aligned}-\frac{1}{8}(R_{i_1 j_1}\dot{R}_{i_2 j_2} &+ R_{i_2 j_2}\dot{R}_{i_1 j_1}) + \frac{1}{4}(\dot{R}_{i_1 i_2} R_{j_1 j_2} + R_{j_1 j_2}\dot{R}_{i_1 i_2}) \\ &= -\frac{1}{8}\frac{d}{dt}(R_{i_1 j_1} R_{i_2 j_2} - 2R_{i_1 i_2} R_{j_1 j_2}) \\ &= -\frac{1}{8}\frac{d}{dt}(R_{i_1 j_1} R_{i_2 j_2} + R_{i_1 i_2} R_{j_2 j_1} + R_{i_1 j_2} R_{j_1 j_2}) \\ &= -\frac{3}{8}\frac{d}{dt} R_{[i_1 j_1} R_{i_2 j_2]}.\end{aligned}$$

The terms corresponding to ∂ are proportional to

$$\partial_{i_1}\partial_{i_2}\partial_{i_3}\dot{\Gamma}_{j_1 j_2 j_3} - \partial_{i_4}(\dot{\Gamma}_{i_1 i_2 i_3} R_{j_1 j_2 j_3 j_4}). \tag{4.10}$$

We transform the first term using the relation

$$\begin{aligned}\partial_{i_3}\dot{\Gamma}_{j_1 j_2 j_3}\omega^{i_3 j_3} &= \frac{1}{2}(\partial_{i_3}\dot{\Gamma}_{j_1 j_2 j_3} - \partial_{j_3}\Gamma_{j_1 j_2 i_3})\omega^{i_3 j_3} \\ &= \frac{1}{2}\dot{R}_{j_1 j_2 i_3 j_3}.\end{aligned}$$

Thus,

$$\partial_{i_1}\partial_{i_2}\partial_{i_3}\dot{\Gamma}_{j_1 j_2 j_3} = \frac{1}{2}\frac{d}{dt}(\partial_{i_1}\partial_{i_2} R_{j_1 j_2 i_3 j_3}) - \frac{1}{2}\partial_{i_1}\dot{\partial}_{i_2} R_{j_1 j_2 i_3 j_3} - \frac{1}{2}\dot{\partial}_{i_1}\partial_{i_2} R_{j_1 j_2 i_3 j_3}.$$

The last summand is equal to zero. Indeed, the only free index in $\partial_{i_2} R_{j_1 j_2 i_3 j_3}$ is j_1, so

$$\dot{\partial}_{i_1}\partial_{i_2} R_{j_1 j_2 i_3 j_3} = \dot{\Gamma}_{i_1 j_1 i_4}\partial_{i_2} R_{j_4 j_2 i_3 j_3} = 0$$

since Γ_{pqr} is symmetric in all the indices, so that $\dot{\Gamma}_{i_1 j_1 i_4}\omega^{i_1 j_1} \equiv 0$. Now,

$$\dot{\partial}_{i_2} R_{j_1 j_2 i_3 j_3} = \dot{\Gamma}_{i_2 j_1 i_4} R_{j_4 j_2 i_3 j_3} + \dot{\Gamma}_{i_2 j_2 i_4} R_{j_1 j_4 i_3 j_3},$$

and again the last summand is equal to 0, so (4.10) becomes

$$\frac{1}{2}\frac{d}{dt}(\partial_{i_1}\partial_{i_2}R_{j_1j_2i_3j_3}) - \frac{1}{2}\partial_{i_4}(\dot{\Gamma}_{i_1i_2j_4}R_{j_1j_2i_3j_3}) - \partial_{i_4}(\dot{\Gamma}_{i_1i_2i_3}R_{j_1j_2j_3j_4}).$$

The second and the third terms considered as tensors in i_3, j_3, i_4, j_4 (the indices i_1, j_1 and i_2, j_2 in both terms are contracted and we consider them as dumb ones) are treated similarly to the ∂^E-case. Namely,

$$\begin{aligned}
&-\frac{1}{2}\partial_{i_4}\dot{\Gamma}_{i_1i_2j_4}R_{j_1j_2i_3j_3} - \partial_{i_4}\dot{\Gamma}_{i_1i_2i_3}R_{j_1j_2j_3j_4} \\
&\quad = -\frac{1}{4}\dot{R}_{i_1i_2i_4j_4}R_{j_1j_2i_3j_3} - \frac{1}{2}\dot{R}_{i_1i_2i_4i_3}R_{j_1j_2j_3j_4} \\
&\quad = -\frac{1}{8}\frac{d}{dt}(R_{i_1i_2i_4j_4}R_{j_1j_2i_3j_3} + R_{i_1i_2i_4i_3}R_{j_1j_2j_3j_4} + R_{i_1i_2i_4j_3}R_{j_1j_2j_4i_3}) \\
&\quad = -\frac{3}{8}\frac{d}{dt}R_{i_1i_2[i_3j_3}R_{|j_1j_2|i_4j_4]},
\end{aligned}$$

while the remaining terms

$$-\frac{1}{2}\Gamma_{i_1i_2j_4}\partial_{i_4}R_{j_1j_2i_3j_3} - \dot{\Gamma}_{i_1i_2i_3}\partial_{i_4}R_{j_1j_2j_3j_4}$$

vanish in virtue of the Bianchi identity.

Let us formulate the final result as a proposition.

Proposition 4.1. *The trace density up to degree 4 is given by the expression*

$$\begin{aligned}
c = 1 &+ \frac{ih}{2}R_{ij} - \frac{3}{8}h^2R_{[i_1j_1}R_{i_2j_2]} \\
&- \frac{h^2}{64}R_{i_1i_2[i_3j_3}R_{|j_1j_2|i_4j_4]} + \frac{h^2}{48}\partial_{i_1}\partial_{i_2}R_{j_1j_2i_3j_3}.
\end{aligned}$$

After multiplication by the volume form $\omega^n/(2\pi h)^n n!$ one can easily recognize characteristic forms of connections ∂^E and ∂^M. So, the first three terms give

$$1 + \left(-\frac{1}{2\pi i}\frac{1}{2}R^E_{ij}dx^i \wedge dx^j\right) + \frac{1}{2}\left(-\frac{1}{2\pi i}\frac{1}{2}R^E_{ij}dx^i \wedge dx^j\right)^2$$

times $\exp(\omega/2\pi h)$, the forth term yields

$$\frac{1}{12}\operatorname{tr}\left(\frac{1}{2\pi i}\frac{1}{4}R_{ij}dx^i \wedge dx^j\right)^2 \frac{\omega^{n-2}}{(2\pi h)^{n-2}(n-2)!}.$$

References

[1] Fedosov, B., A simple geometrical construction of deformation quantization, *J. Diff. Geom.* 40 (1994), 213–238.

[2] Fedosov, B., *Deformation Quantization and Index Theory*, Math. Top. 9, Akademie Verlag, Berlin 1996.

[3] Connes, A., Flato, M., Sternheimer, D., Closed star-products and cyclic cohomology, *Lett. Math. Phys.* 24 (1992), 1–12.

Deformed double Yangians and quasi-Hopf algebras

Daniel Arnaudon[a], *Jean Avan*[b], *Luc Frappat*[a]*, *and Eric Ragoucy*[a]

[a] *Laboratoire d'Annecy-le-Vieux de Physique Théorique,*
LAPTH, CNRS, UMR 5108, Université de Savoie
B.P. 110, F-74941 Annecy-le-Vieux Cedex, France
e-mail: arnaudon@lapp.in2p3.fr
e-mail: frappat@lapp.in2p3.fr; frappat@univ-savoie.fr
e-mail: ragoucy@lapp.in2p3.fr

[b] *Laboratoire de Physique Théorique et Hautes Énergies*
LPTHE, CNRS, UMR 7589, Universités Paris VI/VII
4, place Jussieu, B.P. 126, F-75252 Paris Cedex 05, France
e-mail: avan@lpthe.jussieu.fr

Abstract. We emphasize the relations between the deformed double Yangian type algebras and the quantum affine elliptic algebras. We show that the notion of Drinfel'd twist is the relevant framework to describe these algebras. The deformed double Yangian type algebras appear as quasi-triangular quasi-Hopf algebras.

Résumé. Nous dérivons les relations entre les algèbres de double Yangien déformées et les algèbres quantiques elliptiques affines. Nous montrons que la notion de "twist" de Drinfel'd constitue le cadre adapté pour décrire ces algèbres et que les algèbres de double Yangien déformées apparaissent comme des algèbres de quasi-Hopf quasi-triangulaires.

2000 Mathematics Subject Classification: 17B37, 81R50

1 Introduction

The study of elliptic quantum algebras, defined with the help of elliptic R-matrices, has yielded a number of algebraic structures relevant to certain integrable systems in quantum mechanics and statistical mechanics (noticeably the XYZ model [1], RSOS models [2, 3] and Sine–Gordon theory [4, 5]). More recently the definition and construction of some scaling limits of these algebras has led to the notion of deformed double Yangian algebras. Two classes of elliptic solutions to the Yang–Baxter equation have been identified, respectively associated with the vertex-type statistical models [6, 7] and the face-type statistical models [2, 8, 9]. The vertex elliptic R-matrix for

*Membre de l'Institut Universitaire de France

$sl(2)$ was first used by Sklyanin [10] to construct a two-parameter deformation of the enveloping algebra $\mathcal{U}(sl(2))$. The central extension of this structure was proposed in [11] for $sl(2)$. The face-type R-matrices, depending on the extra parameters λ belonging to the dual of the Cartan algebra in the underlying algebra, were first used by Felder [12] to define the algebra $\mathcal{B}_{q,p,\lambda}(\widehat{sl}_N)$ in the R-matrix approach.

The degeneracy (scaling) limits of elliptic R-matrices (see below), whether vertex-type [4, 13, 14] or face-type [15] give rise to algebraic structures which have been variously characterized as scaled elliptic algebras [13, 14], or double Yangian type algebras [15, 16, 17]. Although represented by formally identical Yang–Baxter relations ($RLL = LLR$), these two classes of objects differ fundamentally in their structure: in the context of scaled elliptic algebras, the expansion is done in terms of continuous-index Fourier modes of the spectral parameter [4, 14], while in the context of double Yangian type algebras the expansion is done in terms of powers of the spectral parameter [17, 18]. Hence these structures must be considered separately.

It was shown recently that both vertex-type algebras $\mathcal{A}_{q,p}(\hat{g})$ and face-type algebras $\mathcal{B}_{q,p,\lambda}(\hat{g})$ were in fact Drinfel'd twists [19] of the quantum group $\mathcal{U}_q(\hat{g})$ (where $\hat{g}$ is a simple affine Lie algebra, g being restricted to sl_N in the vertex case). Originating with the proposition of [20] on face-type algebras, the construction of the twist operators was undertaken in both cases by C. Frønsdal [21] and finally achieved at the level of formal universal twists in [22, 23]. In [23], the universal twist was obtained by solving a linear equation introduced in [24], and it was also proved in the case of finite (super)algebras that the infinite products defining the twists were convergent. This led to a formal construction of universal R-matrices for the elliptic algebras $\mathcal{A}_{q,p}(\widehat{sl}_2)$ and $\mathcal{B}_{q,p,\lambda}(\widehat{sl}_2)$, of which the BB and ABF (4×4)-matrices are respectively spin $\frac{1}{2}$ evaluation representations [22].

The scaling limit procedure yields R-matrices which are only interpreted as conjectural evaluation representations of hypothetical universal R-matrices for (quasi)-Hopf type algebraic structures and one must establish its actual existence. This will be achieved by identification of these limits as evaluation representations of universal R-matrices for the deformations by particular Drinfel'd twists, known as "shifted-cocycle" twists [25], of Hopf algebra structures. This construction systematically endows these deformations with a Gervais–Neveu–Felder type quasi-triangular quasi-Hopf algebra structure. It is characterized by a particular form of the universal Yang–Baxter equation, called Gervais–Neveu–Felder or dynamical Yang–Baxter equation, to be explicited below.

2 Hopf and quasi-Hopf algebras – Drinfel'd twist

2.1 Hopf algebras

Let us recall some basic definitions about Hopf and quasi-Hopf algebras.

Definition 2.1. A unital associative $\mathbb{C}$-algebra $\mathfrak{A}$ is called a *Hopf algebra* if it is endowed with a coalgebra structure: the coproduct $\Delta : \mathfrak{A} \to \mathfrak{A} \otimes \mathfrak{A}$ and counit

$\epsilon : \mathfrak{A} \to \mathbb{C}$ are algebra homomorphisms, the product $m : \mathfrak{A} \otimes \mathfrak{A} \to \mathfrak{A}$ and unit $\iota : \mathbb{C} \to \mathfrak{A}$ are coalgebra homomorphisms, and $\mathfrak{A}$ is equipped with an antihomomorphism $S : \mathfrak{A} \to \mathfrak{A}$ (antipode), with

$$\begin{aligned} &(\mathrm{id} \otimes \Delta)\Delta = (\Delta \otimes \mathrm{id})\Delta \qquad \text{(coassociativity)} \\ &(\mathrm{id} \otimes \epsilon) \circ \Delta = (\epsilon \otimes \mathrm{id}) \circ \Delta = \mathrm{id} \\ &m \circ (S \otimes \mathrm{id}) \circ \Delta = m \circ (\mathrm{id} \otimes S) \circ \Delta = \iota \circ \epsilon. \end{aligned}$$

If σ denotes the flip map, $\sigma(x \otimes y) = y \otimes x$ for $x, y \in \mathfrak{A}$, $\Delta' = \sigma \circ \Delta$ is the reversed coproduct. Whenever $\Delta = \Delta'$, the coproduct is said cocommutative.

Definition 2.2. A Hopf algebra $\mathfrak{A}$ is said to be *quasi-triangular* if it exists an invertible element $\mathcal{R} \in \mathfrak{A} \otimes \mathfrak{A}$, called the universal R-matrix, such that

$$\begin{aligned} &\Delta'(x) = \mathcal{R}\Delta(x)\mathcal{R}^{-1} \qquad (\forall x \in \mathfrak{A}) \\ &(\Delta \otimes \mathrm{id})\mathcal{R} = \mathcal{R}_{13}\,\mathcal{R}_{23} \\ &(\mathrm{id} \otimes \Delta)\mathcal{R} = \mathcal{R}_{13}\,\mathcal{R}_{12}. \end{aligned}$$

It follows that $\mathcal{R}$ satisfies the *Yang–Baxter equation* (in $\mathfrak{A} \otimes \mathfrak{A} \otimes \mathfrak{A}$):

$$\mathcal{R}_{12}\,\mathcal{R}_{13}\,\mathcal{R}_{23} = \mathcal{R}_{23}\,\mathcal{R}_{13}\,\mathcal{R}_{12}. \tag{2.1}$$

2.2 Quasi-Hopf algebras

Definition 2.3. A unital associative $\mathbb{C}$-algebra $\mathfrak{A}$ is called a *quasi-Hopf algebra* if it is endowed with a coalgebra structure: the coproduct $\Delta : \mathfrak{A} \to \mathfrak{A} \otimes \mathfrak{A}$ and counit $\epsilon : \mathfrak{A} \to \mathbb{C}$ are algebra homomorphisms, the product $m : \mathfrak{A}\otimes\mathfrak{A} \to \mathfrak{A}$ and unit $\iota : \mathbb{C} \to \mathfrak{A}$ are coalgebra homomorphisms, and $\mathfrak{A}$ is equipped with an antihomomorphism $S : \mathfrak{A} \to \mathfrak{A}$ (antipode) and elements $\alpha, \beta \in \mathfrak{A}$, and an invertible element $\Phi \in \mathfrak{A} \otimes \mathfrak{A} \otimes \mathfrak{A}$ (coassociator), with

$$\begin{aligned} &(\mathrm{id} \otimes \Delta)\Delta = \Phi(\Delta \otimes \mathrm{id})\Delta\Phi^{-1} \\ &(\mathrm{id} \otimes \epsilon) \circ \Delta = (\epsilon \otimes \mathrm{id}) \circ \Delta = \mathrm{id} \\ &(\mathrm{id} \otimes \mathrm{id} \otimes \Delta)\Phi \cdot (\Delta \otimes \mathrm{id} \otimes \mathrm{id})\Phi = (1 \otimes \Phi) \cdot (\mathrm{id} \otimes \Delta \otimes \mathrm{id})\Phi \cdot (\Phi \otimes 1) \\ &(\mathrm{id} \otimes \epsilon \otimes \mathrm{id})\Phi = 1, \end{aligned}$$

and for the antipode

$$\begin{aligned} &\sum_i S(x_i^{(1)})\alpha x_i^{(2)} = \epsilon(x)\alpha, \qquad \sum_i x_i^{(1)}\beta S(x_i^{(2)}) = \epsilon(x)\beta, \\ &\sum_i S(\varphi_i^{(1)})\alpha\varphi_i^{(2)}\beta S(\varphi_i^{(3)}) = 1, \qquad \sum_i \psi_i^{(1)}\beta S(\psi_i^{(2)})\alpha\psi_i^{(3)} = 1, \end{aligned}$$

where $x \in \mathfrak{A}$ with $\Delta(x) = \sum_i x_i^{(1)} \otimes x_i^{(2)}$ and

$$\Phi^{-1} = \sum_i \varphi_i^{(1)} \otimes \varphi_i^{(2)} \otimes \varphi_i^{(3)}, \qquad \Phi = \sum_i \psi_i^{(1)} \otimes \psi_i^{(2)} \otimes \psi_i^{(3)}.$$

The element Φ measures the lack of coassociativity of the coproduct.

Definition 2.4. A quasi-Hopf algebra $\mathfrak{A}$ is said to be *quasi-triangular* if it exists an invertible element $\mathcal{R} \in \mathfrak{A} \otimes \mathfrak{A}$, called the universal R-matrix, such that

$$\begin{aligned} \Delta'(x) &= \mathcal{R}\Delta(x)\mathcal{R}^{-1} \qquad (\forall x \in \mathfrak{A}) \\ (\Delta \otimes \mathrm{id})\mathcal{R} &= \Phi^{(312)} \mathcal{R}_{13} \Phi^{(132)^{-1}} \mathcal{R}_{23} \Phi^{(123)} \\ (\mathrm{id} \otimes \Delta)\mathcal{R} &= \Phi^{(231)^{-1}} \mathcal{R}_{13} \Phi^{(213)} \mathcal{R}_{12} \Phi^{(123)^{-1}}. \end{aligned}$$

It follows that $\mathcal{R}$ satisfies the generalized Yang–Baxter equation (in $\mathfrak{A} \otimes \mathfrak{A} \otimes \mathfrak{A}$):

$$\mathcal{R}_{12} \Phi^{(312)} \mathcal{R}_{13} \Phi^{(132)^{-1}} \mathcal{R}_{23} \Phi^{(123)} = \Phi^{(321)} \mathcal{R}_{23} \Phi^{(231)^{-1}} \mathcal{R}_{13} \Phi^{(213)} \mathcal{R}_{12}. \tag{2.2}$$

The notation $\Phi^{(312)}$ means that if $\Phi^{(123)} = \sum_i \varphi_i^{(1)} \otimes \varphi_i^{(2)} \otimes \varphi_i^{(3)}$, then $\Phi^{(312)} = \sum_i \varphi_i^{(3)} \otimes \varphi_i^{(1)} \otimes \varphi_i^{(2)}$, and so on.

Obviously, a quasi-Hopf algebra with $\Phi = 1$ is a Hopf algebra.

2.3 Drinfel'd twist

The notion of Drinfel'd twist allows us to associate to a given quasi-triangular quasi-Hopf algebra another quasi-triangular quasi-Hopf algebra in the following way. Consider an invertible element $\mathcal{F} \in \mathfrak{A} \otimes \mathfrak{A}$ such that $(\mathrm{id} \otimes \epsilon)\mathcal{F} = (\epsilon \otimes \mathrm{id})\mathcal{F} = 1$ (when $\mathfrak{A}$ is a quantum universal enveloping algebra, this means that the "leading" term in $\mathcal{F}$ is $1 \otimes 1$). One sets

$$\widetilde{\Delta}(x) = \mathcal{F}_{12}\, \Delta(x)\, \mathcal{F}_{12}^{-1} \tag{2.3}$$

$$\widetilde{\mathcal{R}} = \mathcal{F}_{21} \mathcal{R}_{12} \mathcal{F}_{12}^{-1} \tag{2.4}$$

$$\widetilde{\Phi} = (\mathcal{F}_{23}(\mathrm{id} \otimes \Delta)\mathcal{F})\, \Phi\, (\mathcal{F}_{12}(\Delta \otimes \mathrm{id})\mathcal{F})^{-1} \tag{2.5}$$

$$\widetilde{\alpha} = \sum_i S(w_i^{(1)})\, \alpha\, w_i^{(2)} \quad \text{and} \quad \widetilde{\beta} = \sum_i v_i^{(1)}\, \beta\, S(v_i^{(2)}), \tag{2.6}$$

where

$$\mathcal{F}_{12} = \sum_i v_i^{(1)} \otimes v_i^{(2)} \quad \text{and} \quad \mathcal{F}_{12}^{-1} = \sum_i w_i^{(1)} \otimes w_i^{(2)}. \tag{2.7}$$

Proposition 2.5 (Drinfel'd). *If $(\mathfrak{A}, \Phi, \Delta, \epsilon, S, \alpha, \beta, \mathcal{R})$ is a quasi-triangular quasi-Hopf algebra (QTQHA), then $(\mathfrak{A}, \widetilde{\Phi}, \widetilde{\Delta}, \epsilon, S, \widetilde{\alpha}, \widetilde{\beta}, \widetilde{\mathcal{R}})$ is also a QTQHA (denoted $\widetilde{\mathfrak{A}}$ for short).*

$\mathcal{F}$ is called a Drinfel'd twist.

In the following, we will mainly be concerned with twists of Hopf algebras. From now on, we consider the case where $\mathfrak{A}$ is a Hopf algebra ($\Phi = 1$) and $\mathcal{F}$ depends on parameters $\lambda \in \mathfrak{H}$ where $\mathfrak{H}$ is an Abelian subalgebra of $\mathfrak{A}$.

If $\mathcal{F}$ satisfies the so-called *shifted cocycle condition* ($h \in \mathfrak{H}$)

$$\mathcal{F}_{12}(\lambda)\,(\Delta \otimes \mathrm{id})\mathcal{F}(\lambda) = \mathcal{F}_{23}(\lambda + h^{(1)})\,(\mathrm{id} \otimes \Delta)\mathcal{F}(\lambda) \tag{2.8}$$

then the associator $\widetilde{\Phi}$ is given by $\widetilde{\Phi}^{(123)} = \mathcal{F}_{23}(\lambda)\,\mathcal{F}_{23}(\lambda + h^{(1)})^{-1}$ and the universal R-matrix $\widetilde{\mathcal{R}}$ satisfies the so-called *Gervais–Neveu–Felder* or *dynamical Yang–Baxter equation*:

$$\widetilde{\mathcal{R}}_{12}(\lambda + h^{(3)})\,\widetilde{\mathcal{R}}_{13}(\lambda)\,\widetilde{\mathcal{R}}_{23}(\lambda + h^{(1)}) = \widetilde{\mathcal{R}}_{23}(\lambda)\,\widetilde{\mathcal{R}}_{13}(\lambda + h^{(2)})\,\widetilde{\mathcal{R}}_{12}(\lambda). \tag{2.9}$$

Denoting by $\pi_{ev}(x)$ an evaluation representation of $\mathfrak{A}$ with evaluation parameter x, the Lax matrix $\widetilde{L}(x) = (\pi_{ev}(x) \otimes \mathrm{id})\,\widetilde{\mathcal{R}}$ realizes an FRT-type formalism of $\widetilde{\mathfrak{A}}$ with an evaluated R-matrix defined by $\widetilde{R}(x_1 - x_2) = (\pi_{ev}(x_1) \otimes \pi_{ev}(x_2))\,\widetilde{\mathcal{R}}$. The RLL relations take the form

$$\begin{aligned}\widetilde{R}_{12}(x_1 - x_2, \lambda + h)\,\widetilde{L}_1(x_1, \lambda)\,\widetilde{L}_2(x_2, \lambda + h^{(1)})\\ = \widetilde{L}_2(x_2, \lambda)\,\widetilde{L}_1(x_1, \lambda + h^{(2)})\,\widetilde{R}_{12}(x_1 - x_2, \lambda).\end{aligned} \tag{2.10}$$

Denoting by $\{h_i\}$ a basis of $\mathfrak{H}$ and $\{h_i^\vee\}$ the dual basis, the notation $\lambda + h^{(k)}$ means $\sum_i (\lambda_i + h_i{}^{(k)}) h_i^\vee$ if $\lambda = \sum_i \lambda_i h_i^\vee$ and the superscript (k) labels the space where h acts.

3 Elliptic algebras

3.1 The quantum affine elliptic algebra $\mathcal{A}_{q,p}(\widehat{sl}_N)$

The quantum affine elliptic algebra $\mathcal{A}_{q,p}(\widehat{sl}_2)$ was introduced in [11], generalized to the sl_N case in [26].

The quantum affine elliptic algebra $\mathcal{A}_{q,p}(\widehat{sl}_N)$ is defined in the RLL formalism. The R-matrix of the quantum affine elliptic algebra $\mathcal{A}_{q,p}(\widehat{sl}_N)$, associated to the $\mathbb{Z}_N$-vertex model, is defined as follows [27, 28]:

$$R[\mathcal{A}_{q,p}(\widehat{sl}_N)] = \rho(z, q, p) \sum_{(\alpha_1,\alpha_2) \in \mathbb{Z}_N \times \mathbb{Z}_N} W_{(\alpha_1,\alpha_2)}(\xi, \zeta, \tau)\; I_{(\alpha_1,\alpha_2)} \otimes I^{-1}_{(\alpha_1,\alpha_2)}, \tag{3.1}$$

where the variables z, q, p are related to the variables ξ, ζ, τ by

$$z = e^{i\pi\xi}, \quad q = e^{i\pi\zeta}, \quad p = e^{2i\pi\tau}.$$

The functions $W_{(\alpha_1,\alpha_2)}$ are given by

$$W_{(\alpha_1,\alpha_2)}(\xi,\zeta,\tau) = \frac{\vartheta\begin{bmatrix}\frac{1}{2}+\alpha_1/N\\ \frac{1}{2}+\alpha_2/N\end{bmatrix}(\xi+\zeta/N,\tau)}{N\vartheta\begin{bmatrix}\frac{1}{2}+\alpha_1/N\\ \frac{1}{2}+\alpha_2/N\end{bmatrix}(\zeta/N,\tau)}, \tag{3.2}$$

where the Jacobi theta functions with rational characteristics $(\gamma_1,\gamma_2) \in \frac{1}{N}\mathbb{Z} \times \frac{1}{N}\mathbb{Z}$ are defined by

$$\vartheta\begin{bmatrix}\gamma_1\\ \gamma_2\end{bmatrix}(\xi,\tau) = \sum_{m\in\mathbb{Z}} \exp\left(i\pi(m+\gamma_1)^2\tau + 2i\pi(m+\gamma_1)(\xi+\gamma_2)\right) \tag{3.3}$$

and the matrices $I_{(\alpha_1,\alpha_2)}$ by

$$I_{(\alpha_1,\alpha_2)} = g^{\alpha_2}\,h^{\alpha_1}, \tag{3.4}$$

the $(N\times N)$-matrices g and h being given by $g_{ij} = \omega^i\delta_{ij}$ and $h_{ij} = \delta_{i+1,j}$, the addition of indices being understood modulo N and $\omega = e^{2i\pi/N}$.

Finally, $\rho(z,q,p)$ is a suitable normalization factor.

The matrix R is $\mathbb{Z}_N$-symmetric, that is

$$R^{c+s,d+s}_{a+s,b+s} = R^{c,d}_{a,b} \tag{3.5}$$

for any indices $a,b,c,d,s \in \mathbb{Z}_N$.

In the particular case of $\mathcal{A}_{q,p}(\widehat{sl}_2)$, the R-matrix reduces, up to a gauge transformation by $g^{\frac{1}{2}} \otimes g^{\frac{1}{2}}$, to the R-matrix of the eight vertex model [29]:

$$R[\mathcal{A}_{q,p}(\widehat{sl}_2)] = \rho(z,q,p)\begin{pmatrix} a(z) & 0 & 0 & d(z)\\ 0 & b(z) & c(z) & 0\\ 0 & c(z) & b(z) & 0\\ d(z) & 0 & 0 & a(z)\end{pmatrix}, \tag{3.6}$$

where

$$a(z) = \frac{\Theta_{p^2}(q^2z^2)\,\Theta_{p^2}(pq^2)}{\Theta_{p^2}(pq^2z^2)\,\Theta_{p^2}(q^2)}, \quad d(z) = -\frac{p^{1/2}}{qz^2}\,\frac{\Theta_{p^2}(z^2)\,\Theta_{p^2}(q^2z^2)}{\Theta_{p^2}(pz^2)\,\Theta_{p^2}(pq^2z^2)},$$

$$b(z) = \frac{\Theta_{p^2}(z^2)\,\Theta_{p^2}(pq^2)}{\Theta_{p^2}(pz^2)\,\Theta_{p^2}(q^2)}, \qquad c(z) = 1 \tag{3.7}$$

with $\Theta_a(x) = (x;a)_\infty\,(ax^{-1};a)_\infty\,(a;a)_\infty$ and $(x;a)_\infty = \prod_{n\ge 0}(1-xa^n)$.

The quantum affine elliptic algebra $\mathcal{A}_{q,p}(\widehat{sl}_N)$ is defined as an algebra of operators $L_{ij}(z) = \sum_{n\in\mathbb{Z}} L_{ij}(n)\,z^n$ where $i,j \in \mathbb{Z}_N$, encapsulated into a $(N\times N)$-matrix. One first defines $\mathcal{A}_{q,p}(\widehat{gl}_N)$ by imposing the following constraints on the $L_{ij}(z)$:

$$R_{12}(z/w,q,p)\,L_1(z)\,L_2(w) = L_2(w)\,L_1(z)\,R_{12}(z/w,q,pq^{-2c}) \tag{3.8}$$

with $L_1(z) = L(z) \otimes 1$, $L_2(z) = 1 \otimes L(z)$.

The q-determinant of $L(z)$ given by

$$q\text{-}\det L(z) \equiv \sum_{\sigma \in \mathfrak{S}_N} \varepsilon(\sigma) \prod_{i=1}^{N} L_{i,\sigma(i)}(zq^{i-N-1}) \tag{3.9}$$

($\varepsilon(\sigma)$ being the signature of the permutation σ) is in the center of $\mathcal{A}_{q,p}(\widehat{gl}_N)$. It can be factored out, and set to the value $q^{\frac{c}{2}}$ (c being the central charge) so as to get

$$\mathcal{A}_{q,p}(\widehat{sl}_N) = \mathcal{A}_{q,p}(\widehat{gl}_N)/\langle q\text{-}\det L - q^{\frac{c}{2}} \rangle. \tag{3.10}$$

It was proved in [22] that $\mathcal{A}_{q,p}(\widehat{sl}_N)$ was a QTQHA through a Drinfel'd twist applied to the Hopf algebra $\mathcal{U}_q(\widehat{sl}_N)$. The sketch of the proof is as follows.

Consider the affine Lie algebra $\widehat{sl}_N$ with Cartan basis $\mathfrak{H} = \{h_0, \dots, h_{N-1}, d\}$ and dual basis $\mathfrak{H}^* = \{\Lambda_0, \dots, \Lambda_{N-1}, c\}$ where Λ_i are the fundamental weights, d the derivation and c the central charge. Let τ be the automorphism of order N related to the cyclic symmetry of the extended Dynkin diagram of $\widehat{sl}_N$: $\tau(x_i) = x_{i+1 \bmod N}$ where the x_i's denote the Chevalley generators of $\widehat{sl}_N$. One defines

$$\varphi = \tau \circ \operatorname{Ad}\left(q^{2(r+c)\rho/N}\right), \tag{3.11}$$

where $\rho = \sum_{i=0}^{N-1} \Lambda_i$ gives the principal grading and r is some complex number.

The vertex-type Drinfel'd twist is then given by

$$\mathcal{F}(r) = \prod_{k \geq 1}^{\curvearrowleft} (\varphi^k \otimes \mathrm{id})(\widehat{\mathcal{R}}^{-1}), \tag{3.12}$$

where $\widehat{\mathcal{R}} = q^T \mathcal{R}$ and $T = \frac{1}{N}(\rho \otimes c + c \otimes \rho - \frac{N^2-1}{12} c \otimes c)$.

Theorem 3.1 (JKOS). *The vertex-type Drinfel'd twist* (3.12) *satisfies the shifted cocycle condition*

$$\mathcal{F}_{12}(r)\,(\Delta \otimes \mathrm{id})\mathcal{F}(r) = \mathcal{F}_{23}(r + c^{(1)})\,(\mathrm{id} \otimes \Delta)\mathcal{F}(r).$$

The quantum affine elliptic algebra $\mathcal{A}_{q,p}(\widehat{sl}_N)$ (where $p = q^{2r}$) is a quasi-triangular quasi-Hopf algebra with the universal R-matrix $\widetilde{\mathcal{R}}(r) = \mathcal{F}_{21}(r)\,\mathcal{R}_{12}\,\mathcal{F}_{12}^{-1}(r)$ where $\mathcal{R}$ is the universal R-matrix of $\mathcal{U}_q(\widehat{sl}_N)$. $\widetilde{\mathcal{R}}(r)$ satisfies the shifted Yang–Baxter equation:

$$\widetilde{\mathcal{R}}_{12}(r + c^{(3)})\,\widetilde{\mathcal{R}}_{13}(r)\,\widetilde{\mathcal{R}}_{23}(r + c^{(1)}) = \widetilde{\mathcal{R}}_{23}(r)\,\widetilde{\mathcal{R}}_{13}(r + c^{(2)})\,\widetilde{\mathcal{R}}_{12}(r).$$

3.2 The quantum affine elliptic algebra $\mathcal{B}_{q,p,\lambda}(\widehat{sl}_N)$

The face-type R-matrices depend on the extra parameters λ belonging to the dual of the Cartan algebra in the underlying algebra. Let $\mathfrak{H}$ be the Cartan subalgebra of the affine Lie algebra $\widehat{sl}_N$, with basis $\{h_i\}$ and dual basis $\{h_i^\vee\}$, and $\Pi^0 = \{\alpha_i\}$ be the corresponding simple root system of $\widehat{sl}_N$. We set $\lambda = \rho + s_1 h_1 + \cdots + s_{N-1} h_{N-1} + rd + s'c$ where $\rho \in \mathfrak{H}$ is such that $(\rho, \alpha_i) = 1$.

The R-matrix of the quantum affine elliptic algebra $\mathcal{B}_{q,p,\lambda}(\widehat{sl}_N)$ takes the following form $(1 \le a, b \le N)$ [12]:

$$R(z) = \rho(z)\Big(\sum_a E_{aa} \otimes E_{aa} + \sum_{a \neq b} q^2 \frac{\Theta_p(q^{-2} w_{ab})}{\Theta_p(w_{ab})} \frac{\Theta_p(z)}{\Theta_p(q^2 z)} E_{aa} \otimes E_{bb}$$

$$+ \sum_{a \neq b} \frac{\Theta_p(w_{ab} z)}{\Theta_p(w_{ab})} \frac{\Theta_p(q^2)}{\Theta_p(q^2 z)} E_{ab} \otimes E_{ba}\Big), \tag{3.13}$$

where the E_{ab} are the elementary $(N \times N)$-matrices with entries $(E_{ab})_i^j = \delta_{ai}\delta_{jb}$, $w_{ab} = q^{x_a - x_b}$, $x_a = 2s_a - 2s_{a-1}$ with $s_0 = s_N = 0$, and $\rho(z)$ is a suitable normalization factor.

It is proved in [22] that the quantum affine elliptic algebra $\mathcal{B}_{q,p,\lambda}(\widehat{sl}_N)$ was a QTQHA obtained by a Drinfel'd twist of shifted-cocycle type from $\mathcal{U}_q(\widehat{sl}_N)$. The R-matrix then obtained is actually a gauge transform of (3.13) where the coefficients of $E_{aa} \otimes E_{bb}$ become

$$R_{ab}^{ab} = \rho(z) \begin{cases} q \, \dfrac{(p w_{ab}^{-1} q^2; p)_\infty \, (p w_{ab}^{-1} q^{-2}; p)_\infty}{(p w_{ab}^{-1}; p)_\infty^2} \, \dfrac{\Theta_p(z)}{\Theta_p(q^2 z)} & \text{if } b > a, \\[2ex] q \, \dfrac{(w_{ab}^{-1} q^2; p)_\infty \, (w_{ab}^{-1} q^{-2}; p)_\infty}{(w_{ab}^{-1}; p)_\infty^2} \, \dfrac{\Theta_p(z)}{\Theta_p(q^2 z)} & \text{if } b < a. \end{cases} \tag{3.14}$$

Defining

$$\varphi = \mathrm{Ad}\left(q^{\sum_i h_i h_i^\vee + 2(\lambda - \rho)}\right), \tag{3.15}$$

the face-type Drinfel'd twist is then given by

$$\mathcal{F}(\lambda) = \prod_{k \ge 1}^{\curvearrowleft} (\varphi^k \otimes \mathrm{id})(\widehat{\mathcal{R}}^{-1}), \tag{3.16}$$

where $\widehat{\mathcal{R}} = q^T \mathcal{R}$ and $T = \sum_i h_i \otimes h_i^\vee$.

Theorem 3.2 (JKOS). *The face-type Drinfel'd twist* (3.16) *satisfies the shifted cocycle condition*

$$\mathcal{F}_{12}(\lambda)\,(\Delta \otimes \mathrm{id})\mathcal{F}(\lambda) = \mathcal{F}_{23}(\lambda + h^{(1)})\,(\mathrm{id} \otimes \Delta)\mathcal{F}(\lambda).$$

The quantum affine elliptic algebra $\mathcal{B}_{q,p,\lambda}(\widehat{sl}_N)$ is a quasi-triangular quasi-Hopf algebra with the universal R-matrix $\widetilde{\mathcal{R}}(\lambda) = \mathcal{F}_{21}(\lambda)\,\mathcal{R}_{12}\,\mathcal{F}_{12}^{-1}(\lambda)$ where $\mathcal{R}$ is the universal R-matrix of $\mathcal{U}_q(\widehat{sl}_N)$. $\widetilde{\mathcal{R}}(\lambda)$ satisfies the dynamical Yang–Baxter equation:

$$\widetilde{\mathcal{R}}_{12}(\lambda + h^{(3)})\,\widetilde{\mathcal{R}}_{13}(\lambda)\,\widetilde{\mathcal{R}}_{23}(\lambda + h^{(1)}) = \widetilde{\mathcal{R}}_{23}(\lambda)\,\widetilde{\mathcal{R}}_{13}(\lambda + h^{(2)})\,\widetilde{\mathcal{R}}_{12}(\lambda).$$

4 Deformed double Yangian

4.1 Double Yangian $\mathcal{D}Y(sl_2)$

The centrally extended double Yangian $\mathcal{D}Y(sl_2)$ is defined in the RLL formalism by introducing the following R-matrix:

$$R[\mathcal{D}Y(sl_2)] = \rho(u)\begin{pmatrix} 1 & 0 & 0 & 0 \\ 0 & \frac{u}{u+1} & \frac{1}{u+1} & 0 \\ 0 & \frac{1}{u+1} & \frac{u}{u+1} & 0 \\ 0 & 0 & 0 & 1 \end{pmatrix} \tag{4.1}$$

the normalization factor being $\rho(u) = \Gamma_1(u|2)\Gamma_1(2+u|2)\Gamma_1(1+u|2)^{-2}$. This matrix can be obtained as the scaling limit $q \to 1$ with $z = q^u$, u being kept fixed, of the R-matrix of $\mathcal{U}_q(\widehat{sl}_2)$. $\mathcal{D}Y(sl_2)$ is then defined by the relations

$$\begin{aligned} R_{12}(u_1 - u_2)\,L_1^{\pm}(u_1)\,L_2^{\pm}(u_2) &= L_2^{\pm}(u_2)\,L_1^{\pm}(u_2)\,R_{12}(u_1 - u_2) \\ R_{12}(u_1 - u_2 - c)\,L_1^{-}(u_1)\,L_2^{+}(u_2) &= L_2^{+}(u_2)\,L_1^{-}(u_2)\,R_{12}(u_1 - u_2), \end{aligned} \tag{4.2}$$

where $L^+(u) = \sum_{k\geq 0} L_k^+ u^{-k}$ and $L^-(u) = \sum_{k\leq 0} L_k^- u^{-k}$ are two 2×2 independent matrices. There exists in this case a Gauss decomposition of the Lax matrices allowing for an alternative Drinfeld presentation [18]. Indeed, $L^{\pm}$ are decomposed as

$$L^{\pm}(u) = \begin{pmatrix} 1 & f^{\pm}(u^{\mp}) \\ 0 & 1 \end{pmatrix}\begin{pmatrix} k_1^{\pm}(u) & 0 \\ 0 & k_2^{\pm}(u) \end{pmatrix}\begin{pmatrix} 1 & 0 \\ e^{\pm}(u) & 1 \end{pmatrix}$$

with $u^+ = u$ and $u^- = u - c$. Furthermore, $k_1^{\pm}(u)k_2^{\pm}(u-1) = 1$ and one defines $h^{\pm}(u) \equiv k_2^{\pm}(u)^{-1}k_1^{\pm}(u)$.

The generators of $\mathcal{D}Y(sl_2)$ being expanded as

$$e^{\pm}(u) = \pm\sum_{\substack{k\geq 0\\ k<0}} e_k u^{-k-1}, \quad f^{\pm}(u) = \pm\sum_{\substack{k\geq 0\\ k<0}} f_k u^{-k-1}, \quad h^{\pm}(u) = 1 \pm \sum_{\substack{k\geq 0\\ k<0}} h_k u^{-k-1}, \tag{4.3}$$

the universal R-matrix of $\mathcal{D}Y(sl_2)$ is given by [17]

$$\mathcal{R}[\mathcal{D}Y(sl_2)] = \mathcal{R}_E\,\mathcal{R}_H\,\mathcal{R}_F, \tag{4.4}$$

where

$$\begin{aligned}
\mathcal{R}_E &= \prod_{k\geq 0}^{\curvearrowright} \exp(-e_k \otimes f_{-k-1}) \,, \qquad \mathcal{R}_F = \prod_{k\geq 0}^{\curvearrowleft} \exp(-f_k \otimes e_{-k-1}) \,, \\
\mathcal{R}_H &= \prod_{k\geq 0} \exp\Big(\sum_{n\geq 0} \Big(-\frac{d}{du} \ln h_+(u)\Big)_n \otimes \big(\ln h_-(x+2k+1)\big)_{-n-1}\Big).
\end{aligned} \tag{4.5}$$

4.2 Deformed double Yangian $\mathcal{D}Y_r(sl_2)$

4.2.1 Definition of the deformed double Yangian $\mathcal{D}Y_r(sl_2)$. Consider the R-matrix of $\mathcal{A}_{q,p}(\widehat{sl}_2)$, and perform the scaling limit $q \to 1$, with $z = q^u$, $p = q^{2r}$, keeping u and r fixed. One obtains, up to a gauge transformation, the following R-matrix:

$$R[\mathcal{D}Y_r(sl_2)] = \rho(u,r) \begin{pmatrix} 1 & 0 & 0 & 0 \\ 0 & \frac{\sin \pi u/r}{\sin \pi(u+1)/r} & \frac{\sin \pi/r}{\sin \pi(u+1)/r} & 0 \\ 0 & \frac{\sin \pi/r}{\sin \pi(u+1)/r} & \frac{\sin \pi u/r}{\sin \pi(u+1)/r} & 0 \\ 0 & 0 & 0 & 1 \end{pmatrix}, \tag{4.6}$$

the normalization factor being

$$\rho(u,r) = \frac{\Gamma_2(r+1-u|r,2)^2\, \Gamma_2(2+u|r,2)\, \Gamma_2(u|r,2)}{\Gamma_2(u+1|r,2)^2\, \Gamma_2(r-u|r,2)\, \Gamma_2(r+2-u|r,2)}, \tag{4.7}$$

where Γ_2 is the multiple Gamma function of order 2 (see Appendix).

Taking now $L(u) = \sum_{n\in\mathbb{Z}} L_n u^{-n}$, the deformed double Yangian $\mathcal{D}Y_r(sl_2)$ is defined by

$$R_{12}(u_1-u_2,r)\, L_1(u_1)\, L_2(u_2) = L_2(u_2)\, L_1(u_2)\, R_{12}(u_1-u_2, r-c). \tag{4.8}$$

4.2.2 The deformed double Yangian $\mathcal{D}Y_r(sl_2)$ is a QTQHA. What is the status of this algebra? In fact, in the same way the quantum affine elliptic algebra $\mathcal{A}_{q,p}(\widehat{sl}_2)$ appears as a Drinfel'd twist of the universal quantum affine algebra $\mathcal{U}_q(\widehat{sl}_2)$, we sketch below how the deformed double Yangian $\mathcal{D}Y_r(sl_2)$ can be obtained as a Drinfel'd twist of the double Yangian $\mathcal{D}Y(sl_2)$, promoting $\mathcal{D}Y_r(sl_2)$ as a QTQHA. For this purpose, one must needs prove that the matrix (4.6) is indeed an evaluation representation of a universal R-matrix obtained as a Drinfel'd twist of the universal R-matrix of $\mathcal{D}Y(sl_2)$.

Introducing the notation

$$M(b^+,b^-) = \begin{pmatrix} 1 & 0 & 0 & 0 \\ 0 & \frac{1}{2}(b^+ + b^-) & \frac{1}{2}(b^+ - b^-) & 0 \\ 0 & \frac{1}{2}(b^+ - b^-) & \frac{1}{2}(b^+ + b^-) & 0 \\ 0 & 0 & 0 & 1 \end{pmatrix}, \tag{4.9}$$

one has obviously $M(a,b)M(a',b') = M(aa',bb')$ and $M(a,b)^{-1} = M(a^{-1},b^{-1})$. The R-matrix of $\mathcal{D}Y(sl_2)$ can be written as $R[\mathcal{D}Y(sl_2)](u) = \rho(u)M\left(1,\frac{u-1}{u+1}\right)$, and the R-matrix of $\mathcal{D}Y_r(sl_2)$ takes the form $R[\mathcal{D}Y_r(sl_2)](u) = \rho_r(u)M(b_r^+,b_r^-)$ with

$$b_r^+ = \frac{\cos\frac{u-1}{2r}}{\cos\frac{u+1}{2r}} = \frac{\Gamma_1(r+u+1|2r)\Gamma_1(r-u-1|2r)}{\Gamma_1(r+u-1|2r)\Gamma_1(r-u+1|2r)},$$

$$\begin{aligned} b_r^- &= \frac{\sin\frac{u-1}{2r}}{\sin\frac{u+1}{2r}} = \frac{\Gamma_1(u+1|2r)\Gamma_1(2r-u-1|2r)}{\Gamma_1(u-1|2r)\Gamma_1(2r-u+1|2r)} \\ &= \frac{\Gamma_1(2r+u+1|2r)\Gamma_1(2r-u-1|2r)}{\Gamma_1(2r+u-1|2r)\Gamma_1(2r-u+1|2r)}\cdot\frac{u-1}{u+1}. \end{aligned}$$

The normalization factor of $\mathcal{D}Y_r(sl_2)$ being rewritten as $\rho_r(u) = \rho_F(-u;r)\rho(u)$ $\rho_F(u;r)^{-1}$ with $\rho_F(u) = \dfrac{\Gamma_2(u+1+r|2,r)^2}{\Gamma_2(u+r|2,r)\,\Gamma_2(u+2+r|2,r)}$, one obtains

$$R[\mathcal{D}Y_r(sl_2)] = F_{21}(-u)R[\mathcal{D}Y(sl_2)]F_{12}(u)^{-1}. \tag{4.10}$$

Using the notation (4.9), $F_{12}(u)$ is given by

$$F_{12}(u) = \rho_F(u)\,M\left(\frac{\Gamma_1(u+r-1|2r)}{\Gamma_1(u+r+1|2r)},\frac{\Gamma_1(u+2r-1|2r)}{\Gamma_1(u+2r+1|2r)}\right). \tag{4.11}$$

The main point is that $F_{12}(u)$ satisfies a *difference equation*:

$$F_{12}(u) = \mathrm{Ad}\left((-1)^{-\frac{1}{2}h_0}\otimes 1\right)F_{12}(u-r)\cdot R_{12}(u). \tag{4.12}$$

One imposes then the difference equation *at the universal level*:

$$\mathcal{F}_{12}(u) = \mathrm{Ad}(\phi\otimes 1)[\mathcal{F}_{12}(u)]\,\mathcal{R}_{12}(u), \tag{4.13}$$

where $\mathcal{F}_{12}(u) = \mathrm{Ad}(e^{ud}\otimes\mathrm{id})\mathcal{F}_{12}$ and $\phi = (-1)^{-\frac{1}{2}h_0}e^{-(r+c)d}$, h_0 being the Cartan generator of $sl_2 \subset \mathcal{D}Y(sl_2)$.

A solution of (4.13) is given by an infinite product

$$\mathcal{F}_{21}(u) = \prod_{n\geq 1}^{\curvearrowleft} \mathrm{Ad}(\phi^n\otimes 1)[\mathcal{R}_{12}^{-1}(u)]. \tag{4.14}$$

Finally, one can prove the following theorem [30]:

Theorem 4.1. *The Yangian-type Drinfel'd twist* (4.14) *satisfies the shifted cocycle condition*

$$\mathcal{F}_{12}(r)\,(\Delta\otimes\mathrm{id})\mathcal{F}(r) = \mathcal{F}_{23}(r+c^{(1)})\,(\mathrm{id}\otimes\Delta)\mathcal{F}(r).$$

The deformed double Yangian $\mathcal{D}Y_r(sl_2)$ is a quasi-triangular quasi-Hopf algebra with the universal R-matrix $\tilde{\mathcal{R}}(r) = \mathcal{F}_{21}(r)\,\mathcal{R}_{12}\,\mathcal{F}_{12}^{-1}(r)$ where $\mathcal{R}$ is the universal

R-matrix of $\mathcal{D}Y(sl_2)$. $\widetilde{\mathcal{R}}(r)$ satisfies the shifted Yang–Baxter equation:

$$\widetilde{\mathcal{R}}_{12}(r+c^{(3)})\,\widetilde{\mathcal{R}}_{13}(r)\,\widetilde{\mathcal{R}}_{23}(r+c^{(1)}) = \widetilde{\mathcal{R}}_{23}(r)\,\widetilde{\mathcal{R}}_{13}(r+c^{(2)})\,\widetilde{\mathcal{R}}_{12}(r).$$

Remark 4.2. It may happen that the infinite product (4.14) is divergent in representation and therefore needs to be regularized. For example, in the case of $\mathcal{D}Y_r(sl_2)$ illustrated here, one can show that the product (4.14) in representation is at most logarithmically divergent. The Gamma functions are precisely a consistent regularization of that product.

4.2.3 Other presentation of $\mathcal{D}Y_r(sl_2)$. As it was said before, the R-matrix $\overline{R}[\mathcal{D}Y_r(sl_2)]$ obtained from the scaling limit of the R-matrix of $\mathcal{A}_{q,p}(\widehat{sl_2})$ differs from the R-matrix $R[\mathcal{D}Y_r(sl_2)]$ given by Eq. (4.6) by a gauge transformation. Indeed, one has

$$\overline{R}[\mathcal{D}Y_r(sl_2)] = K_{21}\; R[\mathcal{D}Y_r(sl_2)]\; K_{12}^{-1}\,, \tag{4.15}$$

where

$$K = V\otimes V \quad \text{with } V = \frac{1}{\sqrt{2}}\begin{pmatrix} 1 & 1 \\ -1 & 1 \end{pmatrix}. \tag{4.16}$$

The Lax operators L and $\overline{L}$ associated respectively to the R-matrices $R[\mathcal{D}Y_r(sl_2)]$ and $\overline{R}[\mathcal{D}Y_r(sl_2)]$ are connected by $\overline{L} = VLV^{-1}$, implying an isomorphism between the two corresponding Yangian structures.

One can identify V with an evaluation representation of an element g: $V = \pi_{ev}(g)$ with $g = \exp\left(\frac{-\pi}{4}(f_0 - e_0)\right)$. Since e_0 and f_0 lie in the undeformed Hopf subalgebra $\mathcal{U}(sl(2))$ of $\mathcal{D}Y(sl_2)$ [18], the coproduct of g reads $\Delta(g) = g\otimes g$, so that

$$g_1 g_2\, \Delta^{\mathcal{F}}(g^{-1})\mathcal{F} = g_1 g_2\, \mathcal{F} g_1^{-1} g_2^{-1}, \tag{4.17}$$

where Δ is the coproduct of $\mathcal{D}Y(sl_2)$ while $\Delta^{\mathcal{F}}$ is the coproduct of $\mathcal{D}Y_r(sl_2)$.

The two-cocycle $g_1 g_2 \Delta^{\mathcal{F}}(g^{-1})$ is a coboundary (with respect to the coproduct $\Delta^{\mathcal{F}}$). In representation, (4.17) is equal to the scaling limit of the represented twist from $\mathcal{U}_q(\widehat{sl_2})$ to $\mathcal{A}_{q,p}(\widehat{sl_2})$.

It follows that

$$\begin{aligned}\overline{\mathcal{R}}[\mathcal{D}Y_r(sl_2)] &\equiv g_1 g_2\, \Delta_{21}^{\mathcal{F}}(g^{-1})\; \mathcal{R}[\mathcal{D}Y_r(sl_2)]\; \Delta_{12}^{\mathcal{F}}(g)\, g_1^{-1} g_2^{-1} \\ &= g_1 g_2\; \mathcal{R}[\mathcal{D}Y_r(sl_2)]\; g_1^{-1} g_2^{-1}\end{aligned} \tag{4.18}$$

satisfies the shifted Yang–Baxter equation of Theorem 4.1.

To recover (4.15), it suffices to use (4.17) and to remark that $(\pi_{ev}\otimes\pi_{ev})(g\otimes g)$ commutes with $R[\mathcal{D}Y_r(sl_2)]$.

4.3 Generalization to sl_N

4.3.1 Definition of the double Yangian $\mathcal{D}Y(sl_N)$. The definition of the double Yangian $\mathcal{D}Y(sl_2)$ can be generalized to any simple Lie algebra. However in this case, the universal R-matrix has only been conjectured in [17]. In what follows, we restrict the generalization to the sl_N case.

The double Yangian $\mathcal{D}Y(sl_N)$ is a Hopf algebra generated by the elements $e_{i,k}$, $f_{i,k}$ and $h_{i,k}$ with $i = 1, \dots, r$ and $k \in \mathbb{Z}$, the elements $e_{i,0}$, $f_{i,0}$ and $h_{i,0}$ corresponding to the Chevalley generators of the Lie algebra sl_N. In [17], it was conjectured that the universal R-matrix of $\mathcal{D}Y(sl_N)$ was given as follows:[1]

$$\mathcal{R}[\mathcal{D}Y(sl_N)] = \mathcal{R}_E\, \mathcal{R}_H\, \mathcal{R}_F, \tag{4.19}$$

where

$$\mathcal{R}_H = \prod_{k\geq 0} \exp\Big(\sum_{i,j=1,\dots,r} \sum_{n\geq 0} \Big(-\frac{d}{du} \ln h_{i,+}(u)\Big)_n \otimes \big(C_{ji}(T^{\frac{1}{2}}) \ln h_{j,-}(x + N(k+\tfrac{1}{2}))\big)_{-n-1}\Big). \tag{4.20}$$

The matrix $\frac{1}{[N]_q} C_{ij}$ is the inverse of the quantum analogue of the Cartan matrix of sl_N, and T is the shift operator $(Tf(x) = f(x+1))$.

In order to write the expression of the factors $\mathcal{R}_E$ and $\mathcal{R}_F$, one needs to fix some notations. Let E and F be the unital subalgebras of $\mathcal{D}Y(sl_N)$ generated respectively by the generators $e_{i,k}$ $(k \geq 0)$ and $f_{i,k}$ $(k \geq 0)$. Let Π and $\Pi^+ \subset \Pi$ be the root and positive root systems of sl_N and $\widehat{\Pi} = \{\gamma + n\delta,\ m\delta \mid \gamma \in \Pi,\ n, m \in \mathbb{Z},\ m \neq 0\}$ be the root system of $\widehat{sl}_N$. One defines the following two subsets of $\widehat{\Pi}$: $\Pi_E = \{\gamma + n\delta \mid \gamma \in \Pi^+,\ n \in \mathbb{Z}_+\}$ and $\Pi_F = \{-\gamma + n\delta \mid \gamma \in \Pi^+,\ n \in \mathbb{Z}_+\}$. Π_E and Π_F are equipped with two orderings $<_E$ and $<_F$ such that

$$\begin{aligned} &\xi_1 <_E \xi_1 + \xi_2 <_E \xi_2 \text{ if } \xi_1, \xi_2, \xi_1 + \xi_2 \in \Pi_E \quad \text{and} \quad \gamma + n\delta <_E \gamma + m\delta, \\ &\xi_2 <_F \xi_1 + \xi_2 <_F \xi_1 \text{ if } \xi_1, \xi_2, \xi_1 + \xi_2 \in \Pi_F \quad \text{and} \quad -\gamma + n\delta >_F -\gamma + m\delta, \end{aligned}$$

where $\gamma \in \Pi^+$ and $n > m$.

By induction, for any root $\xi = \gamma + n\delta \in \Pi_E$ (resp. $\xi = -\gamma + n\delta \in \Pi_F$), one constructs the roots generators $e_\xi \equiv e_{\gamma,n}$ (resp. $f_{\gamma,n}$) and $e_{-\xi} \equiv f_{\gamma,-n}$ (resp. $e_{\gamma,-n}$) by

$$e_\xi = [e_{\xi_1}, e_{\xi_2}] \quad \text{and} \quad e_{-\xi} = [e_{-\xi_2}, e_{-\xi_1}], \tag{4.21}$$

where $\xi_1 <_E \xi <_E \xi_2$ (resp. $\xi_1 <_F \xi <_F \xi_2$) and (ξ_1, ξ_2) is a minimal segment in the chosen ordering.

[1] To be more precise, the factors $\mathcal{R}_E$ and $\mathcal{R}_F$ were only conjectured, but the factor $\mathcal{R}_H$ was rigorously derived in the construction of the quantum double.

Then the factors $\mathcal{R}_E$ and $\mathcal{R}_F$ of the universal R-matrix of $\mathcal{D}Y(sl_N)$ have been conjectured to be [17]:

$$\mathcal{R}_E = \prod_{\gamma\in\Delta_+, n\in\mathbb{Z}_+}^{\curvearrowright} \exp(-e_{\gamma,n} \otimes f_{\gamma,-n-1}) \,, \tag{4.22}$$

$$\mathcal{R}_F = \prod_{\gamma\in\Delta_+, n\in\mathbb{Z}_+}^{\curvearrowright} \exp(-f_{\gamma,n} \otimes e_{\gamma,-n-1}) \,, \tag{4.23}$$

where the arrows correspond to ordered products according to $<_E$ and $<_F$ respectively.

Now, using the following evaluation representation

$$\begin{aligned} \pi_{ev}(e_{i,k}) &= E_{i,i+1}\ (z + \tfrac{1}{2}(i-1))^k \\ \pi_{ev}(f_{i,k}) &= E_{i+1,i}\ (z + \tfrac{1}{2}(i-1))^k \\ \pi_{ev}(h_{i,k}) &= (E_{i,i} - E_{i+1,i+1})\ (z + \tfrac{1}{2}(i-1))^k, \end{aligned} \tag{4.24}$$

the represented R-matrix of $\mathcal{D}Y(sl_N)$ takes the form $(1 \le a, b \le N)$

$$R[\mathcal{D}Y(sl_N)](u) = \rho(u)\Big(\sum_a E_{aa} \otimes E_{aa} + \frac{1}{u+1} \sum_{a\neq b} (u\, E_{aa} \otimes E_{bb} + E_{ab} \otimes E_{ba})\Big) \tag{4.25}$$

with

$$\rho(u) = \frac{\Gamma_1(u|N)\ \Gamma_1(u+N|N)}{\Gamma_1(u+1|N)\ \Gamma_1(u+N-1|N)}. \tag{4.26}$$

4.3.2 Definition of the deformed double Yangian $\mathcal{D}Y_r(sl_N)$. We now construct the deformed double Yangian $\mathcal{D}Y_r(sl_N)$ [31]. The R-matrix of the deformed double Yangian $\mathcal{D}Y_r(sl_N)$ is obtained by taking the scaling limit ($q \to 1$ with $z = q^u$ and $p = q^{2r}$ and keeping u, r fixed) of the R-matrix of the quantum affine elliptic algebra $\mathcal{A}_{q,p}(\widehat{sl}_N)$. One gets

$$\overline{R}(u,r) = \rho_{DYr}(u) \times \sum_{a,b,c=1}^{N} \frac{\sin\frac{\pi u}{r}\ \sin\frac{\pi}{r}\ \sin\frac{\pi}{Nr}(u+1+(b-a)r)}{N \sin\frac{\pi(u+1)}{r}\ \sin\frac{\pi}{Nr}(u+(b-c)r)\ \sin\frac{\pi}{Nr}(1-(a-c)r)}\ E_{ac} \otimes E_{b,a+b-c}, \tag{4.27}$$

the normalization factor $\rho_{DYr}(u)$ being defined by

$$\rho_{DYr}(u) = \frac{S_2(-u|r,N)\, S_2(1+u|r,N)}{S_2(u|r,N)\, S_2(1-u|r,N)}, \tag{4.28}$$

where $S_2(x|\omega_1, \omega_2)$ is Barnes' double sine function of periods ω_1 and ω_2 (see Appendix).

In fact, it is possible to simplify this matrix by a gauge transformation. Define $V_i^j = 1/\sqrt{N}\ \omega^{(i-1)j}$ where $\omega = \exp(2i\pi/N)$. The gauge transformed matrix $R = (V \otimes V)\overline{R}(V \otimes V)^{-1}$ has the following non zero entries:

$$\begin{aligned} S_{aa}^{aa}(u) &= \cot\tfrac{\pi u}{r} + \cot\tfrac{\pi}{r}, \\ S_{ab}^{ab}(u) &= \frac{e^{i\pi/r}}{\sin\frac{\pi}{r}} e^{-2i\pi(b-a)/Nr} \quad \text{for } b-a \in \{1,\dots,N-1\}, \\ S_{ab}^{ba}(u) &= \frac{e^{i\pi u/r}}{\sin\frac{\pi u}{r}} e^{-2i\pi(b-a)u/Nr} \quad \text{for } b-a \in \{1,\dots,N-1\}, \end{aligned} \tag{4.29}$$

where the matrix elements R_{ij}^{kl} and S_{ij}^{kl} are related by

$$R_{ij}^{kl} = -\rho_{DY_r}(u)\, \frac{\sin\frac{\pi u}{r}\,\sin\frac{\pi}{r}}{\sin\frac{\pi(u+1)}{r}}\, S_{ij}^{kl}. \tag{4.30}$$

4.3.3 The deformed double Yangian $\mathcal{D}Y_r(sl_N)$ is a QTQHA. Inspired by the expressions obtained in the sl_2 case, one postulates the following linear equation for the twist $\mathcal{F}$ [32]:

$$\mathcal{F} \equiv \mathcal{F}(r) = (\phi^{-1} \otimes \mathrm{id})(\mathcal{F}) \cdot \mathcal{C} \tag{4.31}$$

with

$$\phi = \mathrm{Ad}\left(\omega^{h_{0,\rho}} e^{(r+c)d}\right), \tag{4.32}$$

$$\mathcal{C} = e^{\frac{1}{2}(c\otimes d + d\otimes c)}\mathcal{R}. \tag{4.33}$$

Equation (4.31) can be solved by

$$\mathcal{F}(r) = \prod_k^{\curvearrowleft} (\phi^k \otimes \mathrm{id})\ \mathcal{C}^{-1}. \tag{4.34}$$

Using the following properties:

- the operator d in the double Yangian $\mathcal{D}Y(sl_N)$ is defined by $[d, e_\alpha(u)] = \frac{d}{du}e_\alpha(u)$ for any root α of sl_N;

- the operator d satisfies $\Delta(d) = d \otimes 1 + 1 \otimes d$;

- the generator $h_{0,\rho}$ of $\mathcal{D}Y(sl_N)$ is such that

$$\begin{aligned} h_{0,\rho} e_\alpha(u) &= e_\alpha(u)(h_{0,\rho} + (\rho|\alpha)) \\ h_{0,\rho} f_\alpha(u) &= f_\alpha(u)(h_{0,\rho} - (\rho|\alpha)) \end{aligned} \qquad [h_{0,\rho}, h_\alpha(u)] = 0$$

  and hence $\tau = \mathrm{Ad}\left(\omega^{h_{0,\rho}} \otimes \mathrm{id}\right)$ is idempotent;

it can be shown that the solution of the linear equation (4.31) given by the infinite product (4.34) satisfies the shifted cocycle relation

$$\mathcal{F}^{(12)}(r)(\Delta \otimes \mathrm{id})(\mathcal{F}(r)) = \mathcal{F}^{(23)}(r + c^{(1)})(\mathrm{id} \otimes \Delta)(\mathcal{F}(r)). \tag{4.35}$$

The twist $\mathcal{F}(r)$ defines a QTQHA denoted $\mathcal{D}Y_r(sl_N)$ with universal R-matrix

$$\mathcal{R}[\mathcal{D}Y_r(sl_N)](r) = \mathcal{F}_{21}(r)\ \mathcal{R}[\mathcal{D}Y(sl_N)]\ \mathcal{F}_{12}(r)^{-1}.$$

It remains to show that the corresponding evaluated R-matrix indeed coincides with the expression (4.30). In the evaluation representation (4.24), the linear equation (4.31) takes the form

$$F(u) = (\phi \otimes 1)^{-1} F(u - r)(\phi \otimes 1) R(u), \tag{4.36}$$

where $\phi = \mathrm{diag}(\omega^{N-1}, \omega^{N-2}, \ldots, \omega, 1)$.

The solution of (4.36) is expressed in term of hypergeometric functions ${}_2F_1$:

$$F_{ab}^{ab}(u) = \frac{\Gamma\left(\frac{u-1}{r}+1\right)}{\Gamma\left(\frac{u}{r}+1\right)}\ {}_2F_1\left(\begin{matrix} -\frac{1}{r}\ ,\ \frac{u-1}{r}+1 \\ \frac{u}{r}+1 \end{matrix}\ ; \omega^{b-a}\right) \tag{4.37}$$

$$F_{ab}^{ba}(u) = -\frac{\omega^{b-a}}{r}\ \frac{\Gamma\left(\frac{u-1}{r}+1\right)}{\Gamma\left(\frac{u}{r}+2\right)}\ {}_2F_1\left(\begin{matrix} -\frac{1}{r}+1\ ,\ \frac{u-1}{r}+1 \\ \frac{u}{r}+2 \end{matrix}\ ; \omega^{b-a}\right) \tag{4.38}$$

and $F_{aa}^{aa}(u) = 1$ (the indices $a \neq b$ run from 1 to N).

The twist $F(u)$ applied to the R-matrix of $\mathcal{D}Y(sl_N)$, Eq. (4.25), provides the R-matrix $\widetilde{R}(u)$ of the deformed double Yangian $\mathcal{D}Y_r(sl_N)$, Eq. (4.30), the non vanishing entries of which are expressed in terms of (4.29). The proof follows by a direct computation using the properties of the hypergeometric functions ${}_2F_1$.

5 Dynamical double Yangian $\mathcal{D}Y_s(sl_N)$

5.1 The case of $\mathcal{D}Y_s(sl_2)$

Using the same kind of argument, it is possible to construct another deformation of the Yangian $\mathcal{D}Y(sl_2)$, which can be considered as a dynamical deformation, denoted by $\mathcal{D}Y_s(sl_2)$. The corresponding R-matrix reads:

$$R[\mathcal{D}Y_s(sl_2)] = \rho(u) \begin{pmatrix} 1 & 0 & 0 & 0 \\ 0 & \frac{u}{u+1} & \frac{s+u}{s(u+1)} & 0 \\ 0 & \frac{s-u}{s(u+1)} & \frac{s^2-1}{s^2}\frac{u}{u+1} & 0 \\ 0 & 0 & 0 & 1 \end{pmatrix}, \tag{5.1}$$

$\rho(u)$ being the normalization factor of the R-matrix of $\mathcal{D}Y(sl_2)$ and s is a complex number. Actually this matrix can be obtained as a scaling limit of the quantum affine elliptic algebra $\mathcal{B}_{q,p,\lambda}(\widehat{sl_2})$ with $p = 0$.

The algebra $\mathcal{D}Y_s(sl_2)$ is then defined by the RLL-type relations

$$\begin{aligned} &R_{12}(u_1-u_2,\lambda+h)\,L_1(u_1,\lambda)\,L_2(u_2,\lambda+h^{(1)})\\ &\qquad = L_2(u_2,\lambda)\,L_1(u_1,\lambda+h^{(2)})\,R_{12}(u_1-u_2,\lambda). \end{aligned} \tag{5.2}$$

Let us prove that $\mathcal{D}Y_s(sl_2)$ is a QTQHA obtained from $\mathcal{D}Y(sl_2)$ by a Drinfel'd twist. To this aim, one uses the fact that $\mathcal{U}(sl_2)$ is a Hopf subalgebra of $\mathcal{D}Y(sl_2)$.

One first constructs a twist from $\mathcal{U}_q(sl_2)$ to $\mathcal{B}_{q,\lambda}(sl_2)$ [22, 23, 25] as follows. Introduce $\phi = q^\chi = q^{\frac{1}{2}h^2+sh}$ where $s \in \mathbb{C}$, and consider

$$\mathcal{F} = \prod_{k\geq 1}^{\curvearrowleft} \mathrm{Ad}(\phi^k \otimes 1)\,(\widehat{\mathcal{R}}^{-1}), \tag{5.3}$$

where $\mathcal{R} = q^{-\frac{1}{2}h\otimes h}\,\widehat{\mathcal{R}}$ is the universal R-matrix of $\mathcal{U}_q(sl_2)$. The twisted R-matrix $\widetilde{\mathcal{R}}_{12} = \mathcal{F}_{21}\,\mathcal{R}_{12}\,\mathcal{F}_{12}^{-1}$ defines the algebra $\mathcal{B}_{q,\lambda}(sl_2)$. The twist $\mathcal{F}$ satisfies the following difference equation

$$\mathcal{F} = \mathrm{Ad}(\phi^{-1}\otimes 1)(\mathcal{F})\cdot\widehat{\mathcal{R}} \tag{5.4}$$

the solution of which is precisely the infinite product (5.3).

Consider now the twist $\mathcal{F}'$ obtained from (5.3) by taking the limit $q \to 1$ and keeping s fixed. The consistency of this procedure follows from the Hopf algebra identification $\mathcal{U}_\hbar(sl_2)/(\hbar\,\mathcal{U}_\hbar(sl_2)) \simeq \mathcal{U}(sl_2)$ with $q = e^\hbar$. In this quotient, the twist $\mathcal{F}'$ is given by the formula

$$\mathcal{F}' = \sum_{n=0}^{\infty}\frac{1}{n!}\Big(\prod_{k=0}^{n-1}[(1+k-s)1\otimes h]\otimes 1\Big)^{-1} e^n\otimes f^n. \tag{5.5}$$

In fact, the scaling limit of the difference equation (5.4) of the twist $\mathcal{F}$ leads to the following equation for the twist $\mathcal{F}'$, obtained as the first non trivial order in $\hbar$ in (5.4):

$$[\chi\otimes 1, \mathcal{F}'] = \mathcal{F}'\hat{r}, \tag{5.6}$$

where $\hat{r}$ is given by $\widehat{\mathcal{R}} = 1\otimes 1 + \hbar\hat{r} + o(\hbar)$. This equation has a unique solution expressed either by Eq. (5.5) or by the infinite product (it can be checked directly, order by order, that formulae (5.5) and (5.7) lead to the same expression in terms of the generators)

$$\mathcal{F}' = \prod_{k\geq 1}^{\curvearrowleft} \mathrm{Ad}(\chi^{-k}\otimes 1)\Big(1\otimes 1 + (\chi\otimes 1)^{-1}\hat{r}\Big). \tag{5.7}$$

Finally, the following theorem holds [15, 32]:

Theorem 5.1. *The dynamical Yangian-type Drinfel'd twist* (5.7) *satisfies the shifted cocycle condition*

$$\mathcal{F}'_{12}(s)\,(\Delta\otimes\mathrm{id})\mathcal{F}'(s) = \mathcal{F}'_{23}(s+h^{\vee(1)})\,(\mathrm{id}\otimes\Delta)\mathcal{F}'(s).$$

The dynamical double Yangian $\mathcal{D}Y_s(sl_2)$ is a quasi-triangular quasi-Hopf algebra with the universal R-matrix $\widetilde{\mathcal{R}}(s) = \mathcal{F}'_{21}(s)\,\mathcal{R}\,\mathcal{F}'_{12}(s)$ where $\mathcal{R}$ is the universal R-matrix of $\mathcal{D}Y(sl_2)$. $\widetilde{\mathcal{R}}(s)$ satisfies the dynamical Yang–Baxter equation:

$$\widetilde{\mathcal{R}}_{12}(s+h^{\vee(3)})\,\widetilde{\mathcal{R}}_{13}(s)\,\widetilde{\mathcal{R}}_{23}(s+h^{\vee(1)}) = \widetilde{\mathcal{R}}_{23}(s)\,\widetilde{\mathcal{R}}_{13}(s+h^{\vee(2)})\,\widetilde{\mathcal{R}}_{12}(s).$$

In that case, although the computation is rather tedious, the proof of the shifted cocycle condition can be done directly by using the sum formula (5.5).

5.2 Generalization to sl_N

The previous construction can be generalized to the case of sl_N without difficulty [32]. It follows the same lines:

- construction of the twist $\mathcal{F}$ from $\mathcal{U}_q(sl_N)$ to $\mathcal{B}_{q,\lambda}(sl_N)$,
- one takes the scaling limit of this twist consistently from the Hopf algebra identification $\mathcal{U}_\hbar(sl_N)/(\hbar\ \mathcal{U}_\hbar(sl_N)) \simeq \mathcal{U}(sl_N)$ with $q = e^\hbar$,
- the scaled twist $\mathcal{F}'$ can also be obtained in its universal form as an infinite product given by (5.7), the matrix $\hat{r}$ being the classical limit of the matrix $\widehat{\mathcal{R}}$ where $\mathcal{R} = q^{-d_{ij}h_i\otimes h_j}\ \widehat{\mathcal{R}}$ is the universal R-matrix of $\mathcal{U}_q(sl_N)$ and d_{ij} is the inverse of the Cartan matrix of sl_N.

In the fundamental representation for sl_N, the evaluated infinite product expression for $\mathcal{F}'$ reads

$$F' = 1\otimes 1 - \sum_{a<b} \frac{2}{x_a - x_b}\, E_{ab}\otimes E_{ba}. \tag{5.8}$$

Applied to the R-matrix of $\mathcal{D}Y(sl_N)$ given in (4.25), this twist leads to the evaluated R-matrix of $\mathcal{D}Y_s(sl_N)$:

$$\begin{aligned} R[\mathcal{D}Y_s(sl_N)](u) = \rho[\mathcal{D}Y_s(sl_N)](u)\Bigg(&\sum_a E_{aa}\otimes E_{aa} + \sum_{a<b}\frac{u}{u+1}\,E_{aa}\otimes E_{bb} \\ &+ \sum_{a>b}\left(1 - \frac{4}{(x_a-x_b)^2}\right)\frac{u}{u+1}\,E_{aa}\otimes E_{bb} \\ &+ \sum_{a,b}\left(1 + \frac{2u}{x_a-x_b}\right)\frac{1}{u+1}\,E_{ab}\otimes E_{ba}\Bigg), \end{aligned} \tag{5.9}$$

the normalization factor being $\rho[\mathcal{D}Y_s(sl_N)](u) = \rho[\mathcal{D}Y(sl_N)](u)$, see Eq. (4.26).

5.3 Deformed dynamical double Yangian $\mathcal{D}Y_{r,s}(sl_2)$

Considering now the scaling limit of the quantum affine elliptic algebra $\mathcal{B}_{q,p,\lambda}(\widehat{sl_2})$, it is possible to construct a deformed dynamical double Yangian $\mathcal{D}Y_{r,s}(sl_2)$ with the following R-matrix[15]:

$$R[\mathcal{D}Y_{r,s}(sl_2)] = \rho(u,r) \begin{pmatrix} 1 & 0 & 0 & 0 \\ 0 & b(u) & c(u) & 0 \\ 0 & c'(u) & b'(u) & 0 \\ 0 & 0 & 0 & 1 \end{pmatrix}, \tag{5.10}$$

where

$$b(u) = \frac{\Gamma_1(r-s|r)^2}{\Gamma_1(r-s+1|r)\,\Gamma_1(r-s-1|r)} \frac{\sin\frac{\pi u}{r}}{\sin\frac{\pi(1+u)}{r}},$$

$$c(u) = \frac{\sin\frac{\pi(s+u)}{r}}{\sin\frac{\pi s}{r}} \frac{\sin\frac{\pi}{r}}{\sin\frac{\pi(1+u)}{r}},$$

$$b'(u) = \frac{\Gamma_1(s|r)^2}{\Gamma_1(s+1|r)\,\Gamma_1(s-1|r)} \frac{\sin\frac{\pi u}{r}}{\sin\frac{\pi(1+u)}{r}},$$

$$c'(u) = \frac{\sin\frac{\pi(s-u)}{r}}{\sin\frac{\pi s}{r}} \frac{\sin\frac{\pi}{r}}{\sin\frac{\pi(1+u)}{r}},$$

and the normalization factor $\rho(u,r)$ is given by Eq. (4.7).

The algebra $\mathcal{D}Y_{r,s}(sl_2)$ is then defined by the relations

$$\begin{aligned} &R_{12}(u_1-u_2,\lambda+h)\,L_1(u_1,\lambda)\,L_2(u_2,\lambda+h^{(1)}) \\ &\qquad = L_2(u_2,\lambda)\,L_1(u_1,\lambda+h^{(2)})\,R_{12}(u_1-u_2,\lambda). \end{aligned} \tag{5.11}$$

In that case, although the algebra can be defined through the RLL relations, the status of this algebra as a QTQHA, obtained by a Drinfel'd twist of $\mathcal{D}Y(sl_2)$, remains still open. However the interest of this algebra is enforced by the following remark.

Remark 5.2. Considering the R-matrix of the $\mathcal{D}Y_{r,s}(sl_2)$ algebra, Eq. (5.10), and taking the limit $s \to i\infty$, one gets a non dynamical R-matrix given by

$$R = \rho(u,r) \begin{pmatrix} 1 & 0 & 0 & 0 \\ 0 & \frac{\sin \pi u/r}{\sin \pi(u+1)/r} & e^{-i\pi u} \frac{\sin \pi/r}{\sin \pi(u+1)/r} & 0 \\ 0 & e^{i\pi u} \frac{\sin \pi/r}{\sin \pi(u+1)/r} & \frac{\sin \pi u/r}{\sin \pi(u+1)/r} & 0 \\ 0 & 0 & 0 & 1 \end{pmatrix}, \tag{5.12}$$

where $\rho(u,r)$ is the same as above.

In fact, this matrix is a gauge-transformed version of the R-matrix (4.6). This gauge transformation can be constructed as a Drinfel'd twist at the universal level.

Considering $g' = \exp(h_1/2r)$, one defines the following shifted coboundary

$$\mathcal{K}_{12}(r) = g'(r) \otimes g'(r + c^{(1)})\ \Delta^{\mathcal{F}}(g'^{-1}), \tag{5.13}$$

where $\Delta^{\mathcal{F}}$ is the coproduct of $\mathcal{D}Y_r(sl_2)$. It obeys a shifted cocycle condition

$$\mathcal{K}_{12}(r)\ (\Delta^{\mathcal{F}} \otimes \mathrm{id})\mathcal{K}(r) = \mathcal{K}_{23}(r + c^{(1)})\ (\mathrm{id} \otimes \Delta^{\mathcal{F}'})\mathcal{K}(r) \tag{5.14}$$

as a consequence of

$$(\Delta^{\mathcal{F}} \otimes \mathrm{id})\ \Delta^{\mathcal{F}}(g'^{-1}) = (\mathrm{id} \otimes \Delta^{\mathcal{F}'})\ \Delta^{\mathcal{F}}(g'^{-1}) \tag{5.15}$$

with $\mathcal{F}'_{23}(r) = \mathcal{F}_{23}(r + c^{(1)})$. This last equation is the quasi-coassociativity property for the coproduct $\Delta^{\mathcal{F}}$.

Finally, the matrix $\mathcal{K}_{21}(r)\,\mathcal{R}[\mathcal{D}Y_r(sl_2)]\,\mathcal{K}_{12}^{-1}(r)$ satisfies the shifted Yang–Baxter equation, and the corresponding evaluated R-matrix coincides with (5.12).

Acknowledgements. L.F. would like to thank the organizers of the "68ème Rencontre entre Physiciens Théoriciens et Mathématiciens" for their kind invitation.

Appendix

Multiple Gamma functions are defined by

$$\Gamma_r(x|\omega_1,\ldots,\omega_r) = \exp\left(\frac{\partial}{\partial s} \sum_{n_1,\ldots,n_r \geq 0} (x + n_1\omega_1 + \cdots + n_r\omega_2)^{-s}\Big|_{s=0}\right).$$

Barnes' multiple sine function $S_r(x|\omega_1,\ldots,\omega_r)$ of periods $\omega_1,\ldots,\omega_r$ is defined by [33]

$$S_r(x|\omega_1,\omega_2) = \Gamma_r(\omega_1 + \cdots + \omega_r - x|\omega_1,\ldots,\omega_r)^{(-1)^r}\,\Gamma_r(x|\omega_1,\ldots,\omega_r)^{-1}.$$

In particular,

$$\Gamma_1(x|\omega_1) = \frac{\omega_1^{x/\omega_1}}{\sqrt{2\pi\omega_1}}\,\Gamma\left(\frac{x}{\omega_1}\right) \quad \text{and} \quad S_1(x|\omega_1) = 2\sin\frac{\pi x}{\omega_1}.$$

They satisfy for each $i \in [1,\ldots,r]$

$$\frac{\Gamma_r(x+\omega_i|\omega_1,\ldots,\omega_r)}{\Gamma_r(x|\omega_1,\ldots,\omega_r)} = \frac{1}{\Gamma_{r-1}(x|\omega_1,\ldots,\omega_{i-1},\omega_{i+1},\ldots,\omega_r)},$$

$$\frac{S_r(x+\omega_i|\omega_1,\ldots,\omega_r)}{S_r(x|\omega_1,\ldots,\omega_r)} = \frac{1}{S_{r-1}(x|\omega_1,\ldots,\omega_{i-1},\omega_{i+1},\ldots,\omega_r)}.$$

References

[1] M. Jimbo, R. Kedem, H. Konno, T. Miwa and R. Weston, Difference equations in spin chains with a boundary, *Nucl. Phys.* B 448 (1995), 429–456.

[2] G. Andrews, R. Baxter, P. J. Forrester, Eight-vertex SOS model and generalized Rogers-Ramanujan type identities, *J. Statist. Phys.* 35 (1984), 193–266.

[3] H. Konno, An elliptic algebra $U_{q,p}(\widehat{sl}(2))$ and the fusion RSOS model, *Comm. Math. Phys.* 195 (1998), 373–403.

[4] H. Konno, Degeneration of the elliptic algebra $\mathcal{A}_{q,p}(\widehat{sl}_2)$ and form factors in the Sine-Gordon theory, Proceedings of the Nankai-CRM joint meeting on *Extended and Quantum Algebras and their Applications to Physics*, Tianjin, China, 1996, to appear in CRM Ser. Math. Phys., Springer Verlag, hep-th/9701034.

[5] S. Khoroshkin, A. LeClair, S. Pakuliak, Angular quantization of the Sine–Gordon model at the free fermion point, *Adv. Theor. Math. Phys.* 3 (1999), 1227–1287.

[6] R. J. Baxter, Partition function of the eight-vertex lattice model, *Ann. Phys.* 70 (1972), 193–228.

[7] A. A. Belavin, Dynamical symmetry of integrable quantum systems, *Nucl. Phys.* B 180 (1981), 189–200.

[8] E. Date, M. Jimbo, T. Miwa, M. Okado, Fusion of the eight-vertex SOS model, *Lett. Math. Phys.* 12 (1986), 209–215.

[9] M. Jimbo, T. Miwa, M. Okado, *Lett. Math. Phys.* 14 (1987), 123; Solvable lattice models related to the vector representation of classical simple Lie algebras, *Comm. Math. Phys.* 116 (1988), 507–525.

[10] E. K. Sklyanin, Some algebraic structures connected with the Yang–Baxter equation, *Funct. Anal. Appl.* 16 (1982), 263–270; Some algebraic structures connected with the Yang–Baxter equation. Representations of quantum algebras, *Funct. Anal. Appl.* 17 (1983), 273–284.

[11] O. Foda, K. Iohara, M. Jimbo, R. Kedem, T. Miwa, H. Yan, An elliptic quantum algebra for $\widehat{sl}_2$, *Lett. Math. Phys.* 32 (1994), 259–268.

[12] G. Felder, Elliptic quantum groups, in *Proceedings of the XIth International Congress on Mathematical Physics* (Iagolnitzer, D., ed.), Paris, France, July 18–23, 1994, International Press, Cambridge, MA, 1995, 211–218.

[13] M. Jimbo, H. Konno, T. Miwa, Massless XXZ model and degeneration of the elliptic algebra $\mathcal{A}_{q,p}(\widehat{sl}_2)$, in *Deformation theory and symplectic geometry, Ascona 1996* (D. Sternheimer, J. Rawnsley and S. Gutt, eds.), Math. Phys. Stud. 20, Kluwer Academic Publishers, Dordrecht 1997, 117–138.

[14] S. Khoroshkin, D. Lebedev, S. Pakuliak, Elliptic algebra $A_{q,p}(\hat{sl}_2)$ in the scaling limit, *Comm. Math. Phys.* 190 (1998), 597–627.

[15] D. Arnaudon, J. Avan, L. Frappat, E. Ragoucy, M. Rossi, Towards a cladistics of double Yangians and elliptic algebras, *J. Phys.* A 33 (2000), 6279–6309.

[16] D. Bernard, A. LeClair, The quantum double in integrable quantum field theory, *Nucl. Phys.* B 399 (1993), 709–748.

Definition 4.2. A *-algebra $\mathcal{A}$ over C has *sufficiently many positive linear functionals* if for every non-zero Hermitian element $H \in \mathcal{A}$ there exists a positive linear functional $\omega : \mathcal{A} \to \mathsf{C}$ such that $\omega(H) \neq 0$.

There are simple examples of *-algebras *without* sufficiently many positive linear functionals, see e.g. [12, Sect. 2].

Definition 4.3. Let $\mathcal{A}$ be a *-algebra over C and $\mathcal{J} \subseteq \mathcal{A}$ a *-ideal. Then $\mathcal{J}$ is called *closed* if it is the kernel of a *-representation. Moreover, the *minimal ideal* of $\mathcal{A}$ is defined by

$$\mathcal{J}_{\mathrm{min}}(\mathcal{A}) = \bigcap_{\mathcal{J} \text{ closed}} \mathcal{J}. \tag{4.5}$$

As expected from the C^*-algebra case the closed *-ideals form a lattice with minimal element $\mathcal{J}_{\mathrm{min}}(\mathcal{A})$. The following theorem relates the existence of positive linear functionals to the 'size' of $\mathcal{J}_{\mathrm{min}}(\mathcal{A})$ and the existence of *-representations.

Theorem 4.4. *Let $\mathcal{A}$ be a unital *-algebra over C. Then the following statements are equivalent:*

1. *$\mathcal{A}$ has sufficiently many positive linear functionals.*
2. *$\mathcal{A}$ has a faithful *-representation.*
3. *$\mathcal{J}_{\mathrm{min}}(\mathcal{A}) = \{0\}$.*

In general one has the characterization

$$\mathcal{J}_{\mathrm{min}}(\mathcal{A}) = \bigcap_{\omega} \ker \pi_\omega = \bigcap_{\omega} \mathcal{J}_\omega = \bigcap_{\omega} \ker \omega, \tag{4.6}$$

where the intersection runs over all positive functionals of $\mathcal{A}$. Moreover, if $A^*A = 0$, or if A is normal and nilpotent, or if $zA = 0$ for $0 \neq z \in \mathsf{C}$ then $A \in \mathcal{J}_{\mathrm{min}}(\mathcal{A})$.

The proof relies on the GNS construction, see [15, Prop. 2.8] and [12, Thm. 4.7]. In particular, if there exists a faithful *-representation then one can already obtain a faithful *-representation by direct sums of GNS representations. Moreover, since all the unpleasant and 'non-C^*-algebra like' elements of $\mathcal{A}$ are contained in $\mathcal{J}_{\mathrm{min}}(\mathcal{A})$ we can pass to the quotient

$$\varrho : \mathcal{A} \to \mathcal{A} \big/ \mathcal{J}_{\mathrm{min}}(\mathcal{A}). \tag{4.7}$$

This procedure is clearly functorial and can be seen as an algebraic analog of the construction of the *C^*-enveloping algebra* of a Banach *-algebra, see e.g. [22, Sect. II.7]. In particular, the pull-back ϱ^* of *-representations of $\mathcal{A}/\mathcal{J}_{\mathrm{min}}(\mathcal{A})$ to $\mathcal{A}$ implements an equivalence of the categories *-$\mathsf{Rep}(\mathcal{A})$ and *-$\mathsf{Rep}(\mathcal{A}/\mathcal{J}_{\mathrm{min}}(\mathcal{A}))$. Thus elements in $\mathcal{J}_{\mathrm{min}}(\mathcal{A})$ are 'invisible' in *-representations of $\mathcal{A}$, see [12, Prop. 5.6].

5 Deformation of *-algebras

Now we shall consider formal deformations of *-algebras. Since $\mathsf{R}[[\lambda]]$ is again ordered one stays in the same framework of *-algebras over ordered rings.

Definition 5.1. Let $\mathcal{A}$ be a *-algebra over C and let $\boldsymbol{\mathcal{A}} = (\mathcal{A}[[\lambda]], \star)$ be a formal associative deformation of $\mathcal{A}$, viewed as an algebra over $\mathsf{C}[[\lambda]]$. Then $\boldsymbol{\mathcal{A}}$ is called a *Hermitian deformation* if

$$(A \star B)^* = B^* \star A^*. \tag{5.1}$$

In principle one can also deform the *-involution of $\mathcal{A}$ in order to obtain a *-involution for $\star$ but we shall only consider the case of Hermitian deformations here, see also [15, Sect. 8]. Hermitian star products are of course our main example. For later use we shall recall the following lemma [13, Lem. 2.1].

Lemma 5.2. *Let $B_0 \in M_n(\mathcal{A})$ be invertible and let $A = B_0^* B_0 + \sum_{r=1}^{\infty} \lambda^r A_r \in M_n(\mathcal{A})[[\lambda]]$. Let $\star$ be a Hermitian deformation of $\mathcal{A}$. Then there exists $B_1, \ldots \in M_n(\mathcal{A})$ such that for $B = \sum_{r=0}^{\infty} \lambda^r B_r$ one has $A = B^* \star B$.*

In order to compare *-representations of $\mathcal{A}$ and $\boldsymbol{\mathcal{A}}$ we first need the notion of the *classical limit* for pre-Hilbert spaces, see [15, Lem. 8.2 and 8.3].

Lemma 5.3. *Let $\boldsymbol{\mathfrak{H}}$ be a pre-Hilbert space over* $\mathsf{C}[[\lambda]]$. *Then*

$$\boldsymbol{\mathfrak{H}}_0 = \{\phi \in \boldsymbol{\mathfrak{H}} \mid \langle \phi, \phi \rangle |_{\lambda=0} = 0\} \tag{5.2}$$

is a $\mathsf{C}[[\lambda]]$ *sub-module and $\mathfrak{H} := \mathfrak{C}\boldsymbol{\mathfrak{H}} = \boldsymbol{\mathfrak{H}} / \boldsymbol{\mathfrak{H}}_0$ becomes a pre-Hilbert space over* C *by setting $\langle \mathfrak{C}\phi, \mathfrak{C}\psi \rangle = \langle \phi, \psi \rangle |_{\lambda=0}$. Moreover, if $A \in \mathfrak{B}(\boldsymbol{\mathfrak{H}}, \boldsymbol{\mathfrak{K}})$ then $\mathfrak{C}A : \mathfrak{C}\boldsymbol{\mathfrak{H}} \to \mathfrak{C}\boldsymbol{\mathfrak{K}}$ defined by*

$$\mathfrak{C}A(\mathfrak{C}\phi) = \mathfrak{C}(A\phi) \tag{5.3}$$

is again adjointable $\mathfrak{C}A \in \mathfrak{B}(\mathfrak{C}\boldsymbol{\mathfrak{H}}, \mathfrak{C}\boldsymbol{\mathfrak{K}})$.

It follows immediately that $\mathfrak{C}$ defines a functor, the so-called *classical limit*

$$\mathfrak{C} : \text{pre-Hilbert}(\mathsf{C}[[\lambda]]) \longrightarrow \text{pre-Hilbert}(\mathsf{C}). \tag{5.4}$$

Example 5.4. Let $\mathfrak{H}$ be a pre-Hilbert space over C. Then $\boldsymbol{\mathfrak{H}} = \mathfrak{H}[[\lambda]]$ is a pre-Hilbert space over $\mathsf{C}[[\lambda]]$ by λ-linear extension of the inner product such that canonically $\mathfrak{C}\boldsymbol{\mathfrak{H}} \cong \mathfrak{H}$. Moreover, $\mathfrak{B}(\boldsymbol{\mathfrak{H}}) = \mathfrak{B}(\mathfrak{H})[[\lambda]]$.

This example shows that the classical limit functor $\mathfrak{C}$ is essentially surjective. On the other hand there are pre-Hilbert spaces $\boldsymbol{\mathfrak{H}}$ over $\mathsf{C}[[\lambda]]$ which are not of the form $\mathfrak{H}[[\lambda]]$ as in Example 5.4, even if $\boldsymbol{\mathfrak{H}} \cong \mathfrak{H}[[\lambda]]$ as $\mathsf{C}[[\lambda]]$-modules, since the inner products may depend on λ in a non-trivial way.

One also can compute the classical limit of *-representations of Hermitian deformations in a functorial way, see [15, Prop. 8.5].

Lemma 5.5. *Let $\boldsymbol{\mathcal{A}}$ be a Hermitian deformation of a *-algebra $\mathcal{A}$ over* C *and let $\boldsymbol{\pi} : \boldsymbol{\mathcal{A}} \to \mathfrak{B}(\boldsymbol{\mathfrak{H}})$ be a *-representation. Then $(\mathfrak{C}\boldsymbol{\pi})(A) = \mathfrak{C}(\boldsymbol{\pi}(A))$ defines a *-representation of $\mathcal{A}$ on $\mathfrak{H} = \mathfrak{C}\boldsymbol{\mathfrak{H}}$ and*

$$\mathfrak{C} : {}^*\text{-rep}(\boldsymbol{\mathcal{A}}) \longrightarrow {}^*\text{-rep}(\mathcal{A}) \tag{5.5}$$

is functorial such that $\mathfrak{C}({}^*\text{-Rep}(\boldsymbol{\mathcal{A}})) \subseteq {}^*\text{-Rep}(\mathcal{A})$.

Definition 5.6. Let $(\pi, \mathfrak{H})$ be a *-representation of $\mathcal{A}$ and $\boldsymbol{\mathcal{A}}$ a Hermitian deformation of $\mathcal{A}$. A *deformation* of $(\pi, \mathfrak{H})$ with respect to $\boldsymbol{\mathcal{A}}$ is a *-representation $(\boldsymbol{\pi}, \boldsymbol{\mathfrak{H}})$ of $\boldsymbol{\mathcal{A}}$ such that $(\pi, \mathfrak{H})$ is unitarily equivalent to $\mathfrak{C}(\boldsymbol{\pi}, \boldsymbol{\mathfrak{H}})$.

Turning this definition into a question we arrive at the problem of whether a given *-representation of $\mathcal{A}$ can be *deformed* into a *-representation of $\boldsymbol{\mathcal{A}}$. In general this problem seems to be quite hard to solve. In particular, some easy examples like the Bargmann–Fock representation show that it is *not* enough to consider deformations with $\boldsymbol{\mathfrak{H}} = \mathfrak{H}[[\lambda]]$, see [48].

Definition 5.7. Let $\omega_0 : \mathcal{A} \to \mathsf{C}$ be a positive C-linear functional. Then the $\mathsf{C}[[\lambda]]$-linear functional $\boldsymbol{\omega} = \sum_{r=0}^{\infty} \lambda^r \omega_r : \boldsymbol{\mathcal{A}} \to \mathsf{C}[[\lambda]]$ is called a *deformation* of ω_0 if $\boldsymbol{\omega}$ is positive with respect to $\star$. If any positive linear functional of $\mathcal{A}$ can be deformed then $\boldsymbol{\mathcal{A}}$ is called a *positive deformation.*

The physical interpretation of this definition is that *every classical state is the classical limit of a quantum state.* Obviously, from the physical intuition this should be a feature of any reasonable quantization scheme. In fact, Hermitian star products on symplectic manifolds are always positive deformations [14]. From [48] we have:

Lemma 5.8. *Let $\boldsymbol{\omega} : \boldsymbol{\mathcal{A}} \to \mathsf{C}[[\lambda]]$ be a deformation of a positive* C*-linear functional $\omega_0 : \mathcal{A} \to \mathsf{C}$. Then the GNS representation $(\boldsymbol{\pi}_{\boldsymbol{\omega}}, \boldsymbol{\mathfrak{H}}_{\boldsymbol{\omega}})$ is a deformation of the GNS representation $(\pi_{\omega_0}, \mathfrak{H}_{\omega_0})$.*

From this fact one easily obtains that a Hermitian deformation $\boldsymbol{\mathcal{A}}$ of a unital *-algebra $\mathcal{A}$ is positive if and only if any GNS representation of $\mathcal{A}$ can be deformed. In particular, if $\mathfrak{C}\boldsymbol{\Omega} = \psi_1$ for some $\boldsymbol{\Omega} \in \boldsymbol{\mathfrak{H}}$ with $\mathfrak{C}(\boldsymbol{\pi}, \boldsymbol{\mathfrak{H}}) = (\pi_{\omega_0}, \mathfrak{H}_{\omega_0})$ then $\boldsymbol{\omega}(A) = \langle \boldsymbol{\Omega}, \boldsymbol{\pi}(A)\boldsymbol{\Omega} \rangle$ is a deformation of ω_0. This again emphasizes the importance of GNS representations.

6 Rieffel induction

Given two *-algebras $\mathcal{A}$, $\mathcal{B}$ over C we want to learn something about their relation by comparing *-Rep($\mathcal{A}$) and *-Rep($\mathcal{B}$) in a functorial way. This wish can be made precise by considering a $\mathcal{B}$-$\mathcal{A}$ bimodule $\mathcal{E}$. As usual $\mathcal{E}$ shall also possess an underlying C-module structure. If $\mathcal{A}$ and $\mathcal{B}$ are unital we always assume that the units act as units on $\mathcal{E}$. In the non-unital case we require that $\mathcal{E}$ is strongly non-degenerate for both actions.

Given such a bimodule $\mathcal{E}$ and given a *-representation $(\pi, \mathfrak{H})$ of $\mathcal{A}$,

$$\widetilde{\mathfrak{K}} = \mathcal{E} \otimes_{\mathcal{A}} \mathfrak{H} \tag{6.1}$$

clearly becomes a $\mathcal{B}$-left module. The main idea of *Rieffel induction*, as known from C^*-algebra theory [38, 37], is to impose additional structures on $\mathcal{E}$ such that (6.1) eventually gives a *-representation of $\mathcal{B}$. We shall now sketch how the construction for C^*-algebras can be extended to this purely algebraic framework of *-algebras over $\mathbb{C}$.

Definition 6.1. A *$\mathcal{A}$-valued, $\mathcal{A}$-right linear Hermitian inner product* on a $\mathcal{A}$-right module $\mathcal{E}$ is a map $\langle \cdot\,, \cdot \rangle : \mathcal{E} \times \mathcal{E} \to \mathcal{A}$ such that for all $x, y \in \mathcal{E}$ and $A \in \mathcal{A}$

1. $\langle \cdot\,, \cdot \rangle$ is $\mathbb{C}$-linear in the second argument,
2. $\langle x, y \rangle = \langle y, x \rangle^*$,
3. $\langle x, y \cdot A \rangle = \langle x, y \rangle A$,
4. $\langle x, x \rangle \in \mathcal{A}^+$.

If $\mathcal{E}$ is even a $\mathcal{B}$-$\mathcal{A}$ bimodule then $\langle \cdot\,, \cdot \rangle$ is *compatible* with the $\mathcal{B}$-left action if in addition

5. $\langle B \cdot x, y \rangle = \langle x, B^* \cdot y \rangle$.

Moreover, $\langle \cdot\,, \cdot \rangle$ satisfies the *positivity condition* **P** if

$$\langle x \otimes \phi, y \otimes \psi \rangle := \langle \phi, \pi(\langle x, y \rangle)\psi \rangle \quad \text{for} \quad x \otimes \phi, y \otimes \psi \in \widetilde{\mathfrak{K}} \tag{6.2}$$

extends to a positive semi-definite Hermitian inner product on $\widetilde{\mathfrak{K}}$ for *all* *-representations $(\pi, \mathfrak{H})$ of $\mathcal{A}$.

Actually, the positivity condition **P** turns out to be rather mild. It follows 'almost' from $\langle x, x \rangle \in \mathcal{A}^+$. In case of a C^*-algebra (6.2) is necessarily positive semi-definite and one can give various sufficient conditions in the general case, too, see [15] for a detailed discussion.

Now let us assume we have a $\mathcal{B}$-$\mathcal{A}$ bimodule with compatible $\mathcal{A}$-valued $\mathcal{A}$-right linear inner product $\langle \cdot\,, \cdot \rangle$ such that **P** holds. Then we consider the quotient

$$\mathfrak{K} = \widetilde{\mathfrak{K}} \,/\, \widetilde{\mathfrak{K}}^{\perp} \tag{6.3}$$

of $\widetilde{\mathfrak{K}}$ divided by the vectors of length zero and obtain a pre-Hilbert space $\mathfrak{K}$ over $\mathbb{C}$. Moreover, we observe that the $\mathcal{B}$-left action on $\widetilde{\mathfrak{K}}$ passes to the quotient $\mathfrak{K}$ and yields a *-representation of $\mathcal{B}$, thanks to the compatibility condition. A final check shows that this construction is functorial: indeed, let $(\pi_1, \mathfrak{H}_1)$ and $(\pi_2, \mathfrak{H}_2)$ be *-representations of $\mathcal{A}$ and let $U : \mathfrak{H}_1 \to \mathfrak{H}_2$ be an isometric intertwiner. Then

$$\widetilde{V}(x \otimes \phi) = x \otimes U(\phi) \tag{6.4}$$

Let us now discuss a few examples for the unital case. Since here $\mathcal{E}$ has to be projective we consider a *Hermitian* projection $P_0 \in M_n(\mathcal{A})$, i.e. $P_0^2 = P_0 = P_0^*$ and $\mathcal{E} = P_0\mathcal{A}^n \subseteq \mathcal{A}^n$ as $\mathcal{A}$-right module. In principle, this is an extra requirement. But for C^*-algebras as well as for $C^\infty(M)$ any projection turns out to be equivalent to a Hermitian projection. Thus we shall always assume this in the following. Then we have the canonical $\mathcal{A}$-valued inner product

$$\langle x, y\rangle = \sum_{i=1}^{n} x_i^* y_i \tag{7.6}$$

for $x, y \in \mathcal{E}$, which is the restriction of the canonical $\mathcal{A}$-valued $\mathcal{A}$-right linear Hermitian inner product on the free $\mathcal{A}$-module $\mathcal{A}^n$. Next, consider $\mathcal{B} = \mathsf{End}_{\mathcal{A}}(\mathcal{E}) = P_0 M_n(\mathcal{A}) P_0$, which clearly is a *-algebra acting on $\mathcal{E}$ such that $\mathcal{E}$ becomes a $\mathcal{B}$-$\mathcal{A}$ bimodule. Moreover, this left action is clearly compatible with (7.6). Finally, we have the canonical $\mathcal{B}$-valued inner product $\Theta_{x,y} \in \mathcal{B}$ *defined* by

$$\Theta_{x,y} \cdot z = x \cdot \langle y, z\rangle \, . \tag{7.7}$$

Thus the compatibility conditions will all be satisfied automatically. However, in general $\mathcal{E}$ will be not a Morita equivalence bimodule since the fullness conditions and the positivity conditions are not fulfilled in general. Nevertheless there is a simple *sufficient* condition, see [13, Def. 4.1 and Thm. 4.2]:

Definition 7.5. A Hermitian projection $P_0 \in M_n(\mathcal{A})$ is called *strongly full* if there exists an invertible $\tau \in \mathcal{A}$ such that $\mathsf{tr} P_0 = \tau^*\tau$.

Theorem 7.6. *Let $P_0 \in M_n(\mathcal{A})$ be a strongly full Hermitian projection. Then $\mathcal{E} = P_0\mathcal{A}^n$ is a Morita equivalence bimodule for $\mathsf{End}_{\mathcal{A}}(\mathcal{E}) = P_0 M_n(\mathcal{A}) P_0$ and $\mathcal{A}$.*

Let us now discuss a more concrete example bringing us back to deformation quantization: If $E \to M$ is a complex vector bundle of fiber dimension $k > 0$ with Hermitian fiber metric h_0 then the classical theorem of Serre and Swan, see e.g. [46], says that $\mathcal{E} = \Gamma^\infty(E)$ is isomorphic to some $P_0 C^\infty(M)^n$ as $C^\infty(M)$-module. Moreover, since any two Hermitian fiber metrics are isometric, we can assume P_0 to be Hermitian. In this case $\mathsf{tr} P_0$ equals the fiber dimension of E and thus P_0 is strongly full whence $\mathcal{E}$ is a Morita equivalence bimodule for $\mathsf{End}_{C^\infty(M)}(\Gamma^\infty(E)) = \Gamma^\infty(\mathsf{End}(E))$ and $C^\infty(M)$, see also [12, Prop. 6.7] for a alternative proof.

Lemma 7.7. *Let $E \to M$ be a non-zero Hermitian vector bundle over M. Then the algebras $\Gamma^\infty(\mathsf{End}(E))$ and $C^\infty(M)$ are strongly Morita equivalent as *-algebras over $\mathbb{C}$ via the equivalence bimodule $\Gamma^\infty(E)$. In particular, writing $\Gamma^\infty(E)$ as $P_0 C^\infty(M)^n$ the Hermitian projection P_0 is strongly full.*

8 Deformation of projective modules

Since isomorphism classes of Hermitian vector bundles $E \to M$ are in one-to-one correspondence to isomorphism classes of Morita equivalence bimodules for $C^\infty(M)$ we shall now deform the module structure of $\Gamma^\infty(E)$ as well as the fiber metric h_0 in order to obtain equivalence bimodules for star products. Note that there are other motivations for deforming the (Hermitian) vector bundle structure like e.g. the index theorems of deformation quantization [40, 24] or the global description of non-commutative field theories arising from string theory [31, 32, 30, 49].

Again we consider the general situation where $\mathcal{A}$ is a unital *-algebra over C and $\boldsymbol{\mathcal{A}} = (\mathcal{A}[[\lambda]], \star)$ is a Hermitian deformation of $\mathcal{A}$. Here and in the following we shall assume that $\mathbb{Q} \subseteq \mathsf{R}$. Let $P_0 \in M_n(\mathcal{A})$ be a Hermitian projection and $\mathcal{E} = P_0\mathcal{A}^n$ as before. Since the matrices $M_n(\boldsymbol{\mathcal{A}})$ inherit the deformed product $\star$ we can consider

$$\boldsymbol{P} = \frac{1}{2} + \left(P_0 - \frac{1}{2}\right) \star \frac{1}{\sqrt[\star]{1 + 4(P_0 \star P_0 - P_0)}}, \tag{8.1}$$

which gives a well-defined formal power series $\boldsymbol{P} \in M_n(\boldsymbol{\mathcal{A}})$. It is an easy computation to show that $\boldsymbol{P} \star \boldsymbol{P} = \boldsymbol{P} = \boldsymbol{P}^*$ is a Hermitian projection with respect to $\star$, see [24, Eq. (6.1.4)] or [26]. Thus one always can deform Hermitian projections since clearly $\boldsymbol{P}$ coincides with P_0 in zeroth order. Moreover,

$$I : \mathcal{E}[[\lambda]] \ni x \mapsto \boldsymbol{P} \star x \in \boldsymbol{P} \star \boldsymbol{\mathcal{A}}^n \tag{8.2}$$

turns out to be an isomorphism of $\mathsf{C}[[\lambda]]$-modules. Since the right hand side is obviously a projective $\star$-right module we can pull-back this module structure via I to obtain a finitely generated projective $\star$-right module $(\mathcal{E}[[\lambda]], \bullet)$ which is a *deformation* of the original module $\mathcal{E}$. Moreover, the restriction of the canonical $\boldsymbol{\mathcal{A}}$-valued $\boldsymbol{\mathcal{A}}$-right linear Hermitian inner product on $\boldsymbol{\mathcal{A}}^n$ to $\boldsymbol{P} \star \boldsymbol{\mathcal{A}}^n$ induces via (8.2) a Hermitian inner product $\boldsymbol{h}$ on $(\mathcal{E}[[\lambda]], \bullet)$, which is again a deformation of the original h_0. Analogously to (8.2) the map

$$P_0 M_n(\mathcal{A}[[\lambda]]) P_0 \ni B \mapsto \boldsymbol{P} \star B \star \boldsymbol{P} \in \boldsymbol{P} \star M_n(\boldsymbol{\mathcal{A}}) \star \boldsymbol{P} \cong \mathsf{End}_{\boldsymbol{\mathcal{A}}}(\boldsymbol{P} \star \boldsymbol{\mathcal{A}}^n) \tag{8.3}$$

is an isomorphism of $\mathsf{C}[[\lambda]]$-modules. Hence the pull-back of the operator product to the left hand side gives a deformation $\star'$ of $\mathsf{End}_{\mathcal{A}}(\mathcal{E})$. Since (8.3) is compatible with the *-involutions, $\star'$ is Hermitian whence $(\mathsf{End}_{\mathcal{A}}(\mathcal{E})[[\lambda]], \star') \cong \mathsf{End}_{\boldsymbol{\mathcal{A}}}(\mathcal{E}[[\lambda]], \bullet, \boldsymbol{h})$ as *-algebras. Clearly, any choice of an isomorphism of the algebras $(\mathsf{End}_{\mathcal{A}}(\mathcal{E})[[\lambda]], \star')$ and $\mathsf{End}_{\boldsymbol{\mathcal{A}}}(\mathcal{E}[[\lambda]], \bullet, \boldsymbol{h})$ gives automatically a left action of the first algebra on $\mathcal{E}[[\lambda]]$ by $\bullet$-right linear endomorphisms. Using the particular isomorphism (8.3) this left action is obviously a *deformation* of the original left action of $\mathsf{End}_{\mathcal{A}}(\mathcal{E})$ on $\mathcal{E}$.

Finally we remark that any other deformations $(\bullet', \boldsymbol{h}')$ of the original module structure and inner product are necessarily equivalent to the ones above. This is a direct consequence of $(\mathcal{E}[[\lambda]], \bullet)$ being *projective* over $\boldsymbol{\mathcal{A}}$ as well as Lemma 5.2. Hence all choices lead to isomorphic deformations $\star'$. The above choice is distinguished in so far that it gives a deformation of the bimodule structure of $\mathcal{E}$. It can be shown easily that any other $\star'$ admitting a deformation of the left action is actually *equivalent* to the above choice (and not only isomorphic). In the following we shall always use such a $\star'$.

Theorem 8.1. *Let $\mathcal{A}$ be a *-algebra over* C *and $P_0 = P_0^2 = P_0^* \in M_n(\mathcal{A})$ and $\mathcal{E} = P_0\mathcal{A}^n$ endowed with its $\mathcal{A}$-valued $\mathcal{A}$-right linear Hermitian inner product h_0 inherited from $\mathcal{A}^n$. Moreover, let $\boldsymbol{\mathcal{A}} = (\mathcal{A}[[\lambda]], \star)$ be a Hermitian deformation of $\mathcal{A}$. Then there exists a deformation $\bullet$ of the right module structure of $\mathcal{E}$ with respect to $\star$ as well as a deformation $\boldsymbol{h}$ of h_0. Any two such deformations are (isometrically) equivalent. Moreover, $\bullet$ yields a finitely generated projective $\boldsymbol{\mathcal{A}}$-module. Finally, $\bullet$ induces a Hermitian deformation $\star'$ of $\mathcal{B} = \mathsf{End}_{\mathcal{A}}(\mathcal{E})$ such that $(\mathcal{B}[[\lambda]], \star') \cong \mathsf{End}_{\boldsymbol{\mathcal{A}}}(\mathcal{E}[[\lambda]], \bullet)$ with a deformed bimodule structure, and $\star'$ is uniquely determined up to equivalence.*

Remark 8.2.

1. Since for any Hermitian vector bundle $E \to M$ the sections $\mathcal{E} = \Gamma^\infty(E)$ are exactly of this form we obtain the result that *Hermitian vector bundles can always be deformed in a unique way up to equivalence.* See also [39] for the case of a trivial vector bundle.

2. Since clearly any projection $\boldsymbol{P} \in M_n(\boldsymbol{\mathcal{A}})$ has a projection $P_0 \in M_n(\mathcal{A})$ as classical limit it follows that all finitely generated projective modules over $\boldsymbol{\mathcal{A}}$ are deformations of finitely generated projective modules over $\mathcal{A}$. Thus the *K-theory does not change under deformation*, i.e. $K_0(\mathcal{A}) \cong K_0(\boldsymbol{\mathcal{A}})$, see e.g. [43].

The use of strongly full projections is illustrated by the following lemma, which follows directly form Lemma 5.2.

Lemma 8.3. *Let $P_0 \in M_n(\mathcal{A})$ be a strongly full projection then any deformation $\boldsymbol{P} \in M_n(\boldsymbol{\mathcal{A}})$ is again strongly full.*

Thus if $\mathcal{E} = P_0\mathcal{A}^n$ is a Morita equivalence bimodule for $\mathcal{A}$ and $\mathcal{B} = \mathsf{End}_{\mathcal{A}}(\mathcal{E})$ with a strongly full P_0 then $\boldsymbol{\mathcal{A}}$ and $\boldsymbol{\mathcal{B}} = (\mathcal{B}[[\lambda]], \star')$, with $\star'$ as above, are again strongly Morita equivalent as *-algebras over C[[λ]] via $(\mathcal{E}[[\lambda]], \bullet, \boldsymbol{h})$ by Theorem 8.1.

9 Local description of deformed vector bundles

Before we shall investigate the Morita equivalence of star products we shall give a local description of the deformed Hermitian vector bundles in case where $\mathcal{A} = C^\infty(M)$ and $\mathcal{E} = \Gamma^\infty(E)$, see also [31] for the case of a line bundle. Let $\star$ be a Hermitian star product for M and $\bullet$, $\boldsymbol{h}$ some fixed choices of deformations for the Hermitian vector bundle. In particular, $\bullet$ and $\boldsymbol{h}$ can be chosen to be *local* thanks to the explicit map (8.2).

Now we choose a good open and locally finite cover $\{\mathcal{O}_\alpha\}$ of M and denote by $e_\alpha = (e_{\alpha,1}, \ldots, e_{\alpha,k})$ a local frame on $\mathcal{O}_\alpha$ of the undeformed vector bundle, i.e. the $e_{\alpha,i}$ give local base sections and any local section s can be written as $s = \sum_{i=1}^k e_{\alpha,i} s_\alpha^i$ with unique coefficient functions $s_\alpha^i \in C^\infty(\mathcal{O}_\alpha)$. For abbreviation we simply write $s_\alpha = (s_\alpha^1, \ldots, s_\alpha^k)$ and $s = e_\alpha s_\alpha$. A simple induction shows the quantum analog of this:

Lemma 9.1. $\boldsymbol{e}_{\alpha,i} = e_{\alpha,i} + \lambda e_{\alpha,i}^{(1)} + \cdots \in \Gamma^\infty(E)[[\lambda]]$ *be deformations of the classical base sections* $e_{\alpha,i}$. *Then any local section* $\boldsymbol{s} \in \Gamma^\infty(E|_{\mathcal{O}_\alpha})[[\lambda]]$ *can uniquely be written as*

$$\boldsymbol{s} = \sum_{i-1}^{k} \boldsymbol{e}_{\alpha,i} \bullet \boldsymbol{s}_\alpha^i \qquad (\boldsymbol{s}_\alpha^i \in C^\infty(\mathcal{O}_\alpha)[[\lambda]]) \tag{9.1}$$

Moreover, the $\mathbb{C}[[\lambda]]$*-linear map* $\boldsymbol{\Psi}_\alpha : \Gamma^\infty(E|_{\mathcal{O}_\alpha})[[\lambda]] \to C^\infty(\mathcal{O}_\alpha)^k[[\lambda]]$ *defined by*

$$\boldsymbol{\Psi}_\alpha(\boldsymbol{s}) = \boldsymbol{s}_\alpha = (\boldsymbol{s}_\alpha^1, \ldots, \boldsymbol{s}_\alpha^k) \tag{9.2}$$

is an isomorphism of $\star$*-right modules, i.e.* $\boldsymbol{\Psi}_\alpha(s \bullet f) = \boldsymbol{\Psi}_\alpha(s) \star f$.

Clearly the $\boldsymbol{\Psi}_\alpha$ can be viewed as deformations of the classical trivialization isomorphisms ψ_α determined by the classical frame e_α. We shall call $\boldsymbol{e}_\alpha = (\boldsymbol{e}_{\alpha,1}, \ldots, \boldsymbol{e}_{\alpha,k})$ a local frame for the deformed vector bundle and write $\boldsymbol{s} = \boldsymbol{e}_\alpha \bullet \boldsymbol{s}_\alpha$ for short. In the case where we have also deformed the Hermitian fiber metric we can apply again Lemma 5.2 to obtain orthonormal frames:

Lemma 9.2. *Let* $\widetilde{\boldsymbol{e}}_\alpha$ *be a local frame such that the zeroth orders constitute an orthonormal frame with respect to* h_0. *Then there exists a matrix* $\boldsymbol{V} = 1 + \sum_{r=1}^\infty \lambda^r V_r \in M_k(C^\infty(\mathcal{O}_\alpha))[[\lambda]]$ *such that* $\boldsymbol{e}_\alpha = \widetilde{\boldsymbol{e}}_\alpha \bullet \boldsymbol{V}$ *is an orthonormal frame with respect to* $\boldsymbol{h}$, *i.e.*

$$\boldsymbol{h}(\boldsymbol{e}_{\alpha,i}, \boldsymbol{e}_{\alpha,j}) = \delta_{ij}. \tag{9.3}$$

As in the undeformed case the choice of local frames gives local *transition matrices* which satisfy the cocycle identity:

Lemma 9.3. *Let* $\{\boldsymbol{e}_\alpha\}$ *be local frames then there exist unique transition matrices* $\boldsymbol{\Phi}_{\alpha\beta} \in M_k(C^\infty(\mathcal{O}_\alpha \cap \mathcal{O}_\beta))[[\lambda]]$ *on every non-empty overlap such that*

$$\boldsymbol{e}_\beta = \boldsymbol{e}_\alpha \bullet \boldsymbol{\Phi}_{\alpha\beta} \quad \textit{and} \quad \boldsymbol{s}_\alpha = \boldsymbol{\Phi}_{\alpha\beta} \star \boldsymbol{s}_\beta = \boldsymbol{\Psi}_\alpha \circ \boldsymbol{\Psi}_\beta^{-1}(\boldsymbol{s}_\beta). \tag{9.4}$$

Moreover, the cocycle condition

$$\boldsymbol{\Phi}_{\alpha\beta} \star \boldsymbol{\Phi}_{\beta\alpha} = 1 \quad \textit{and} \quad \boldsymbol{\Phi}_{\alpha\beta} \star \boldsymbol{\Phi}_{\beta\gamma} \star \boldsymbol{\Phi}_{\gamma\alpha} = 1 \tag{9.5}$$

is fulfilled on the corresponding non-empty intersections. If the $\{\boldsymbol{e}_\alpha\}$ *are orthonormal we have unitary transition matrices* $\boldsymbol{\Phi}_{\alpha\beta}^* = \boldsymbol{\Phi}_{\beta\alpha}$.

Again the $\boldsymbol{\Phi}_{\alpha\beta}$ are deformations of the transition matrices $\phi_{\alpha\beta}$ corresponding to the classical limit of the frames $\{\boldsymbol{e}_\alpha\}$.

Let us now discuss the $\bullet$-right linear endomorphisms of the deformed vector bundle. If $A(s \bullet f) = A(s) \bullet f$ holds globally then locally we have a matrix $A_\alpha \in M_k(C^\infty(\mathcal{O}_\alpha))[[\lambda]]$ such that

$$(A(s))_\alpha = A_\alpha \star s_\alpha \quad \text{and} \quad A(e_\alpha) = e_\alpha \bullet A_\alpha. \tag{9.6}$$

On overlaps $\mathcal{O}_\alpha \cap \mathcal{O}_\beta$ we clearly have

$$A_\beta = \Phi_{\beta\alpha} \star A_\alpha \star \Phi_{\alpha\beta}. \tag{9.7}$$

Conversely, if a collection of matrices $\{A_\alpha\}$ satisfies (9.7) then it defines a global $\bullet$-right linear endomorphism A by (9.6). The composition of A and B is locally given by the deformed matrix multiplication

$$(A \circ B)_\alpha = A_\alpha \star B_\alpha, \tag{9.8}$$

and if the frames $\{e_\alpha\}$ are orthonormal we have $(A^*)_\alpha = (A_\alpha)^*$.

10 From local to global deformations

In this section we present some new material, namely a detailed comparison between the approach to deformed vector bundles by deformed local transition matrices as used in [31, 32] with our approach of globally deforming a projection. As we already know from Theorem 8.1 both approaches are necessarily equivalent. The local approach is in some sense more familiar and more suited for explicit constructions while in the global approach questions of uniqueness up to equivalence can be handled more easily.

The starting point is an open cover $\{\mathcal{U}_\alpha\}$ of M such that $E|_{\mathcal{U}_\alpha}$ is trivial. It will be necessary to assume that $\{\mathcal{U}_\alpha\}$ is actually a *finite* cover whence we shall use a different symbol $\mathcal{U}_\alpha$ here, see e.g. [51, Prop. 4.1, Chap. III] for the existence of such covers. We assume that we have found deformations $\Phi_{\alpha\beta}$ of the classical transition matrices $\phi_{\alpha\beta}$ corresponding to some local frames $\{e_\alpha\}$, i.e. we have $\Phi_{\alpha\beta} = \phi_{\alpha\beta} + \sum_{r=1}^\infty \lambda^r \Phi_{\alpha\beta}^{(r)}$ satisfying (9.5). We shall mainly be interested in the Hermitian case whence we assume in addition $\Phi_{\alpha\beta}^* = \Phi_{\beta\alpha}$. Then the aim is to build a deformation $\bullet$ of $\Gamma^\infty(E)$ and $\star'$ of $\Gamma^\infty(\mathsf{End}(E))$ out of this data and to show that these deformations are of the form as obtained in Section 8.

We choose a partition of unity $\{\chi_\alpha\}$ subordinate to $\{\mathcal{U}_\alpha\}$. For $s \in \Gamma^\infty(E)[[\lambda]]$ we define

$$\mathsf{s}_\alpha := \sum_\beta \Phi_{\alpha\beta} \star \chi_\beta \star s_\beta, \tag{10.1}$$

where classically $s = e_\beta s_\beta$ on $\mathcal{U}_\beta$. Clearly $\mathsf{s}_\alpha \in C^\infty(\mathcal{U}_\alpha)^k[[\lambda]]$ and $\Phi_{\alpha\beta} \star \mathsf{s}_\beta = \mathsf{s}_\alpha$. Finally note that $\mathsf{s}_\alpha = s_\alpha + \sum_{r=1}^\infty \lambda^r s_\alpha^{(r)}$. Then the following lemma is a simple induction and can be seen as the 'converse' statement to Lemma 9.1.

Lemma 10.1. *The map $s \mapsto \{s_\alpha\}$ from $\Gamma^\infty(E)[[\lambda]]$ into the set of all collections $\{s_\alpha\}$ with $s_\alpha \in C^\infty(\mathcal{U}_\alpha)^k[[\lambda]]$ satisfying $\boldsymbol{\Phi}_{\alpha\beta} \star s_\beta = s_\alpha$ is a $\mathbb{C}[[\lambda]]$-linear bijection.*

This allows to define a global $\star$-right module structure on the sections by setting

$$(s \bullet f)_\alpha := s_\alpha \star f, \tag{10.2}$$

which is clearly a deformation of the classical module structure. Thus the existence of a deformation quantization of E follows from the existence of deformed transition matrices. A Hermitian inner product is defined by

$$\boldsymbol{h}(s,t)|_{\mathcal{U}_\alpha} = \langle s_\alpha, t_\alpha \rangle = \sum_{l=1}^{k} \overline{s_\alpha^l} \star t_\alpha^l, \tag{10.3}$$

which is globally defined since $\boldsymbol{\Phi}_{\alpha\beta}^* = \boldsymbol{\Phi}_{\beta\alpha}$ and clearly a deformation of a classical Hermitian fiber metric h_0.

Now let us consider a 'local' way to deal with the $\bullet$-right linear endomorphisms. First we choose a *quadratic* partition of unity $\{\chi_\alpha\}$ subordinate to $\{\mathcal{U}_\alpha\}$ and define for $A \in \Gamma^\infty(\mathsf{End}(E))[[\lambda]]$ the matrix $\widetilde{T}_\alpha(A) \in M_k(C^\infty(\mathcal{U}_\alpha))[[\lambda]]$ by

$$\widetilde{T}_\alpha(A) = \sum_\gamma \boldsymbol{\Phi}_{\alpha\gamma} \star \overline{\chi}_\gamma \star A_\gamma \star \chi_\gamma \star \boldsymbol{\Phi}_{\gamma\alpha}, \tag{10.4}$$

where A_γ is the local matrix of A corresponding to the undeformed structures. Then clearly

$$\boldsymbol{\Phi}_{\alpha\beta} \star \widetilde{T}_\beta(A) \star \boldsymbol{\Phi}_{\beta\alpha} = \widetilde{T}_\alpha(A) \quad \text{and} \quad \widetilde{T}_\alpha(A^*) = \widetilde{T}_\alpha(A)^*, \tag{10.5}$$

whence $(\widetilde{T}(A)s)_\alpha = \widetilde{T}_\alpha(A) \star s_\alpha$ defines a global $\bullet$-right linear endomorphism $\widetilde{T}(A)$. In zeroth order we have $\widetilde{T}_\alpha(A) = A_\alpha$ whence a simple induction shows that all $\bullet$-right linear endomorphisms are obtained this way, i.e. $\widetilde{T}$ is a bijection. Thus we obtain a Hermitian deformation $\tilde{\star}$ of $\Gamma^\infty(\mathsf{End}(E))[[\lambda]]$ such that $\widetilde{T}_\alpha(A \tilde{\star} B) = \widetilde{T}_\alpha(A) \star \widetilde{T}_\alpha(B)$. However, $\widetilde{T}_\alpha(1) \neq 1$ in general. Thus the identity endomorphism is no longer the unit for $\tilde{\star}$ but some other $E = \widetilde{T}^{-1}(1) = 1 + \sum_{r=1}^\infty \lambda^r E_r$. Nevertheless this can be 'repaired' in the usual way by passing to an equivalent deformation $\star'$. By Lemma 5.2 we know that $1 = (U^{-1})^* \tilde{\star} U^{-1}$ with some $\tilde{\star}$-invertible U. The equivalence transformation $S(A) := (U^{-1})^* \tilde{\star} A \tilde{\star} U^{-1}$ gives a new deformation $S(A \tilde{\star} B) = S(A) \star' S(B)$, which is still Hermitian and satisfies $1 \star' A = A = A \star' 1$. Moreover, we obtain a new isomorphism $T(A) = \widetilde{T}(S^{-1}(A))$ locally given by

$$T_\alpha(A) = \widetilde{T}_\alpha(U^*) \star \widetilde{T}_\alpha(A) \star \widetilde{T}_\alpha(U) = \sum_\gamma \boldsymbol{U}_\alpha^* \star \boldsymbol{\Phi}_{\alpha\gamma} \star \overline{\chi}_\gamma \star A_\gamma \star \chi_\gamma \star \boldsymbol{\Phi}_{\gamma\alpha} \star \boldsymbol{U}_\alpha, \tag{10.6}$$

where $\boldsymbol{U}_\alpha := \widetilde{T}_\alpha(U)$. Now we indeed have

$$T_\alpha(A \star' B) = T_\alpha(A) \star T_\alpha(B), \quad T_\alpha(A^*) = T_\alpha(A)^*, \quad \text{and} \quad T_\alpha(1) = 1. \tag{10.7}$$

Again T as well as $\star'$ are not uniquely determined and there may be situations where a more concrete description of the isomorphism T is available.

Let us now prove 'intrinsically' that such a deformation $\bullet$ yields a finitely generated projective module. This can be viewed as a sort of *quantum Serre–Swan theorem*. We follow the ideas of [32] where the case of a line bundle was considered. It is crucial to have a finite cover $\{\mathcal{U}_\alpha\}$.

Let χ_α be as before, only viewed as element in $\Gamma^\infty(\mathsf{End}(E))$. Using Lemma 5.2 we find a $\star'$-invertible $V \in \Gamma^\infty(\mathsf{End}(E))[[\lambda]]$ such that

$$\sum_{\alpha=1}^{m} \overline{\chi}_\alpha \star' \chi_\alpha = V^* \star' V, \tag{10.8}$$

Hence the global endomorphisms $\boldsymbol{\chi}_\alpha = \chi_\alpha \star' V^{-1}$ satisfy

$$\sum_{\alpha=1}^{m} \boldsymbol{\chi}_\alpha^* \star' \boldsymbol{\chi}_\alpha = 1 \quad \text{and} \quad \operatorname{supp} \boldsymbol{\chi}_\alpha \subseteq \mathcal{U}_\alpha. \tag{10.9}$$

We define a $\mathbb{C}[[\lambda]]$-linear map $\epsilon : \Gamma^\infty(E)[[\lambda]] \to C^\infty(M)^n[[\lambda]]$, where $n = mk$, by

$$\epsilon(s) = \left(T_1(\boldsymbol{\chi}_1) \star s_1, \ldots, T_m(\boldsymbol{\chi}_m) \star s_m\right), \tag{10.10}$$

which indeed gives *global* functions since T and $\star$ are local and $\operatorname{supp} \boldsymbol{\chi}_\alpha \subseteq \mathcal{U}_\alpha$. Clearly we have $\epsilon(s \bullet f) = \epsilon(s) \star f$, where $C^\infty(M)^n[[\lambda]]$ is endowed with its usual *free* $\star$-right module structure. Next we define $\pi : C^\infty(M)^n[[\lambda]] \to \Gamma^\infty(E)[[\lambda]]$ by

$$(\pi(t))_\alpha := \sum_{\gamma=1}^{m} \boldsymbol{\Phi}_{\alpha\gamma} \star T_\gamma(\boldsymbol{\chi}_\gamma^*) \star t_\gamma\big|_{\mathcal{U}_\alpha}, \tag{10.11}$$

where $t = (t_1, \ldots, t_m)$ with global $t_\gamma \in C^\infty(M)^k[[\lambda]]$. Then $\boldsymbol{\Phi}_{\alpha\beta} \star \pi(t)_\beta = \pi(t)_\alpha$ whence $\pi(t)$ is indeed a section of E. Again π is a $\star$-right module morphism and a straightforward computation yields $\pi \circ \epsilon = \mathsf{id}$, whence $\epsilon \circ \pi$ is a projection. Since it is even $\star$-right linear it is of the form

$$\epsilon(\pi(t)) = \boldsymbol{P} \star t \quad \text{with} \quad \boldsymbol{P} \star \boldsymbol{P} = \boldsymbol{P} \in M_n(C^\infty(M))[[\lambda]]. \tag{10.12}$$

But this shows that $(\Gamma^\infty(E)[[\lambda]], \bullet)$ is indeed a finitely generated projective module as ϵ provides a module isomorphism into the image of $\boldsymbol{P}$. Thus we have an alternative proof of this fact and can now easily use the uniqueness statements of Theorem 8.1 also in this (completely equivalent) approach.

Example 10.2 (Flat bundles). Let $E \to M$ be a *flat* Hermitian vector bundle and let $\{e_\alpha\}$ be covariantly constant local frames with *constant* transition functions $\phi_{\alpha\beta}$. Then $\boldsymbol{\Phi}_{\alpha\beta} = \phi_{\alpha\beta}$ satisfy the cocycle condition (9.5). Hence $\mathsf{s}_\alpha = s_\alpha$ and $s \bullet f = e_\alpha(s_\alpha \star f)$ if classically $s = e_\alpha s_\alpha$. Moreover, in this case $T_\alpha(A) = A_\alpha$ will do the job, i.e.

$$(T(A)s)_\alpha = T_\alpha(A) \star s_\alpha = A_\alpha \star s_\alpha \tag{10.13}$$

gives a simple description of the $\bullet$-right linear endomorphisms and we have

$$(A \star' B)_\alpha = A_\alpha \star A_\beta. \tag{10.14}$$

Surprisingly, in this case the center $C^\infty(M)$ of the undeformed algebra $\Gamma^\infty(\mathsf{End}(E))$ is still a *sub-algebra* with respect to $\star'$. Moreover, from (10.14) it immediately follows that $\star' = \star$ when restricted to $C^\infty(M)[[\lambda]] \subseteq \Gamma^\infty(\mathsf{End}(E))[[\lambda]]$. Thus a flat vector bundle becomes automatically a *$\star$-bimodule*, see also Corollary 11.5.

It seems to be an interesting question whether in the case of a general vector bundle the center is deformed into a sub-algebra with respect to $\star'$. In this case $\star'$ would induce an associative deformation of the center by restricting $\star'$.

Question 10.3. Let $\mathcal{A}$ be an *Abelian* algebra with associative deformation $\star$ and let $\mathcal{E}$ be a finitely generated projective module over $\mathcal{A}$ with deformation $\bullet$ and let $\star'$ be the corresponding (unique up to equivalence) deformation of $\mathsf{End}_{\mathcal{A}}(\mathcal{E})$. Then $\mathcal{A} \subseteq \mathsf{End}_{\mathcal{A}}(\mathcal{E})$ belongs to the center. Is this deformed into a *sub-algebra* with respect to $\star'$ or, if not, what is the algebra generated by $\mathcal{A}$ inside $\mathsf{End}_{\mathcal{A}}(\mathcal{E})[[\lambda]]$ with respect to $\star'$? Moreover, does the assumption of *fullness* of $\mathcal{E}$ simplify the problem?

11 Morita equivalence of star products

In this final section we shall sketch the classification of star products up to Morita equivalence on symplectic manifolds.

First we recall the construction of the *relative characteristic class* of two star products $\star, \star'$ on (M, ω) as it can be found in [28]. The *characteristic class* $c(\star)$ of a symplectic star product can e.g. be defined by means of the Weyl curvature in the Fedosov construction, see e.g. [40, 4], and yields a class

$$c(\star) \in \frac{1}{\mathrm{i}\lambda}[\omega] + \mathrm{H}^2_{\mathrm{dR}}(M, \mathbb{C})[[\lambda]]. \tag{11.1}$$

Then the relative class of two star products is defined by

$$t(\star', \star) = c(\star') - c(\star) \in \mathrm{H}^2_{\mathrm{dR}}(M, \mathbb{C})[[\lambda]], \tag{11.2}$$

and $\star'$ is equivalent to $\star$ if and only if their relative class vanishes. There is a simple Čech cohomological construction of $t(\star', \star)$. We again use the good cover as before. Then on $\mathcal{O}_\alpha$ the star products $\star'$ and $\star$ are equivalent via some $T_\alpha = \mathsf{id} + \sum_{r=1}^\infty \lambda^r T_\alpha^{(r)}$, i.e. $T_\alpha(f \star' g) = T_\alpha f \star T_\alpha g$, since $\mathcal{O}_\alpha$ is contractible. On overlaps it follows that $T_\alpha^{-1} T_\beta$ is a $\star$-automorphism starting with the identity. Hence it is an inner automorphism

$$T_\alpha^{-1} T_\beta(f) = \mathrm{e}^{[t_{\alpha\beta}, \cdot]}(f) = \mathrm{Exp}(t_{\alpha\beta}) \star f \star \mathrm{Exp}(-t_{\alpha\beta}) \tag{11.3}$$

with some function $t_{\alpha\beta} \in C^\infty(\mathcal{O}_\alpha \cap \mathcal{O}_\beta)[[\lambda]]$. Here Exp denotes the $\star$-exponential, see e.g. [8, 16]. Moreover, the $T_\alpha^{-1} T_\beta$ satisfy the 'cocycle condition'

$$T_\alpha^{-1} T_\beta T_\beta^{-1} T_\gamma T_\gamma^{-1} T_\alpha = \mathsf{id} \tag{11.4}$$

whence $\mathrm{Exp}(t_{\alpha\beta}) \star \mathrm{Exp}(t_{\beta\gamma}) \star \mathrm{Exp}(t_{\gamma\alpha})$ has to be a central element and thus *constant* on the triple overlap. Since for the star exponential one has a BCH formula denoted by $\mathrm{Exp}(f) \star \mathrm{Exp}(g) = \mathrm{Exp}(f \circ_\star g)$ we find that

$$t_{\alpha\beta\gamma} = t_{\alpha\beta} \circ_\star t_{\beta\gamma} \circ_\star t_{\gamma\alpha} \in \mathbb{C}[[\lambda]] \tag{11.5}$$

has to be *constant* on $\mathcal{O}_\alpha \cap \mathcal{O}_\beta \cap \mathcal{O}_\gamma$ as well. Thus we have a Čech cochain and it turns out that it is even a cocycle. Thereby it defines a Čech cohomology class $[t_{\alpha\beta\gamma}] \in \check{\mathrm{H}}^2(M, \mathbb{C})[[\lambda]]$. This class coincides under the usual identification of $\check{\mathrm{H}}^2(M, \mathbb{C})$ with $\mathrm{H}^2_{\mathrm{dR}}(M, \mathbb{C})$ with the relative class, i.e. $[t_{\alpha\beta\gamma}] = t(\star', \star)$.

Now let us turn to (strong) Morita equivalence of star products. If $\star$ and $\star'$ are Morita equivalent by some equivalence bimodule, then it has to be a finitely generated and projective $\star$-right module. Thus it is a deformed vector bundle $E \to M$ with deformation $\bullet$ and $\boldsymbol{h}$. But since we want $(C^\infty(M)[[\lambda]], \star')$ to be *-isomorphic to $\mathsf{End}(\Gamma^\infty(E)[[\lambda]], \bullet, \boldsymbol{h})$ it is necessary that E is a *line bundle* $L \to M$ since only in this case $\Gamma^\infty(\mathsf{End}(L)) \cong C^\infty(M)$. Thus let $L \to M$ be a line bundle with Hermitian fiber metric and corresponding deformations quantization $\bullet$ and $\boldsymbol{h}$ with respect to $\star$ such that $\star'$ is the star product for M such that

$$(C^\infty(M)[[\lambda]], \star') \cong \mathsf{End}(\Gamma^\infty(L)[[\lambda]], \bullet, \boldsymbol{h}). \tag{11.6}$$

We already know that it does not matter which deformation $\bullet$ and $\boldsymbol{h}$ we actually choose since all are equivalent anyway. Moreover, we can choose within the *isomorphism* class of $\star'$ a representative such that we have a bimodule deformation. This corresponds to the distinguished *equivalence* class as discussed in Theorem 8.1. The subtle point here is the fact that an isomorphism of star products starts in zeroth order with the pull-back by some diffeomorphism which is not local. At the moment we want to exclude this non-locality and consider this additional freedom later. For such $\star'$ we know from Section 10 that we can find *local* operators

$$T_\alpha = \mathsf{id} + \sum_{r=1}^{\infty} \lambda^r T_\alpha^{(r)} : C^\infty(\mathcal{O}_\alpha)[[\lambda]] \to C^\infty(\mathcal{O}_\alpha)[[\lambda]] \tag{11.7}$$

such that $T_\alpha(f)$ is the local matrix of a global $\bullet$-right linear endomorphism $T(f)$ of the deformed line bundle. Moreover, $\star'$ can be obtained by $T(f \star' g) = T(f) \circ T(g)$. On the other hand, from (10.7) we know

$$T_\alpha(f \star' g) = T_\alpha(f) \star T_\alpha(g). \tag{11.8}$$

Up to now we have not yet used that $\star$ was a symplectic star product. Thus we have the following statement, which is non-trivial in the Poisson case:

Lemma 11.1. *Let $\star$ and $\star'$ be Morita equivalent star products on a Poisson manifold with a deformed equivalence bimodule. Then $\star$ and $\star'$ are locally equivalent. In particular, they deform the same Poisson bracket, see also* [11, Lem. 3.4].

Since $T(f)$ is a global $\bullet$-right linear endomorphism we conclude with (9.7) that

$$T_\alpha^{-1} T_\beta(f) = \boldsymbol{\Phi}_{\alpha\beta} \star f \star \boldsymbol{\Phi}_{\beta\alpha}. \tag{11.9}$$

Since in zeroth order $\mathbf{\Phi}_{\alpha\beta}$ is just $\phi_{\alpha\beta}$ we can write them as star exponentials $\mathbf{\Phi}_{\alpha\beta} = \operatorname{Exp}(t_{\alpha\beta})$ with some $t_{\alpha\beta} = \sum_{r=0}^{\infty} \lambda^r t_{\alpha\beta}^{(r)}$, where $\mathrm{e}^{t_{\alpha\beta}^{(0)}} = \phi_{\alpha\beta}$. Thus in the symplectic case the relative class of $\star'$ and $\star$ is given by the class $t(\star', \star) = [t_{\alpha\beta\gamma}]$ coming from these $t_{\alpha\beta}$ as in (11.5). On the other hand, this class can be computed explicitly. Since the $\mathbf{\Phi}_{\alpha\beta}$ satisfy the cocycle identity (9.5) we have

$$t_{\alpha\beta\gamma} = t_{\alpha\beta} \circ_\star t_{\beta\gamma} \circ_\star t_{\gamma\alpha} = 2\pi\mathrm{i}\, n_{\alpha\beta\gamma} \quad \text{with} \quad n_{\alpha\beta\gamma} \in \mathbb{Z}. \tag{11.10}$$

In particular, $t_{\alpha\beta\gamma}$ does *not* depend on λ at all. But this shows that it coincides with its classical limit which is given by $t_{\alpha\beta}^{(0)} + t_{\beta\gamma}^{(0)} + t_{\gamma\alpha}^{(0)}$. Since the $t_{\alpha\beta}^{(0)}$ where just the logarithms of the transitions functions $\phi_{\alpha\beta}$ of the undeformed line bundle we conclude that $t(\star', \star) = 2\pi\mathrm{i}\, c_1(L)$ yields the *Chern class* of L. On the other hand, if $t(\star', \star)$ is a $2\pi\mathrm{i}$-*integral class* then there exists a line bundle with fiber metric which we can deform. Then Theorem 8.1 together with Lemma 8.3 and Theorem 7.6 shows that $\star$ and $\star'$ are strongly Morita equivalent in this case. Thus we have the following characterization [16, Thm. 3.1 and Cor. 3.2].

Theorem 11.2. *Let (M, ω) be a symplectic manifold with Hermitian star products $\star$ and $\star'$.*

1. *$\star'$ is *-equivalent to the star product induced by the deformation quantization of a line bundle $L \to M$ with respect to $\star$ if and only if*

$$t(\star', \star) = 2\pi\mathrm{i}\, c_1(L). \tag{11.11}$$

2. *Two star products $\star$ and $\tilde{\star}$ are strongly Morita equivalent if and only if there exists a diffeomorphic star product $\star'$ to $\tilde{\star}$ such that*

$$t(\star', \star) \in 2\pi\mathrm{i}\, \mathrm{H}^2_{\mathrm{dR}}(M, \mathbb{Z}). \tag{11.12}$$

Here $\mathrm{H}^2_{\mathrm{dR}}(M, \mathbb{Z})$ denotes the image of the integer cohomology in the complex-valued deRham cohomology under the usual inclusion map. Note also that we had to re-implement the freedom of a diffeomorphism.

Remark 11.3. There exists also a characterization of Morita equivalent star products in the Poisson case. If one formally 'inverts' the condition $\lambda c(\star') = \lambda c(\star) + 2\pi\mathrm{i}\lambda c_1(L)$ one obtains a geometric series for the relation between the corresponding Poisson structures π and π'. Here one has to observe that $c(\star)$ starts in order λ^{-1} with the symplectic form ω. Thus one obtains a relation between the formal deformations of the original Poisson bracket in Kontsevich's sense [35], which are known to classify star products in the Poisson case. That this heuristic reasoning is actually correct is a result of Jurčo, Schupp, and Wess [32].

In the above construction the inner products only played a minor role. Indeed, we can also consider the Morita equivalence of star products from a purely ring-theoretical point of view. It turns out that for Hermitian star products this notion of Morita equivalence *coincides* with the strong Morita equivalence as *-algebras over $\mathbb{C}[[\lambda]]$, see also the discussion in [16]. We have emphasized here the aspect of strong Morita equivalence since this implies the equivalence of the categories of *-representations on pre-Hilbert spaces, which from a physical point of view is the most interesting consequence of Morita equivalence. Nevertheless, there are also several other consequences.

Recall that the *Picard group* $\mathsf{Pic}(\mathcal{A})$ of an algebra is the group of equivalence classes of self-Morita equivalences. In the classical case we have $\mathsf{Pic}(C^\infty(M)) = \check{\mathrm{H}}^2(M, \mathbb{Z}) = \mathsf{Pic}(M)$, where the (geometric) Picard group of a manifold is the group of equivalence classes of line bundles with group structure from the tensor product. From [11] one knows that the Picard group $\mathsf{Pic}(\mathcal{A})$ acts canonically on the moduli space of equivalence classes of deformations of $\mathcal{A}$. Thus Theorem 11.2 can also be understood as an explicit computation of this group action: For the star products on a symplectic manifold the Picard group acts on the equivalence classes just by adding the $2\pi\mathrm{i}$-multiple of the Chern class of the corresponding line bundle in the zeroth order of λ to the characteristic class.

Corollary 11.4. *Let (M, ω) be a symplectic manifold. Then action Φ of the Picard group $\mathsf{Pic}(M)$ on the equivalence classes of star products $\mathsf{Def}(M, \omega)$ is given by*

$$\Phi : \mathsf{Pic}(M) \times \mathsf{Def}(M, \omega) \ni ([L], c(\star)) \mapsto c(\star) + 2\pi\mathrm{i}\, c_1(L) \in \mathsf{Def}(M, \omega), \tag{11.13}$$

with the identification (11.1).

In the case where the characteristic class of $\star$ has a vanishing zeroth order term we can also compute the Picard group of $\star$, since in this case the additional diffeomorphism in (11.12) does not spoil the argument.

Corollary 11.5. *Let (M, ω) be a symplectic manifold with star product $\star$ such that $c(\star)$ has vanishing zeroth order term.*

1. *The Picard group of $(C^\infty(M)[[\lambda]], \star)$ is given by the group of equivalence classes of flat line bundles over M.*

2. *$\Gamma^\infty(L)$ admits a $\star$-bimodule structure deforming the classical one if and only if L is flat, i.e. $c_1(L) = 0$.*

Remark 11.6. A very explicit physical interpretation of the above characterization of Morita equivalent star products can be achieved in the case of star products on cotangent bundles T^*Q. Here one explicitly knows various star products, in particular a star product $\star^B$ as quantization of a charged particle in presence of a magnetic field B on the configuration space Q. As usual B is viewed as a closed two-form. One knows that $c(\star^B)$ equals the class of $\mathrm{i}B$. Then $\star^B$ is Morita equivalent to $\star$ without B if and only if the magnetic field satisfies *Dirac's quantization/integrality condition for the magnetic monopole charges* described by B. Thus Dirac's condition turns out to imply *equivalence* of the whole representation theory and not just the existence of a particular

'Schrödinger-like' representation on section of the line bundle defined by B. In fact, the latter can be understood as Rieffel induced by the ordinary Schrödinger representation. Thus the above developed framework proves to give physically interesting and non-trivial applications, see [16] for a further discussion.

References

[1] Ara, P., Morita equivalence for rings with involution, *Algebr. Represent. Theory* 2 (1999), 227–247.

[2] Basart, H., Flato, M., Lichnerowicz, A., Sternheimer, D., Deformation Theory applied to Quantization and Statistical Mechanics, *Lett. Math. Phys.* 8 (1984), 483–394.

[3] Bayen, F., Flato, M., Frønsdal, C., Lichnerowicz, A., Sternheimer, D., Deformation Theory and Quantization, *Ann. Phys.* 111 (1978), 61–151.

[4] Bertelson, M., Cahen, M., Gutt, S., Equivalence of Star Products, *Classical Quantum Gravity* 14 (1997), A93–A107.

[5] Bordemann, M., Neumaier, N., Pflaum, M. J., Waldmann, S., On representations of star product algebras over cotangent spaces on Hermitian line bundles, Preprint Freiburg FR-THEP-98/24 `math.QA/9811055` (November 1998).

[6] Bordemann, M., Neumaier, N., Waldmann, S., Homogeneous Fedosov Star Products on Cotangent Bundles I: Weyl and Standard Ordering with Differential Operator Representation, *Commun. Math. Phys.* 198 (1998), 363–396.

[7] Bordemann, M., Neumaier, N., Waldmann, S., Homogeneous Fedosov star products on cotangent bundles II: GNS representations, the WKB expansion, traces, and applications, *J. Geom. Phys.* 29 (1999), 199–234.

[8] Bordemann, M., Römer, H., Waldmann, S., A Remark on Formal KMS States in Deformation Quantization, *Lett. Math. Phys.* 45 (1998), 49–61.

[9] Bordemann, M., Waldmann, S., A Fedosov Star Product of Wick Type for Kähler Manifolds, *Lett. Math. Phys.* 41 (1997), 243–253.

[10] Bordemann, M., Waldmann, S., Formal GNS Construction and States in Deformation Quantization, *Commun. Math. Phys.* 195 (1998), 549–583.

[11] Bursztyn, H., Semiclassical Geometry of Quantum Line Bundles and Morita Equivalence of Star Products, Preprint `math.QA/0105001` (May 2001).

[12] Bursztyn, H., Waldmann, S., *-Ideals and Formal Morita Equivalence of *-Algebras, *Internat. J. Math.* 12 (2001), 555–577.

[13] Bursztyn, H., Waldmann, S., Deformation Quantization of Hermitian Vector Bundles, *Lett. Math. Phys.* 53 (2000), 349–365.

[14] Bursztyn, H., Waldmann, S., On Positive Deformations of *-Algebras, in *Quantization, Deformations, and Symmetries* (Dito, G., Sternheimer, D., eds.), Conférence Moshé Flato 1999, Math. Phys. Stud. 22, Kluwer Academic Publishers, Dordrecht, Boston, London 2000, 69–80.

[15] Bursztyn, H., Waldmann, S., Algebraic Rieffel Induction, Formal Morita Equivalence and Applications to Deformation Quantization, *J. Geom. Phys.* 37 (2001), 307–364.

[16] Bursztyn, H., Waldmann, S., The characterisitc classes of Morita equivalent star products on symplectic manifolds, Preprint `math.QA/0106178` (June 2001).

[17] Cattaneo, A., Felder, G., A Path Integral Approach to the Kontsevich Quantization Formula, *Commun. Math. Phys.* 212 (2000), 591–611.

[18] Connes, A., *Noncommutative Geometry*, Academic Press, San Diego, New York, London, 1994.

[19] Connes, A., Douglas, M. R., Schwarz, A., Noncommutative geometry and matrix theory: compactification on tori, *J. High Energy Phys.* 02 (1998), 003.

[20] DeWilde, M., Lecomte, P. B. A., Existence of Star-Products and of Formal Deformations of the Poisson Lie Algebra of Arbitrary Symplectic Manifolds, *Lett. Math. Phys.* 7 (1983), 487–496.

[21] Dito, G., Sternheimer, D. (eds.), *Quantization, deformations, and symmetries*, Conférence Moshé Flato 1999, Math. Phys. Stud. 22, Kluwer Academic Publishers, Dordrecht, Boston, London 2000.

[22] Dixmier, J., *C^*-algebras*, Translated from the French by Francis Jellett, North-Holland Math. Library 15, North-Holland Publishing Co., Amsterdam 1977.

[23] Fedosov, B. V., A Simple Geometrical Construction of Deformation Quantization, *J. Diff. Geom.* 40 (1994), 213–238.

[24] Fedosov, B. V., *Deformation Quantization and Index Theory*, Math. Top. 9. Akademie Verlag, Berlin 1996.

[25] Gerstenhaber, M., On the Deformation of Rings and Algebras, *Ann. Math.* 79 (1964), 59–103.

[26] Gerstenhaber, M., Schack, S. D., Algebraic Cohomology and Deformation Theory, in *Deformation Theory of Algebras and Structures and Applications* (Hazewinkel, M., Gerstenhaber, M., eds.), Kluwer Academic Press, Dordrecht 1988, 13–264.

[27] Gutt, S., Variations on deformation quantization, in *Quantization, Deformations, and Symmetries* (Dito, G., Sternheimer, D., eds.), Conférence Moshé Flato 1999, Math. Phys. Stud. 22, Kluwer Academic Publishers, Dordrecht, Boston, London 2000, 217–254.

[28] Gutt, S., Rawnsley, J., Equivalence of star products on a symplectic manifold; an introduction to Deligne's Čech cohomology classes, *J. Geom. Phys.* 29 (1999), 347–392.

[29] Haag, R., *Local Quantum Physics*, 2. edition, Springer-Verlag, Berlin, Heidelberg, New York 1993.

[30] Jurco, B., Schupp, P., Wess, J., Noncommutative gauge theory for Poisson manifolds, *Nucl. Phys.* B 584 (2000), 784–794.

[31] Jurco, B., Schupp, P., Wess, J., Nonabelian noncommutative gauge theory via noncommutative extra dimensions, Preprint `hep-th/0102129` (February 2001).

[32] Jurco, B., Schupp, P., Wess, J., Noncommutative line bundle and Morita equivalence, Preprint `hep-th/0106110` (June 2001).

[33] Karabegov, A. V., Deformation Quantization with Separation of Variables on a Kähler Manifold, *Commun. Math. Phys.* 180 (1996), 745–755.

[34] Karabegov, A. V., On Fedosov's approach to Deformation Quantization with Separation of Variables, in *Quantization, Deformations, and Symmetries* (Dito, G., Sternheimer, D., eds.), Conférence Moshé Flato 1999, Math. Phys. Stud. 22, Kluwer Academic Publishers, Dordrecht, Boston, London 2000, 167–176.

[35] Kontsevich, M., Deformation Quantization of Poisson Manifolds, I, Preprint q-alg/9709040 (September 1997).

[36] Lam, T. Y., *Lectures on Modules and Rings*, Grad. Texts in Math. 189, Springer-Verlag, Berlin, Heidelberg, New York 1999.

[37] Lance, E. C., *Hilbert C^*-modules. A toolkit for operator algebraists*, London Math. Soc. Lecture Note Ser. 210, Cambridge University Press, Cambridge 1995.

[38] Landsman, N. P., *Mathematical Topics between Classical and Quantum Mechanics*, Springer Monogr. Math., Springer-Verlag, Berlin, Heidelberg, New York 1998.

[39] Lecomte, P., Roger, C., Formal deformations of the associative algebra of smooth matrices, *Lett. Math. Phys.* 15,1 (1988), 55–63.

[40] Nest, R., Tsygan, B., Algebraic Index Theorem, *Commun. Math. Phys.* 172 (1995), 223–262.

[41] Omori, H., Maeda, Y., Yoshioka, A., Weyl Manifolds and Deformation Quantization, *Adv. Math.* 85 (1991), 224–255.

[42] Rieffel, M. A., Morita equivalence for C^*-algebras and W^*-algebras, *J. Pure. Appl. Math.* 5 (1974), 51–96.

[43] Rosenberg, J., Rigidity of K-theory under deformation quantization, Preprint q-alg/9607021 (July 1996).

[44] Rudin, W., *Functional Analysis*, McGraw-Hill Book Company, New York, 2. edition, 1991.

[45] Schomerus, V., D-branes and deformation quantization, J. High Energy Phys. 06 (1999).

[46] Swan, R. G., Vector bundles and projective modules, *Trans. Amer. Math. Soc.* 105 (1962), 264–277.

[47] Waldmann, S., Locality in GNS Representations of Deformation Quantization, *Commun. Math. Phys.* 210 (2000), 467–495.

[48] Waldmann, S., A Remark on the Deformation of GNS Representations of $*$-Algebras, *Rep. Math. Phys.* 48 (2001), 389–396.

[49] Waldmann, S., Deformation of Hermitian Vector Bundles and Non-Commutative Field Theory, Preprint Freiburg (March 2001), Contribution to the *Proceedings for the Keio Workshop on Non-commutative geometry and String Theory*, to appear in *Progr. Theoret. Phys. Suppl.* 144.

[50] Weinstein, A., Deformation Quantization, Séminaire Bourbaki 46ème année 789 (1994).

[51] Wells, R. O., *Differential Analysis on Complex Manifolds*, Grad. Texts in Math. 65, Springer-Verlag, New York, Berlin, Heidelberg 1980.

Unimodular vector fields and deformation quantization

Claude Roger

Institut Girard Desargues - UMR 5028
Université Claude Bernard (Lyon 1)
43 Boulevard du 11 novembre 1918,69622 Villeurbanne Cedex, France
email: `roger@igd.univ-lyon1.fr`

Abstract. We survey various properties of the Lie algebra of vector fields with vanishing divergence, its cohomologies, central extensions, and coadjoint orbits. We recall the proof of its rigidity, and finally try to sketch some ways of introducing the deformation quantization schedule in this framework.

2000 Mathematics Subject Classification: 17B56 (17B66)

This work has been motivated by some recent publications ([MS], [Ho], [A-L-M-Y], [S]) which consider the problem of constructing deformation quantization for unimodular vector fields, seen as a natural generalization of the Hamiltonian case. The physical meaning of unimodular vector fields in hydrodynamics is well known since a long time, but in the present case, these quantization problems come mainly from the theory of "branes", higher dimensional generalization of string theory, in which unimodular vector fields appear as infinitesimal automorphisms of the action functional for branes [M-S]. We shall briefly describe the main properties of the Lie algebra of unimodular vector fields, including some cohomological computations which imply its rigidity. So deformation of this Lie algebra (in the usual sense) is not possible; we shall indicate several tentative methods for overcoming that difficulty, and survey some of them with more details; the approach through "up to homotopy algebras" seems to be the most promising. Before entering the core of our subject, let us fix the terminology: for us "unimodular vector field" means "a vector field with vanishing divergence"; those fields are often called "volume preserving" in the physical literature with some abuse of language. We think that the term unimodular, as well as the notation $\mathrm{SVect}(V)$ are fully justified by analogy with the linear case and the groups $\mathrm{SL}(n)$. The author thanks Boris Pioline for several questions and discussion.

1 Definitions and notations

Let V be an orientable n-dimensional ($n \geq 3$) manifold and $\omega \in \Omega^n(V)$ be a volume form. Let $\mathrm{Vect}(V)$ be the space of vector fields on the manifold V, the Lie bracket of vector fields $X, Y \to [X, Y]$ endows $\mathrm{Vect}(V)$ with a Lie algebra structure. The inner product and Lie derivative of a differential form α with a vector field X will be denoted by $i(X)\alpha$ and $L_X\alpha$ respectively.

The divergence $\mathrm{Div}(X)$ of X is then the function on V defined as: $\mathrm{Div}(X)\omega = L_X\omega = \mathrm{d}(i(X)\omega)$.

Definition 1.1

(1) A vector field X is *unimodular* if $\mathrm{Div}(X) = 0$.

(2) A vector field X is *exact unimodular* if there exists a $(n-2)$ form α such that $i(X)\omega = \mathrm{d}\,\alpha$.

Straightforward computations show that the bracket of two unimodular (resp. exact unimodular) vector fields is again an unimodular (resp. exact unimodular) vector field; so we have obtained two sub-Lie algebras of $\mathrm{Vect}(V)$; they will be denoted by $\mathrm{S}\overline{\mathrm{Vect}}(V)$ and $\mathrm{SVect}(V)$ respectively.

One should recall now the classification of simple primitive transitive Lie subalgebras of $\mathrm{Vect}(V)$ due to E. Cartan (see [S-S]); the Lie algebra $\mathrm{S}\overline{\mathrm{Vect}}(V)$[1] is one of them.

2 Derivations and derived algebra of $\mathrm{S}\overline{\mathrm{Vect}}(V)$

Let us consider the map $p\colon\ \mathrm{S}\overline{\mathrm{Vect}}(V) \to H^{n-1}_{\mathrm{d}R}(V)$, which associates to X the cohomology class of the $(n-1)$ form $[i(X)\omega]$. One then has an exact sequence

$$0 \longrightarrow \mathrm{SVect}(V) \longrightarrow \mathrm{S}\overline{\mathrm{Vect}}(V) \longrightarrow H^{n-1}_{\mathrm{d}R}(V) \longrightarrow 0. \qquad (2.1)$$

This is an exact sequence of Lie algebras if one considers $H^{n-1}_{\mathrm{d}R}(V)$ as an abelian Lie algebra. One has the following results (Lichnerowicz [L]).

Theorem 2.1. (1) $\mathrm{SVect}(V)$ *is the derived subalgebra of* $\mathrm{S}\overline{\mathrm{Vect}}(V)$.

(2) *All derivations of* $\mathrm{S}\overline{\mathrm{Vect}}(V)$ *are inner.*

(3) *All derivations of* $\mathrm{SVect}(V)$ *are given by brackets with vector fields in* $\mathrm{S}\overline{\mathrm{Vect}}(V)$.

In cohomological terms, one has

$$H_1(\mathrm{SVect}(V)) = H^{n-1}_{\mathrm{d}R}(V), \quad H^1(\mathrm{S}\overline{\mathrm{Vect}}(V), \mathrm{S}\overline{\mathrm{Vect}}(V)) = 0$$
$$\text{and} \quad H^1(\mathrm{SVect}(V), \mathrm{SVect}(V)) = \mathrm{S}\overline{\mathrm{Vect}}(V)/\,\mathrm{SVect}(V) = H^{n-1}_{\mathrm{d}R}(V).$$

[1] *Remark:* Rigourously, one should note $\mathrm{S}\overline{\mathrm{Vect}}(V, \omega)$ since this algebra depends on the volume form ω; but it is easy to check the independence from ω, up to isomorphism.

A remarkable fact is that no special hypothesis for the derivations, such as continuity, locality, or differentiability, is needed. For $n = 2$, $\mathrm{S}\overline{\mathrm{Vect}}(V)$ (resp. $\mathrm{SVect}(V)$) is the Lie algebra of locally Hamiltonian (resp. Hamiltonian) vector fields. So it is natural to consider the general case, and more particularly the case $n = 3$, as a generalization of the Hamiltonian case $n = 2$.

3 The central extension of SVect(V)

First of all, let us recall that a Lie algebra $\hat{\mathfrak{g}}$ is a central extension of a Lie algebra $\mathfrak{g}$ if there exists an exact sequence of Lie algebras

$$0 \longrightarrow E \longrightarrow \widehat{\mathfrak{g}} \longrightarrow \mathfrak{g} \longrightarrow 0,$$

the kernel E being abelian. Such extensions are classified up to natural isomorphism by a cohomology class in $H^2(\mathfrak{g}, E)$.

The most natural way to construct central extensions for $\mathrm{SVect}(V)$ will make use of contravariant tensor fields we shall now consider. For each $0 \leq p \leq n$, the space of p contravariant antisymmetric tensor fields will be denoted by $\Omega_p(V)$. Of course for $p = 0$ or 1 one recovers functions and vector fields; from the sheaf theory point of view, $\Omega_p(V)$ is nothing but the space of sections of the pth exterior product of the tangent bundle of V with itself. In a system of local coordinates $(x_1, \ldots, x_n)$ we shall write $A = A^{i_1 \ldots i_p} \partial_{i_1} \wedge \cdots \wedge \partial_{i_p}$, where as usual $\partial_i = \dfrac{\partial}{\partial x_i}$. The exterior product of tensor fields defines on the direct sum $\Omega_*(V) = \bigoplus_{p=0}^{n} \Omega_p(V)$ an associative and graded commutative multiplication. Moreover, the choice of the volume form ω allows a duality isomorphism between covariant and contravariant tensor fields, given by $A \longrightarrow i(A)\omega$. This isomorphism is called sometimes "musical isomorphism", because of lowering and raising indexes, and denoted by $(\sharp)^{-1}$. So for every $0 \leq p \leq n$, one has $\sharp \colon \Omega^{n-p} \longrightarrow \Omega_p$. The De Rham differential for forms can then be transposed to contravariant tensors using this isomorphism, and one obtains the *codifferential* δ. Explicitly $\delta = \sharp \circ \mathrm{d} \circ (\sharp)^{-1}$, and one has a commutative diagram:

$$\begin{array}{ccc} \Omega_p & \xrightarrow{\ (\sharp)^{-1}\ } & \Omega^{n-p} \\ \delta \big\downarrow & & \big\downarrow \mathrm{d} \\ \Omega_{p-1} & \xleftarrow{\ \ \sharp\ \ } & \Omega^{n-p+1} \end{array} \qquad \text{for each } 1 \leq p \leq n.$$

Of course one has $\delta \circ \delta = 0$, and in a system of local coordinates for which $\omega = \mathrm{d}x_1 \wedge \cdots \wedge \mathrm{d}x_n$, one has $(\delta A)^{i_1 \dots i_{p-1}} = \partial_i A^{i\, i_1 \dots i_{p-1}}$. So for $p = 1$, $\delta(X) = \mathrm{Div}(X)$ and it is natural to define the divergence of an arbitrary tensor field by $\mathrm{Div}(A) = \delta(A)$, but some authors prefer the word "rotational". But δ is not a graded derivation and one defines the *Schouten bracket*[2] as the corresponding obstruction; more explicitly, for A and B in $\Omega_*(V)$, one has

$$[A, B] = \delta(A \wedge B) - \delta A \wedge B - (-1)^{|A|} A \wedge \delta B, \tag{3.1}$$

where $|A|$ denotes as usual the degree of A. This bracket endows $\Omega_*(V)$ with a graded Lie algebra structure, but with a shift of order one in the degrees; one has

$$\begin{cases} |\,[A, B]\,| = |A| + |B| - 1 & \\ [B, A] = (-1)^{(|A|-1)(|B|-1)+1}\,[B, A] & \text{(graded antisymmetry)} \\ \sum\limits_{\text{(cycl)}} (-1)^{(|A|-1)(|C|-1)} [\,[A, B], C\,] = 0 & \text{(graded Jacobi identity)} \end{cases} \tag{3.2}$$

One can say briefly that the graded space $\Omega_*(V)[+1]$ where the graduation has been shifted by one is a graded Lie algebra. The relation between associative product and Schouten bracket is given by the following formula:

$$[A \wedge B, C] = A \wedge [B, C] + (-1)^{|B|\,(|C|-1)} [A, C] \wedge B. \tag{3.3}$$

This kind of algebraic structure looks very much like a graded Poisson algebra, except the above shift in graduation. An associative graded commutative algebra with a bracket satisfying relations (3.2)–(3.3) is called *Gerstenhaber algebra*, shortly *G-algebra*. If moreover the bracket is induced by a differential δ satisfying (3.1), one has a *Batalin–Vilkovisky (B-V) algebra*. We have thus associated to every manifold with volume form (V, ω) a B-V algebra $\big(\Omega_*(V), \wedge, \delta\big)$. The reader can check easily how much this construction depends on the choice of ω: change of ω induces a change of δ [3] but the Schouten bracket remains the same.

But what is the relation with unimodular vector fields? Let $X \in \mathrm{SVect}(V)$, there exists $\alpha \in \Omega^{n-2}(V)$ such that $i(X)\omega = \mathrm{d}\alpha$. Using the isomorphism $\sharp$, one obtains

$$i(X)\omega = (\sharp^{-1} \circ \delta \circ \sharp)(\alpha), \quad \text{so } \delta\big(\sharp(\alpha)\big) = \sharp\big(i(X)\omega\big) = X.$$

If one sets $\sharp\alpha = A \in \Omega_2(V)$ one has $X = \delta A$. The converse is easily checked, so one obtains a surjective mapping

$$\delta \colon \Omega_2(V) \longrightarrow \mathrm{SVect}(V) \longrightarrow 0\,.$$

[2] This bracket is well known since a long time; we shall recall only the characterization of Poisson tensors as square vanishing objects in $\Omega_2(V)$.

[3] Precisely, the difference between the two δ's is a coboundary in the sense of graded Hochschild cohomology of $\Omega_*(V)$.

One can define a bracket $\{\ ,\ \}$ on $\Omega_2(V)$:

$$\begin{array}{ccccl} \Omega_2(V) & \times & \Omega_2(V) & \longrightarrow & \Omega_2(V) \\ A & , & B & \longrightarrow & \{A, B\} = \delta A \wedge \delta B. \end{array}$$

This bracket satisfies

Proposition 3.1. $\delta\big(\{A, B\}\big) = [\delta A, \delta B]$.

It follows immediately from relation (3.1).

Let us look for a moment at the case where $n = 2$; we shall denote by Λ the volume tensor dual to ω, so $\omega(\Lambda) = 1$. Any tensor field in $\Omega_2(V)$ can then be canonically written as $f\Lambda$, where f is a function on V, and one has for every f and g the following relation: $\{f\Lambda, g\Lambda\} = \{f, g\}\Lambda$, where $\{f, g\}$ is nothing but the Poisson bracket of functions f and g with respect to the symplectic structure given by ω.

This computation legitimates the notation $\{A, B\}$ for tensors, but also leads to the question whether the mapping $\delta\colon \Omega_2(V) \longrightarrow \mathrm{SVect}(V) \longrightarrow 0$ is the natural generalization to $n > 2$ of the map which associates to a function its Hamiltonian vector field. Unfortunately, $\{\ ,\ \}$ is not a Lie bracket. One has:

Proposition 3.2.

$$\sum_{(\mathrm{cycl})} \big\{\{A, B\}, C\big\} = 2\delta(\delta A \wedge \delta B \wedge \delta C). \tag{3.4}$$

See Lichnerowicz ([L] Formula (15-4))

So Jacobi identity is valid only up to divergence of 3-tensors; one possible solution is to quotient $\Omega_2(V)$ by the image $\delta\big(\Omega_3(V)\big)$, the bracket $\{\ ,\ \}$ is well defined on the quotient and the induced map from $\Omega_2(V)/\delta(\Omega_3(V))$ into $\mathrm{SVect}(V)$ is a Lie algebra morphism.

One then has an exact sequence of Lie algebras

$$0 \longrightarrow \operatorname{Ker}\delta/\,\delta(\Omega_3(V)) \longrightarrow \Omega_2(V)/\delta(\Omega_3(V)) \xrightarrow{\ \bar{\delta}\ } \mathrm{SVect}(V) \longrightarrow 0.$$

This extension is obviously central; $\operatorname{Ker}\delta$ is isomorphic through $\sharp^{-1}$ to $Z^{n-2}(V)$ and $\delta\big(\Omega_3(V)\big)$ to $B^{n-2}(V)$.

One has obtained a central extension

$$0 \longrightarrow H^{n-2}_{DR}(V) \longrightarrow \widehat{\mathrm{SVect}}(V) \xrightarrow{\ \hat{\delta}\ } \mathrm{SVect}(V) \longrightarrow 0. \tag{3.5}$$

It appeared implicitly in ([L]). See also ([R]) for more details.

Remark 3.3 (1) This extension can also be described in a purely covariant way as follows:

$$0 \longrightarrow H^{n-2}_{DR}(V) \longrightarrow \Omega^{n-2}(V)/B^{n-2}(V) \longrightarrow \Omega^{n-2}(V)/Z^{n-2}(V) \longrightarrow 0.$$

(2) One can compute the cohomology of $\mathrm{SVect}(V)$ with scalar coefficients. It can be proved that

$$H^k(\mathrm{SVect}(V), \mathbb{R}) = H_{n-k}(V) \quad \text{for } 0 \leq k \leq n\,.$$

One associates to a $(n-k)$ cycle Δ a k-cocycle $\widetilde{\Delta}$ defined by

$$\widetilde{\Delta}(X_1, \ldots, X_k) = \int_\Delta i(X_1)\cdots i(X_k)\,\omega\,.$$

In particular $H^2(\mathrm{SVect}(V);\mathbb{R}) = H_{n-2}(V) = H^{n-2}_{\mathrm{d}R}(V)^*$ and we recover

$$H^1(\mathrm{SVect}(V);\mathbb{R}) = H_{n-1}(V) = H^{n-1}_{\mathrm{d}R}(V)^*.$$

So the central extension is *universal*: in some sense it classifies all possible central extensions of $\mathrm{SVect}(V)$ (see [HS] for details about extensions of Lie algebras).

(3) The extension (3.5) is not a very satisfactory object since $\widehat{\mathrm{SVect}}(V)$ has no real geometric status: no notions of support or value in some point exist for elements in $\widehat{\mathrm{SVect}}(V)$. Technically speaking, it is not, unlike $\mathrm{SVect}(V)$, the space of sections of a sheaf on V. We shall go back to that difficulty later.

4 The rigidity of SVect(V)

Let us recall some definitions about deformation theory of Lie algebras. If $\mathfrak{g}$ is a Lie algebra whose structure tensor is denoted by $c\colon \mathfrak{g}\times\mathfrak{g}\to\mathfrak{g}$, a deformation of c is given by a Lie algebra bracket

$$c_t\colon \mathfrak{g}\times\mathfrak{g}\longrightarrow\mathfrak{g} \quad \text{such that } c_t = c + tc_1 + t^2c_2 + \cdots\,,$$

where variable t can have several status: formal, analytic, or polynomial up to some order. Equivalence of deformations is naturally defined. For existence and classification problems the cohomology of $\mathfrak{g}$ in the adjoint representation is pertinent; more explicitly, the space $H^2(\mathfrak{g},\mathfrak{g})$ governs classification and $H^3(\mathfrak{g},\mathfrak{g})$ contains potential obstructions to inductive prolongation of deformations.

In particular if $c_t = c + tc_1 + t^2c_2 + \cdots$ is a deformation of $\mathfrak{g}$, then c_1 is a two-cocycle in $Z^2(\mathfrak{g},\mathfrak{g})$ and its cohomology class in $H^2(\mathfrak{g},\mathfrak{g})$ classifies the infinitesimal part of the deformation up to isomorphism. The obstruction to prolongation at order two is then the Richardson–Nijenhuis bracket of $[c_1]$ with itself denoted by $[\![[c_1],[c_1]]\!]$ in $H^3(\mathfrak{g},\mathfrak{g})$.

Formal deformations of Lie algebras of Hamiltonian vector fields appear as by-products of formal deformations of associative products, or $*$-products: they have been extensively studied (cf. e.g. [Ko], [Fe]). For $n=2$, Hamiltonian and exact unimodular vector fields are the same. So it is very natural to look for formal deformations of the Lie algebra of unimodular vector fields for $n\geq 3$. But one has:

Theorem 4.1 (P. Lecomte & C. Roger [L-R]). *The Lie algebra of unimodular vector fields does not admit any non trivial formal deformation.*

As usual we consider only *local* deformations; the deformed bracket must still be local, i.e. compatible with supports, so according to Peetre's theorem, given by differential operators. Therefore one has to study the corresponding cohomology $H^p_{\mathrm{loc}}(\overline{\mathrm{SVect}}(V), \overline{\mathrm{SVect}}(V))$. One has to adapt the techniques of Gelfand and Fuks [F] concerning Lie algebra cohomology which has been so successful for Vect(V); there exists here a supplementary difficulty: $\overline{\mathrm{SVect}}(V)$ is not the space of sections of a vector bundle, technically speaking, the associated sheaf $\underset{\sim}{\overline{\mathrm{SVect}}}(V)$ is not fine. The sequence $\underset{\sim}{\overline{\mathrm{SVect}}}(V) \longrightarrow \underset{\sim}{\Omega}^{n-1} \longrightarrow \underset{\sim}{\Omega}^{n-2}$ defines a fine resolution, suitable for cohomology computations. See [L-R] for more details.

The precise results are:

If $n \geq 4$, then $H^2_{\mathrm{loc}}(\overline{\mathrm{SVect}}(V); \overline{\mathrm{SVect}}(V)) = 0$. So $\overline{\mathrm{SVect}}(V)$ has no infinitesimal deformations, and no other kind of deformations either.

If $n = 3$, then $H^2_{\mathrm{loc}}(\overline{\mathrm{SVect}}(V); \overline{\mathrm{SVect}}(V))$ is 1-dimensional but the corresponding obstruction in $H^3_{\mathrm{loc}}(\overline{\mathrm{SVect}}(V); \overline{\mathrm{SVect}}(V))$ never vanishes. So for $n = 3$, the Lie algebra $\overline{\mathrm{SVect}}(V)$ admits a one-parameter infinitesimal deformation, but this deformation cannot be prolongated to any higher order.

It is interesting to describe the generator of $H^2_{\mathrm{loc}}(\overline{\mathrm{SVect}}(V); \overline{\mathrm{SVect}}(V))$ in details. There exists a class Φ in $H^2_{\mathrm{loc}}(\mathrm{Vect}(V), \Omega^2(V))$ defined as follows: let $P(V) \xrightarrow{\pi} V$ be the frame bundle on V, and ϖ be a connection on it. For X, Y in Vect(V), call $\widetilde{X}$ and $\widetilde{Y}$ the canonical lift in $P(V)$; consider then on $P(V)$ the tensor valued 2-form $L_{\widetilde{X}}\varpi \wedge L_{\widetilde{Y}}\varpi$ and the associated scalar valued form $\mathrm{Tr}(L_{\widetilde{X}}\varpi \wedge L_{\widetilde{Y}}\varpi)$. It is easily checked that the latter is a basic form, so there exists $\varphi(X, Y) \in \Omega^2(V)$ such that $\pi^*(\varphi(X, Y)) = \mathrm{Tr}(L_{\widetilde{X}}\varpi \wedge L_{\widetilde{Y}}\varpi)$. One can check that $X, Y \longrightarrow \varphi(X, Y)$ is a 2-cocycle on Vect(V). The required class Φ is then the cohomology class of φ. This class appeared for the first time in the contribution of I.M. Gelfand at the ICM of Nice in 1970; it has played a fundamental role in deformation theory under the name of Vey class: if Λ is a Poisson tensor and $H_f = i(\mathrm{d}f)\Lambda$, $H_g = i(\mathrm{d}g)\Lambda$ the Hamiltonian vector fields associated to functions f and g, then the Vey class is defined by $S(f, g) = \langle \Phi(H_f, H_g), \Lambda \rangle$. Here, the restriction defines naturally a class in $H^2_{\mathrm{loc}}(\overline{\mathrm{SVect}}(V), \Omega^2(V))$ which again induces a class in $H^2_{\mathrm{loc}}(\overline{\mathrm{SVect}}(V), \overline{\mathrm{SVect}}(V))$ through the exact sequence of sheaves

$$0 \longrightarrow \underset{\sim}{\overline{\mathrm{SVect}}}(V) \longrightarrow \underset{\sim}{\Omega}^2(V) \longrightarrow \underset{\sim}{\Omega}^3(V) \longrightarrow 0$$

(cf. [LR] and more particularly Appendix 3). So one has no chance of finding non trivial formal deformations for the Lie algebra of unimodular vector fields! Before suggesting some tricks to avoid this difficulty, it will be useful to say some words about the group which corresponds to $\overline{\mathrm{SVect}}(V)$ and its coadjoint orbits.

5 The group of volume-preserving diffeomorphisms

We shall consider the group of diffeomorphisms $f: V \to V$ such that $f^*(\omega) = \omega$, and which are isotopic to the identity map through volume preserving maps. This group will be denoted by $S\mathfrak{D}(V)$; it can be endowed with a Fréchet–Lie group structure such that it is a closed Fréchet–Lie subgroup of the group of all diffeomorphisms (see Hamilton [H] Thm 2.5.3, or the recent book [K-M] by Kriegl and Michor for a more extensive theory). Then, $\overline{\mathrm{SVect}}(V)$ can be considered as the Lie algebra of $S\mathfrak{D}(V)$ in the following weak sense: let f_t, $t \in [0,1]$, $f_0 = \mathrm{Id}$ be a 1-parameter group of diffeomorphisms and let $X(x) = \frac{\mathrm{d}}{\mathrm{d}t} f_t(x)|_{t=0}$ be its infinitesimal generator. Then X is a vector field in $\overline{\mathrm{SVect}}(V)$, if f_t belongs to $S\mathfrak{D}(V)$. But, as usual for groups of diffeomorphisms $S\mathfrak{D}(V)$ is not generated by its 1-parameter subgroups. Most of the above results (§2, 3) concerning $\overline{\mathrm{SVect}}(V)$ have their counterpart for $S\mathfrak{D}(V)$ with the usual difficulties coming from non linearity and the topology of V. One can construct the subgroup which corresponds to exact unimodular vector fields, and also the central extension. See the book of Ismagilov ([I]) for details and geometrical aspects of the theory.

We shall now succinctly describe the coadjoint representation of this group. Let us recall the basic principles of coadjoint representation; the famous orbit method, invented by Kirillov (see [Ki], and bibliography inside) consists in classifying the orbits of a Lie group in its adjoint representation, then constructing a canonical symplectic structure on them, and associating to this data a representation of the group. This stuff is now classical for finite dimensional groups, but some infinite dimensional examples have also been developed (see for example [G]).

We shall mainly consider the *regular* dual of Lie algebras of vector fields, i.e. linear forms which are defined by smooth densities. In any reasonable sense, the regular dual $\mathfrak{g}^*_{\mathrm{reg}}$ represents a generic part of the dual $\mathfrak{g}^*$. One has $\mathrm{Vect}(V)^*_{\mathrm{reg}} = \Omega^1(V) \underset{C^\infty(V)}{\otimes} \Omega^n(V)$, since $\alpha \otimes \omega$ where $\alpha \in \Omega^1(V)$ and $\widetilde{\omega} \in \Omega^n(V)$ acts as follows

$$\langle \alpha \otimes \widetilde{\omega}, X \rangle = \int_V \alpha(X)\widetilde{\omega}.$$

For subalgebras $\mathfrak{g} \subset \mathrm{Vect}(V)$, their regular duals $\mathfrak{g}^*_{\mathrm{reg}}$ will naturally be quotients of $\mathrm{Vect}(V)^*_{\mathrm{reg}}$.

For example one finds easily:

$$\overline{\mathrm{SVect}}^*(V)_{\mathrm{reg}} = \Omega^1/B^1 \quad \text{and} \quad \mathrm{SVect}^*(V)_{\mathrm{reg}} = \Omega^1/Z^1.$$

So the exact sequence (2.1) dualizes as follows.

$$\begin{array}{ccccccccc} 0 & \longrightarrow & H^1_{DR}(V) & \longrightarrow & \overline{\mathrm{SVect}}^*(V)_{\mathrm{reg}} & \longrightarrow & \mathrm{SVect}^*(V) & \longrightarrow & 0 \\ & & \| & & \| & & \| & & \\ & & Z^1/B^1 & & \Omega^1/B^1 & & \Omega^1/Z^1 & & \end{array}$$

A class $[\alpha] \in \Omega^1/Z^1$ acts on $X \in \mathrm{SVect}(V)$ as follows:

$$\langle [\alpha], X \rangle = \int_V \alpha(X)\omega.$$

The coadjoint action is then simply the natural action of diffeomorphisms on differential forms. Analogously, one has the dual of the central extension (3.5).

$$\begin{array}{ccccccccc} 0 & \longrightarrow & \mathrm{SVect}^*(V)_{\mathrm{reg}} & \longrightarrow & \widehat{\mathrm{SVect}}^*(V)_{\mathrm{reg}} & \longrightarrow & H^2_{DR}(V) & \longrightarrow & 0 \\ & & \| & & \| & & \| & & \\ 0 & \longrightarrow & \Omega^1/Z^1 & \longrightarrow & Z^2 & \longrightarrow & Z^2/B^2 & \longrightarrow & 0 \end{array}$$

A closed 2-form η acts on $X \in \mathrm{SVect}(V)$ as follows: $\langle \eta, X \rangle = \int_V \eta \wedge \beta$ where β satisfies $\mathrm{d}\beta = i(X)\omega$.

We have written Ω^p instead of $\Omega^p(V)$ and the same for closed and exact forms in order to simplify the formulas. One can also consider non regular momentums [4], i.e. singulars, given by integration over some cycles or submanifolds. Let $\Sigma \subset V$ be a $(n-2)$ dimensional closed submanifold of V; one can define on the space $\mathcal{M}(\Sigma)$ of submanifolds of V, isotopic to Σ, a structure of a Fréchet manifold (see again Hamilton [H]). Besides, any manifold $\widetilde{\Sigma} \in \mathcal{M}(\Sigma)$ induces very naturally a linear form on $\mathrm{SVect}(V)$; simply set

$$\langle \widetilde{\Sigma}, X \rangle = \int_{\widetilde{\Sigma}} \alpha, \quad \text{where } \alpha \text{ satisfies } \mathrm{d}\alpha = i(X)\omega.$$

So one has $\mathcal{M}(\Sigma) \subset \mathrm{SVect}^*(V)$, the coadjoint action of $\mathrm{SDiff}(V)$ is naturally defined on $\mathcal{M}(\Sigma)$ (simply remark that $\widetilde{\Sigma} \to f(\widetilde{\Sigma})$) making $\mathcal{M}(\Sigma)$ a coadjoint orbit. Moreover one can give a very explicit formula for the (weak) symplectic form on this orbit (see Ismagilov [I]) and construct a kind of geometric quantization for it. We must mention also the works of Goldin & al. [G-M-S]) considering some other elements in $\mathrm{SVect}(V)^*$ for various physical applications. But why are those considerations pertinent for the deformation quantization problem? Coadjoint orbits are typical examples of symplectic manifolds, and one could think about generalizing the now classical deformation quantization problem to the infinite-dimensional case. In some papers of physical literature, the distinction between formal deformations of a Lie algebra of vector fields and star product on its coadjoint orbits is not always clearly made.

We shall conclude this part with some words about the particular case $n = 3$. We already saw that infinitesimal deformation exists only in that case. There exists another interesting object when $n = 3$, the *Arnold invariant*. Let X and Y be exact unimodular fields on V and let α and β be 1-forms such that $\mathrm{d}\alpha = i(X)\omega$ and $\mathrm{d}\beta = i(Y)\omega$. The Arnold invariant then reads $I(X, Y) = \int_V \alpha \wedge \mathrm{d}\beta$. It is now an easy exercise to

[4] following along tradition, "momentum" stands for element of the dual

check that this invariant is well defined, and is a bilinear symmetric invariant and non degenerated form on $\mathrm{SVect}(V)$, some kind of a "Killing form".

As a consequence, there exists a canonical isomorphism $\mathrm{SVect}(V) = \mathrm{SVect}(V)^*_{\mathrm{reg}}$ when $n = 3$. Let us finish with some bibliographical indications: the book of J. L. Brylinski [B] contains a very detailed study of the space of knots on a 3-manifold, seen as a symplectic manifold; it corresponds to a special case of the manifold $\mathcal{M}(\Sigma)$ discussed above; for more information about the Arnold invariant, one should read the wonderful book by Arnold and Khesin [A-K] where the applications to hydrodynamics are extensively described.

6 Some prospective approach of the deformation quantization problem

Abruptly speaking, the problem is: how to circumvent rigidity of our Lie algebra? We shall indicate a few possibilities:

a) One way could be to deform the adjoint action of the Lie algebra on itself instead of the bracket. In the philosophy of deformation quantization, one often consider operators (on functions, fields ...) and deform them, so this point of view is relevant. In the framework of algebraic deformation theory, this leads to deformation of homomorphisms, or representations, of Lie algebras; the corresponding cohomology spaces are $H^1\big(\mathfrak{g}; \mathrm{Hom}(\mathfrak{g}, \mathfrak{g})\big)$ and $H^2\big(\mathfrak{g}; \mathrm{Hom}(\mathfrak{g}, \mathfrak{g})\big)$ for classification and obstruction respectively. Recall that $\mathfrak{g} = \mathrm{Vect}(V)$ is rigid, but

$$H^1\big(\mathrm{Vect}(V); C^\infty(V)\big) = H^1_{DR}(V) \oplus \mathbb{R}$$

(the latter class being associated to the divergence), and since $C^\infty(V)$ naturally embeds into $\mathrm{Hom}\big(\mathrm{Vect}(V); \mathrm{Vect}(V)\big)$ one obtains deformations of the action of $\mathrm{Vect}(V)$ onto itself. Here the understanding of $\mathrm{Hom}\big(\mathrm{SVect}(V); \mathrm{SVect}(V)\big)$ as a $\mathrm{SVect}(V)$ module seems to be rather complicated.

b) Another approach is to enlarge a little bit $\mathrm{SVect}(V)$ and consider its counterpart in supergeometry. We shall discuss it in another paper, but as above, the cohomology computations seem to be hard.

c) Another possibility is to consider subalgebras of $\mathrm{SVect}(V)$ naturally associated with some extra geometric structure; for $n = 3$, let us consider the case where $V = S^1 \times \Sigma$ (Σ being a closed oriented surface) and the algebras of vectors which respect the fibration by S^1. Typically, those fields can be arbitrary along Σ, and their term along S^1 is deviced to kill the divergence. This Lie algebra is much smaller than the full $\mathrm{SVect}(V)$ (five independent parameters instead of eight) and does admit non trivial deformations. One can go even further, and consider the volume form on Σ induced by ω and the choice of a vector field along S^1; denote by $\mathrm{Ham}(\Sigma)$ the corresponding Poisson algebra on Σ. Then the loop algebra $L\big(\mathrm{Ham}(\Sigma)\big)$ is naturally embedded into $\mathrm{SVect}(V)$; one can "loop" the well-known deformation of $\mathrm{Ham}(\Sigma)$, or find some extra deformation. The case when $\Sigma = \mathbb{T}^2$ seems to be particularly interesting, since $\mathrm{Ham}(\mathbb{T}^2)$ can be identified as $\mathrm{SU}(\infty)$ (see for example Hoppe [Ho]), so $L(\mathrm{Ham}(\mathbb{T}^2))$ is in turn interpreted as $\widehat{\mathrm{SU}(\infty)}$.

d) We already mentioned the symplectic structures on the co-adjoint orbits, typically on the space of knots when $n = 3$. One could try a construction of $*$-product in that situation, and perhaps it has some links with the hydrodynamical aspects of $\mathrm{SVect}(V)$ or the more recent results about knot theory in physics.

e) But the most popular way (up to now) to introduce deformational type construction for $\mathrm{SVect}(V)$ is to use Nambu structures (see [F-D-S], [D-F-M-T]). We shall not develop this point of view here. Let us say only that it makes use of a n-ary version of Poisson bracket $(f_1, \dots, f_n) \to \{f_1, \dots, f_n\}$.

f) We think that the most promising way is to make use of structures *up to homotopy*. Before defining them, let us go back to the central extension for a while. We saw in §3 that the volume form ω induces a codifferential $\delta\colon \Omega_p(V) \to \Omega_{p-1}(V)\ (p \le n)$; the bracket $\{A, B\} = \delta A \wedge \delta B$ we used for $p = 2$ can be generalized on the whole $\Omega_*(V)$, is graded antisymmetric for the shifted graduation (i.e. $\{A, B\} = (-1)^{(|A|-1)(|B|-1)}\{B, A\}$) and satisfies:

(i) $\delta\{A, B\} = [\delta A, \delta B]$ where $[\delta A, \delta B]$ is the Schouten bracket of δA and δB,

(ii) (3.4′) $$\sum_{(\mathrm{cycl})} (-1)^{(|A|-1)(|C|-1)}\{\{A, B\}, C\} = 2\delta(\delta A \wedge \delta B \wedge \delta C).$$

This generalization of Lichnerowicz's formula (3.4) comes from a straightforward computation.

Now, we can consider, instead of the central extension, the whole complex

$$0 \longrightarrow \Omega_n(V) \longrightarrow \Omega_{n-1}(V) \longrightarrow \cdots \longrightarrow \Omega_2(V) \longrightarrow \mathrm{SVect}(V) \longrightarrow 0.$$

From the point of view of sheaf theory, this is a resolution of $\underset{\sim}{\mathrm{SVect}}(V)$ by fine sheaves: the sequence is of course exact in the category of sheaves, the complex of global sections gives the De Rham cohomology of V.

$(\Omega_*(V)[-2], \delta)_{*\ge 2}$, the truncated differential graded space with graduation shifted by two, can be said to be quasi-isomorphic to the Lie algebra $\mathrm{SVect}(V)$. One can find moreover a richer structure on this space; following the work of G. Barnich and his collaborators ([B-F-L-S]), if one has a Lie algebra $\mathfrak{g}$ and a homological resolution of it, that is to say, a homological complex (E_*, d) such that $H_p(E_*) = 0$ for $p > 0$ and $H_0(E_*) = \mathfrak{g}$, then E_* admits a L^∞-algebra (or Lie algebra up-to-homotopy) structure. Roughly speaking, a L^∞ algebra structure on E_* is given by a family of linear maps $c_p\colon \Lambda^p E^* \to E^*[p-2]$, $p \ge 1$ satisfying *higher order Jacobi identities* (cf. [A]). In the formalism of square vanishing elements in graded Lie algebras (or Maurer–Cartan elements, according to Kontsevich [Ko]) such a structure is given by an element $\underline{c} = \sum_{p=1}^{+\infty} c_p$ in the Richardson–Nijenhuis algebra $A(E^*)$ on the graded space E^* (with $c_p \in A^{p-1}(E)_{p-2}$) and the higher order Jacobi identities are deduced from the single relation $[\![\underline{c}, \underline{c}]\!] = 0$, where $[\![\ ,\ ]\!]$ indicates the Richardson–Nijenhuis bracket. We can adapt the techniques of [B-F-L-S] to our framework, geometric and sheaf-theoretic and one gets:

Theorem 6.1. *One has a L^∞-algebra structure on $\left(\Omega_*(V)[-2]\right)_{*\geq 2}$.*

It is an easy work to compute the first pieces of this structure; one has of course $c_1 = \delta$, and $c_2 = \{\ ,\ \}$ on $\Omega_2(V) \times \Omega_2(V)$; moreover $c_3(A, B, C) = (\delta A \wedge \delta B \wedge \delta C)$ for A, B and C in $\Omega_2(V)$ and relation (5') is then nothing but $[\![c_2, c_2]\!] + 2[\![c_1, c_3]\!] = 0$.

So it is reasonable to consider this L^∞-algebra structure as an equivalent data to the Lie algebra structure on $\mathrm{SVect}(V)$. An L^∞-algebra structure is obviously "softer" than a Lie algebra structure and one can now look for its deformations through the usual formalism for equations like $[\![c, c]\!] = 0$. Such a deformation should begin with deforming the codifferential δ. But this structure is not in our opinion completely satisfactory for two reasons: first, it does not take into account the associative product on $\Omega_*(V)$ (suggestion: try to compute $\{A \wedge B, C\}$) and a generalization could be sought in B-V structures up-to-homotopy; secondly the L^∞-algebra structure as sketched above does not contain the full algebraic information from bracket $\{\ ,\ \}$; for example $c_2 = \{\ ,\ \}$ only on $\Omega_2(V) \times \Omega_2(V)$, it vanishes elsewhere (cf. [B-F-L-S] prop. 4). The right approach might be to consider the Lie superalgebra generalizing naturally $\mathrm{SVect}(V)$, in other words $S\Omega_*(V) = \operatorname{Im}\delta \subset \Omega_*(V)$; one then has $S\Omega_1(V) = \mathrm{SVect}(V)$ (cf. (b) above). One can then consider a doubly graded resolution of $S\Omega_*(V)$ and generalize slightly the construction of ([B-F-L-S]) in order to construct a bigraded bracket c on this resolution which satisfies $[\![c, c]\!] = S$ where S denotes the Schouten bracket on $\Omega_*(V)$. This would lead to "constant curvature" rather than "vanishing curvature" equation, and we hope to discuss it in a forthcoming paper.

References

[A] Akman, F., A master identity for homotopy Gerstenhaber algebras, `q-alg/9710004`.

[A-K] Arnold, V. I., Khesin, B. A., *Topological methods in hydrodynamics*, Appl. Math. Sci. 125, Springer-Verlag, New York 1998.

[A-L-M-Y] Awata, H., Li, M., Minic, D., Yoneya, T., On the quantization of Nambu brackets, `hep-th/9906248`.

[B-F-L-S] Barnich, G., Fulp, R., Lada, T., Stasheff, J., The *sh* Lie structure of Poisson brackets in field theory, *Comm. Math. Phys.* 191 (1998), no. 3, 585–601.

[B] Brylinski, J.-L., *Loop spaces, characteristic classes and geometric quantization*, Progr. Math. 107, Birkhäuser, Boston, MA, 1993.

[D-F-M-T] Dito, G., Flato, M., Sternheimer, D., Takhtajan, L., Deformation quantization and Nambu mechanics, *Comm. Math. Phys.* 183 (1997), no. 1, 1–22.

[Fe] Fedosov, B., *Deformation quantization and index theory*, Math. Top. 9, Akademie Verlag, Berlin 1996.

[F-D-S] Flato, M., Dito, G., Sternheimer, D., Nambu mechanics, n-ary operations and their quantization, in *Deformation theory and symplectic geometry, Ascona, 1996* (D. Sternheimer, J. Rawnsley and S. Gutt, eds.), Math. Phys. Stud. 20, Kluwer Academic Publishers, Dordrecht 1997, 43–66..

[F] Fuks, D. B., *Kogomologii beskonechnomernykh algebr Li* (in Russian), [Cohomology of infinite-dimensional Lie algebras], Nauka, Moscow 1984.

[G-M-S] Goldin, G. A., Menikoff, Ralph, Sharp, David H., Quantum vortex configurations in three dimensions, *Phys. Rev. Lett.* 67 (1991), no. 25, 3499–3502.

[G] Golenistcheva-Kutuzova, M. I., Generic orbits of the diffeomorphism group of a two-manifold in the space $\mathfrak{G}^*_{\mathrm{reg}}$ of regular momenta, Proceedings of the Winter School on Geometry and Physics (Srnì, 1990). *Rend. Circ. Mat.* Palermo 2 Suppl. No. 26 (1991), 171–178.

[H] Hamilton, R. S., The inverse function theorem of Nash and Moser, *Bull. Amer. Math. Soc.* (N.S.) 7 (1982), no. 1, 65–222.

[Ho] Hoppe, J., On M-algebras, the quantisation of Nambu-mechanics, and volume preserving diffeomorphisms, Papers honouring the 60th birthday of Klaus Hepp and of Walter Hunziker, Part II (Zürich, 1995), *Helv. Phys. Acta* 70 (1997), no. 1–2, 302–317.

[H-S] Hilton, P. J., Stammbach, U., A course in homological algebra, Second edition. Grad. Texts in Math. 4, Springer-Verlag, New York 1997.

[I] Ismagilov, R. S., Representations of infinite-dimensional groups, translated from the Russian manuscript by D. Deart., *Transl. Math. Monogr.* 152, Amer. Math. Soc., Providence, RI, 1996.

[K] Kirillov, A. A., Merits and demerits of the orbit method, *Bull. Amer. Math. Soc.* (N.S.) 36 (1999), no. 4, 433–488.

[Ko] Kontsevich, M., Deformation quantization of Poisson manifolds-I, `q-alg/9709040`.

[K-M] Kriegl, A., Michor, P. W., The convenient setting of global analysis, *Math. Surveys Monogr.* 53, Amer. Math. Soc., Providence, RI, 1997.

[L-R] Lecomte, P. B. A., Roger, C., Rigidité de l'algèbre de Lie des champs de vecteurs unimodulaires, *J. Differential Geom.* 44 (1996), no. 3, 529–549.

[L] Lichnerowicz, A., Algèbre de Lie des automorphismes infinitésimaux d'une structure unimodulaire, *Ann. Inst. Fourier* (Grenoble) 24 (1974), no. 3, xiv, 219–266.

[M-S] Matsuo, Y., Shibusa, Y., Volume Preserving Diffeomorphism and Noncommutative Branes, `hep-th/0010040`.

[R] Roger, C., Extensions centrales d'algèbres et de groupes de Lie de dimension infinie, algèbre de Virasoro et généralisations, Mathematics as language and art (Białowieża, 1993), *Rep. Math. Phys.* 35 (1995), no. 2–3, 225–266.

[S] Schomerus, V., D-branes and deformation quantization, hep-th/9903205.

[S-S] Singer, I. M., Sternberg, S., The infinite groups of Lie and Cartan. I. The transitive groups, *J. Analyse Math.* 15 (1965), 1–114.

Harrison cohomology and abelian deformation quantization on algebraic varieties

Christian Frønsdal

Department of Physics & Astronomy
University of California,
Los Angeles, CA 90095-1547, U.S.A.
e-mail: fronsdal@physics.ucla.edu

Abstract. Abelian deformations of ordinary algebras of functions are studied. The role of Harrison cohomology in classifying such deformations is illustrated in the context of simple examples chosen for their relevance to physics. It is well known that Harrison cohomology is trivial on smooth varieties and that, consequently, differentiable abelian $*$-products on such varieties are trivial. The subject is nevertheless not exhausted; first because varieties with singularities appear in the physical context and secondly, because there are some nice deformations that are not differentiable. We investigate cones, to illustrate the situation on algebraic varieties, and we point out that the coordinate algebra on (anti-) de Sitter space is a nontrivial deformation of the coordinate algebra on Minkowski space – although both spaces are smooth manifolds.

2000 Mathematics Subject Classification: 53D55 (81T70, 81-XX, 16E40)

1 Introduction

Deformation quantization has become a dynamic subject in mathematics as well as in physics. The general setting is a (commutative and associative) algebra of functions, and the most important case is the algebra of differentiable functions on a symplectic space, the observables of classical mechanics. The problem is to construct an associative algebra, generally non commutative, on the same space, as a formal or exact deformation of this algebra.

The Poisson bracket of classical mechanics plays an important role. A deformation 'in the direction of the Poisson bracket' is a formal associative product on the algebra of formal power series,

$$f * g = fg + \sum_{n \geq 0} \hbar^n C_n(f, g),$$

such that $C_1(f, g) = \frac{1}{2}\{f, g\}$. The ordinary product fg and the Poisson bracket $\{f, g\}$ are extended to formal power series in $\hbar$. The emphasis on this type deformation can be explained, not only by the physical applications, but by the fact that, on a smooth manifold, every associative differentiable deformation is equivalent to one with C_1 antisymmetric. Then C_1 is linear in df and in dg, and by the associativity of the $*$-product it follows that $f, g \mapsto C_1(f, g)$ is a Poisson bracket.

The existence of a Poisson bracket turns the manifold into a Poisson manifold but not, in general, into a symplectic space. The existence and the construction of $*$-products on an arbitrary Poisson manifold has recently been investigated, with great success [K, T].

Abelian deformations represent another type of generalization. Such deformations are not trivial. In the first place, though it is true that on a smooth manifold every differentiable abelian deformation is trivial, it is easy to give examples of nontrivial, abelian deformations that are not differentiable; that is, the cochains are not differentiable. In the second place, it is interesting to go beyond the context of smooth manifolds; and in particular, and in particular, to attempt the construction of $*$-products on algebraic varieties and manifolds with singularities.

Abelian $*$-products have turned up in at least one physical context: in attempts to quantize Nambu mechanics [D]. First order abelian deformations are classified by Harrison cohomology [B, H], which is trivial on smooth manifolds. Attempts to overcome this difficulty, unrelated to the present paper and generalizing the usual notion of deformation, may be found in [N1] and [P].

Dealing with algebras of functions (in particular, polynomials) on manifolds or varieties, it is natural to ask what types of singularities would be required for the existence of a non trivial, commutative $*$-product. This paper was planned as a study of simple examples of algebraic varieties, notably cones. However, in the course of this study, it became clear that interesting, non trivial deformations exist even on smooth manifolds (including $\mathbb{R}^n$). These are exact deformations: the deformed product is exact to all orders of λ. But the real surprise was the realization that there are exact, abelian, associative $*$-products, on smooth varieties, of the form

$$f * g = fg + \lambda C(f, g)$$

(with no terms of higher order in λ), that are non trivial as exact deformations.

Physical applications, some that are immediate and some that are more remote, include the following:

(a) The conformal anomaly, briefly discussed in Section 2.6.

(b) A cohomological classification of operator product expansions in quantum field theories.

(c) A view of classical (anti-) de Sitter field theory as an abelian deformation of Minkowski field theory; see Section 3. We hope that this may lead to a better understanding of some difficult aspects of anti-de Sitter physics.

(d) Within the program of geometric or $*$-product quantization on co-adjoint orbits, there are unsolved difficulties in the case of conic orbits. Most interesting is the quantization of Keplerian systems.

(e) One may look into the possibility of finding central extensions of Virasoro algebras on algebraic varieties, for example, on cones.

Section 2 initiates our study by detailed calculations on very simple examples. This paper considers only algebraic varieties, $\mathbb{R}^n$ or subvarieties of $\mathbb{R}^n$ defined by the vanishing of a finite set of polynomials. For physical applications it may be that the most interesting algebraic varieties are the cones,

$$\mathbb{R}^n/g, \quad g = \sum g^{ij} x_i x_j, \quad g^{ij} \in \mathbb{R}, \quad i, j = 1, \ldots, n.$$

The simplest case, $n = 2$ and $g = x^2 - y^2 = uv$ is investigated in Section 2.3.

Section 3 deals in great detail with a deformed, abelian algebra of functions on Minkowski space that is isomorphic to an algebra of functions on (anti-) de Sitter space.

The remaining sections study Hochschild (including Harrison) homology and cohomology, on cones of any dimension. The main purpose is to develop an understanding of the relationship between singularities and cohomology and, in particular, the role of singularities within a program of deformation quantization.

2 First examples

We shall investigate the Harrison cohomology for a commutative algebra, in the simplest context.

2.1 Harrison cohomology and abelian $*$-products

Harrison's complex for a commutative and associative algebra $\mathcal{A}$ is a subcomplex of the Hochschild complex. The cochains are valued in the unital augmentation of the algebra itself and familiar formulas for the differential apply. Thus if E is a one-cochain, then

$$dE(f, g) = fE(g) - E(fg) + E(f)g,$$

which vanishes for derivations, and if C is a two-cochain, then

$$dC(f, g, h) = fC(g, h) - C(fg, h) + C(f, gh) - C(f, g)h.$$

A first order, abelian deformation of (the product of) the algebra is a new product on the same space, a $*$-product,

$$f * g = fg + \lambda C(f, g), \quad C(f, g) = C(g, f), \quad f, g \in \mathcal{A},$$

such that, to first order in λ, $(f * g) * h = f * (g * h)$. Here λ is a formal parameter and it is implicit that the original algebra must be extended to the commutative algebra of polynomials in λ. The condition of associativity is equivalent, to first order in λ, to the condition that C be closed, $dC = 0$.

We continue to discuss first order deformations, until further notice. If there is a one-cochain E, such that $C = dE$, then the deformation is said to be trivial. The reason for this is as follows. For $f \in \mathcal{A}$, let $f_\lambda := f + \lambda E(f)$. Then to first order in λ,

$$f_\lambda g_\lambda = (fg)_\lambda + \lambda C(f_\lambda, g_\lambda),$$

so that the first order mapping $f \mapsto f + \lambda E(f)$ is a first order isomorphism from $\mathcal{A}$ to the deformed algebra.

2.2 A trivial $*$-product

Let $\mathcal{A}$ be the coordinate algebra of $\mathbb{R}^n$,

$$A = \mathbb{C}[x_1, \ldots, x_n],$$

and $\mathcal{A}_\lambda$ the same with a deformed product. Notice that this algebra does not have a unit. To prove that every deformation is trivial to all orders (we are here talking about formal deformations by infinite series that may or not converge for any $\lambda \in \mathbb{C}$), let

$$\Phi : \mathcal{A} \to \mathcal{A}, \quad \Phi(f*) = f,$$

where $f*$ is obtained from f by replacing $x_i x_j \ldots$ by $x_i * x_j * \ldots$. This map takes a Poincaré–Birkhoff–Witt basis for $\mathcal{A}_\lambda$ to a Poincaré–Birkhoff–Witt basis for $\mathcal{A}$ and is an algebra isomorphism. To relate this to the foregoing discussion of first order deformations, consider the one-cochain

$$\lambda E(f) = (f*) - f, \quad f* = \Phi^{-1}(f).$$

We have $((fg)*) = (f*) * (g*)$ and to first order in λ,

$$\begin{aligned} \lambda dE(f, g) &= [(f*) * (g*) - fg] - f[(g*) - g] - [(f*) - f]g \\ &= f * g - fg = \lambda C(f, g). \end{aligned}$$

For a more concrete example take $n = 1$, $C(f, g) = f_2 g_2$, where $f = f_1 + x f_2$ is the unique decomposition of f in terms of two even polynomials f_1, f_2. In this case $C = dE$ with

$$E(f) = -\frac{1}{2x}(\partial f_1 + x\partial f_2) = -\frac{1}{2x}\partial f + \frac{1}{2x} f_2.$$

This is a polynomial, though each term separately is not. But the first term is formally closed (a derivation), and

$$dE(f, g) = -\frac{1}{2x}\big(g_2 f_1 + g_2 f_1 - f g_2 - g f_2\big) = f_2 g_2 = C(f, g).$$

To find a nontrivial deformation we must go beyond this example, and one way to go is to introduce relations.

2.3 Algebras defined by relations

Let M be the algebraic variety

$$M = \mathbb{R}^n/R,$$

where R is a set of polynomial relations. Let $\mathcal{A}$ be the coordinate algebra of M, namely

$$\mathcal{A} = \mathbb{C}[x_1, \ldots, x_n]/R.$$

Let $\mathcal{A}_\lambda$ be a deformed algebra and let $\Phi : \mathcal{A} \to \mathcal{A}$ be the map introduced above, that takes $f* \mapsto f$. Let R_* be the relations of $\mathcal{A}_\lambda$ and $R_\lambda = \Phi R_*$; then the two algebras are isomorphic if there is an invertible mapping on $\mathbb{R}^n$ that takes R_λ to R.

2.4 First non trivial example

Let $n = 1$, N a natural number, $r \in \mathbb{R}$, and $R = x^N - r$. The Poincaré–Birkhoff–Witt basis of $\mathcal{A} = \mathbb{C}[x]/R$ is $\{1, x, x^2, \ldots, x^{N-1}\}$. Define a deformed product by setting

$$x^k * x^l = \begin{cases} x^{k+l}, & k+l < N, \\ x^{k+l} + \lambda x^{k+l-N}, & k+l \geq N. \end{cases}$$

Then $R_* = (x*)^N - (r+\lambda)$, $R_\lambda = x^N - (r+\lambda)$ and a trivializing map is given by

$$x \mapsto (1 + \lambda/r)^{1/N} x.$$

Two things can go wrong. (1) This map does not exist if $r = 0$, and in that case the deformation is not trivial. (2) If $r \neq 0$, then the trivializing map is an infinite power series in λ.

We have $f * g = fg + \lambda C(f, g)$, with

$$C(x^k, x^l) = \begin{cases} 0, & k+l < N, \\ x^{k+l-N}, & k+l \geq N. \end{cases}$$

There is no need to add higher order corrections in $f * g$; the product as it stands is associative, to all orders in λ. We have $C = dE$, with $E(x^k) = (k/Nr)x^k$, $k = 1, \ldots, N-1$.

Conclusion: Although C is first order, exact, and the $*$-product is associative to all orders, nevertheless an infinite power series was needed to trivialize it. In this particular case the appearance of infinite series is more or less innocuous, but that is not always the case, as we shall see.

2.5 Cones

Let $\mathcal{A} = \mathbb{C}[x, y]/R$, with $R = y^2 - x^2 - r^2$. There is a unique decomposition

$$f = f_1 + yf_2,$$

where f_1, f_2 are polynomials in x. Take $C(f, g) = f_2 g_2$. This two-form is closed. The deformed relation is $y^2 - x^2 - (r^2 + \lambda)$. These relations define equivalent algebraic varieties if $r^2 \neq 0$ (when $r \neq 0$ there is a neighbourhood of $\lambda = 0$ in which the two varieties are equivalent), so in this case the deformation is trivial. But if $r = 0$, when the initial variety is a cone and has a singularity at $x = y = 0$, the deformation is not trivial. We shall return to cones in Section 4.

The polynomial one-cochain

$$E(f) = \frac{1}{2r^2}[x\partial f_1 + yf_2 + xy\partial f_2] = \frac{x}{2r^2}\partial f + \frac{1}{2y} f_2$$

(in the last expression $\partial y = x/y$) is formally cohomologous to $E'(f) := \frac{1}{2y} f_2$, and

$$dE(f, g) = dE'(f, g) = -\frac{1}{2y}[f_1 g_2 + f_2 g_1 - f g_2 - f_2 g] = f_2 g_2.$$

This shows that the first order deformation is trivial when $r^2 \neq 0$. (The derivation included in the definition of E, to eliminate the non polynomiality of E', does not exist when $r^2 = 0$.) We notice that the obstruction is the constant term in f_2; in other words the value of $C(f, g)$ at the origin.

2.6 Application, conformal anomaly?

We may consider the algebra of polynomials on Dirac's cone. Of interest is the projective cone, one studies functions with a fixed degree of homogeneity, normally a negative integer. In scalar field theory the degree is -1. The projective cone can be covered by two charts. If y is a set of coordinates for $\mathbb{R}^6$ then a pair of charts is given by $x_\mu^\pm = y_\mu/(y_5 \pm y_6)$, $\mu = 1, 2, 3, 4$, each mapping most of the projective cone onto Minkowski space. To define a function on compactified Minkowski space we may take a pair of functions on Minkowski space, with the relation $f_1(x) = f_2(x/x^2)$. The ordinary product would be $fg = (f_1, f_2)(g_1, g_2) = (f_1 g_1, f_2 g_2)$. It can probably be deformed, and such deformations may perhaps relate to the conformal anomaly, but that will be left for another time.

3 Field theory on anti-de Sitter space

Let $\mathcal{A} = \mathbb{C}[x^0, \ldots, x^3]$, the coordinate algebra of polynomial functions on Minkowski space, without constant term. The metric is $g_{ij}dx^i dx^j = (dx^0)^2 - \sum_{i=1}^3 (dx^i)^2$. For every $a \in \mathcal{A}$ let $a = a_+ + a_-$ be the decomposition into an even and an odd part. We regard the algebra as the space of pairs,

$$A = \{(a_+, a_-)\},$$

with the product

$$AB = (a_+ b_+ + a_- b_-, a_+ b_- + a_- b_+),$$

or equivalently as the even subalgebra of $\mathbb{C}[x^0, \dots, x^3, y]/(y^2 - 1)$, with elements

$$a = a_+ + y a_-.$$

We deform the product, setting

$$A * B = AB - \rho x^2 a_- b_-, \quad x^2 := g_{ij} x^i x^j, \quad \rho > 0.$$

The deformed algebra is the even subalgebra of

$$\mathbb{C}[x^0, \dots, x^3, y]/(y^2 - 1 + \rho x^2),$$

which is a coordinate algebra of polynomial functions on anti-de Sitter space.

Referring back to Section 2.3, we notice a great deal of similarity. The above $*$-product is associative to all orders in ρ, no higher order terms are needed. Furthermore, $C(a, b) = \rho x^2 a_- b_-$ defines an exact two-cochain, $C = dE$ with $E(a) = (\rho/2) x^2 a_-$. But the "trivializing" map takes $a \mapsto a_+ + y\sqrt{1 + \rho x^2} a_-$, and here we find an infinite power series in the generators of the algebra, not just in the parameter. And this series is not entire but has singularities, for any ρ different from zero. There is no sensible point of view that would allow us to regard the two algebras as equivalent.

Conclusion. The algebra of coordinate functions on anti-de Sitter space is a nontrivial deformation of that of Minkowski space; this in spite of the fact that both spaces are smooth manifolds. The deformations are not classified by Harrison cohomology, trivial in this case. It is true that Harrison cohomology does classify first order deformations, but it must be understood that, even if $a * b = ab + \lambda C(a, b)$, with no higher order terms, and this is associative to all orders, trivialization is not localized at finite order. Vanishing of Harr^2 merely tells us that the term of first order in λ can be removed, to be replaced by terms of higher orders. Triviality of Harrison cohomology of a regular, commutative algebra is a consequence of a theorem of Hochschild, Kostant and Rosenberg [HKR], according to which Hochschild cohomology is limited to alternating cochains. This result was extended by Vey [V] to the differentiable case (see also [C] and [N2]). The cochain introduced above is not differentiable, and Vey's theorem does not apply.

Abelian $*$-products thus appear in two different ways. In the cases when $\mathrm{Harr}^2 = 0$ the obstructions are singularities of the "trivializing" map. But this does not diminish our interest in trying to understand those abelian $*$-products that owe their non-triviality to the existence of Harrison cohomology, and that is the subject of the rest of this paper.

4 Hochshild homology of the cone $uv = 0$

This section and the next take up the study of the cone algebra introduced in 2.5, with a more convenient choice of coordinates.

Let $M = \{u, v \in \mathbb{R}^2, uv = 0\}$, and let $\mathcal{A}$ be the commutative algebra

$$\mathcal{A} = \mathbb{C}[u, v]/uv.$$

We study the complex

$$0 \overset{\partial}{\leftarrow} C_1 \overset{\partial}{\leftarrow} C_2 \cdots,$$

where $C_n \in \mathcal{A}^{\otimes n}$, and

$$\begin{aligned}\partial(a_1 \otimes a_2 \otimes \cdots \otimes a_n) \\ = a_1a_2 \otimes a_3 \otimes \cdots - a_1 \otimes a_2a_3 \otimes a_4 \otimes \cdots + \cdots + (-1)^n a_1 \otimes \cdots \otimes a_{n-1}a_n.\end{aligned}$$

Let Z_n, B_n denote the subspaces of closed, respectively exact subspaces of C_n:

$$Z_n = \{a \in C_n, \partial a = 0\}, \quad B_n = \partial C_{n+1}.$$

Theorem 4.1. *The Hochschild homology $H_n = Z_n/B_n$ of this complex is of dimension 2 for $n \geq 1$, and is generated as a vector space by $u \otimes v \otimes u \otimes \cdots$ and $v \otimes u \otimes v \otimes \cdots$.*

Proof. For every n-chain a we have

$$\partial(u \otimes a) = ua_1 \otimes a_2 \otimes \cdots - u \otimes da,$$

therefore every n-chain with initial factor ua_1 is homologous to one with initial factor u, and every n-chain is homologous to $u \otimes a + v \otimes b$. Now suppose that $u \otimes a$ is closed. Then

$$u \otimes a = u \otimes vb_1 \otimes b_2 \otimes \cdots \approx u \otimes v \otimes db,$$

and so on. Since no n-chain a with each factor a_i linear in u, v is exact, the theorem is proved. □

5 Hochschild cohomology of the cone $uv = 0$

The cochains are valued in the unital augmentation $\mathcal{A}_1$ of $\mathcal{A}$, and

$$dC(a_1, \ldots, a_n) = a_1C(a_2, \ldots, a_n) - C(\partial a) + (-1)^n C(a_1, \ldots, a_{n-1})a_n.$$

For any n-cochain C, let C' denote the restriction of C to *closed, linear* chains. Thus C' is defined by the values $C'(a) = C(a)$, with $a = u \otimes v \otimes u \otimes \cdots$ and $a = v \otimes u \otimes v \otimes \cdots$. The set of all restricted cochains form a complex with differential

$$d'C'(a) = a_1C'(a_2, \ldots, a_n) + (-1)^n C'(a_1, \ldots, a_{n-1})a_n$$

A restricted $2n$-cochain is closed if

$$C'_{2n}(u, v, \dots, v) = C'_{2n}(v, u, \dots, u),$$

and exact if it is valued in $\mathcal{A}$. A restricted $2n+1$-form is closed if

$$vC'_{2n+1}(u, v, \dots, u) + uC'_{2n+1}(v, u, \dots, v) = 0,$$

which implies that $C(u, \dots, u) = u\alpha(u)$, $C(v, \dots, v) = v\beta(v)$, with $\alpha(u)$, $\beta(v)$ in the unital augmentation of $\mathcal{A}$, and it is exact if $n > 1$ and $\alpha + \beta \in \mathcal{A}$. The cohomology class of a $2n$-cochain is the value of $C(u, v, u, \dots)$ at the origin.

Proposition 5.1. *For every closed n-cochain $\tilde{C}$ of the restricted complex there is a closed Hochschild cochain C such that the restriction C' of C is equal to $\tilde{C}$.*

Proof. To find such C we begin by setting $C' = \tilde{C}$ for linear, closed arguments, then attempt to define C for nonlinear arguments by induction in the polynomial degree, using

$$dC(a) = a_1C(a_2, \dots) - C(\partial a) + (-1)^n C(\dots, a_{n-1})a_n = 0,$$

with the observation that the argument of the middle term is of higher total polynomial degree than the others. The obstruction is $\partial a = 0$ with a not linear; thus $a = \partial b$. For an inductive proof, we must show that, for $a = \partial b$, the last relation holds by virtue of similar relations involving arguments of lower order. In fact,

$$\begin{aligned} dC(db) &= b_1b_2C(b_3, \dots, b_{n+1}) - b_1C(\partial b^+) \\ &\quad + (-1)^n\big(C(\partial b^-)b_{n+1} + (-1)^{n+1}C(b_1, \dots, b_{n-1})b_nb_{n+1}\big), \end{aligned}$$

where $b^+ = b_2 \otimes \cdots \otimes b_{n+1}$ and $b^- = b_1 \otimes \cdots b_n$. By the induction hypothesis the second and third term are

$$\begin{aligned} &- b_1\big(b_2C(b_3, \dots, b_{n+1}) + (-1)^nC(b_2, \dots, b_n)b_{n+1}\big) \\ &+ (-1)^n\big(b_1C(b_2, \dots, b_n) + (-1)^nC(b_1, \dots, b_n)b_{n+1}\big), \end{aligned}$$

and substitution gives the desired result, zero. □

Proposition 5.2. *If C is any closed cochain, and its restriction is exact, then C is exact.*

Proof. We must show that there is E such that

$$C(a) = a_1E(a_2, \dots) - E(\partial a) + (-1)^nE(\dots, a_{n-1})a_n.$$

By hypothesis this is true if a is linear and closed, and that establishes a basis for induction in the total polynomial degree of a. The obstruction is again $a = \partial b$, and it is necessary to show that $C(\partial b)$ is the same as

$$\begin{aligned} &b_1b_2E(b_3, \dots, b_{n+1}) - b_1E(\partial b^+) \\ &\quad + (-1)^n\big(E(\partial b^-)b_{n+1} + (-1)^nE(b_1, \dots, b_{n-1})b_nb_{n+1}\big). \end{aligned}$$

The induction hypothesis gives us $C(b^+)$ and $C(b^-)$; we use that to eliminate $E(\partial b+)$ and $E(\partial b^-)$ to get

$$
\begin{aligned}
&b_1 b_2 E(b_3, \ldots, b_{n+1}) + E(b_1, \ldots, b_{n-1}) b_n b_{n+1} \\
&\quad - b_1\big(b_2 E(b_3, \ldots, b_{n+1}) + (-1)^{n+1} E(b_2, \ldots, b_n) b_{n+1} - C(b^+)\big) \\
&\quad + (-1)^n \big(b_1 E(b_2, \ldots, b_n) + (-1)^{n+1} E(b_1, \ldots, b_{n-1}) b_n - C(b^-)\big) b_{n+1},
\end{aligned}
$$

which is precisely $C(\partial b)$. □

Together, the two propositions give us the following result.

Theorem 5.3. *The cohomology of the Hochshild complex is equivalent to the cohomology of the restricted complex.*

6 Geometric cochains of the cone $uv = 0$

We intend to discover which, if any, geometric properties of the variety are reflected in the Hochschild complex.

6.1 Points

Consider a collection of "geometric n-cochains" $q_1 \otimes \cdots \otimes q_n$, where q_i is a point in M, and the pairing

$$\langle q_1 \otimes \cdots \otimes q_n | a_1 \otimes \cdots \otimes a_n \rangle = \delta_{nm} a_1(q_1) \ldots a_n(q_n),$$

of geometric cochains with Hochschild chains. Here $a_1, \ldots, a_m$ are the same m-chains as before, with $a_i \in \mathcal{A}$, the algebra of coordinate functions on $M = \mathbb{C}[u, v]/uv$.

The differential is $dq_1 = q_1 \otimes q_1$ and

$$d(q_1 \otimes q_2 \otimes \cdots) = (q_1 \otimes q_1) \otimes q_3 \otimes \cdots - q_1 \otimes (q_2 \otimes q_2) \otimes q_3 \otimes \cdots + \cdots ;$$

this ensures the relation of duality,

$$\langle dq | a \rangle = \langle q | \partial a \rangle.$$

In this complex of points there is no cohomology. Notice that these cochains are distributions (delta-functions) valued in $\mathbb{C}$.

The intuitive meaning of this is that the points do not contain any information about those geometric properties of the variety that are related to the existence of deformations. Instead, the tangent vectors do contain such information. The tangent space is two-dimensional at the singular point, one-dimensional elsewhere.

6.2 Points and tangent vectors

Let p_i denote a tangent vector at the point q_i, and consider chains made up of both q's and p's. The pairing is

$$\langle q_1 \otimes p_2 \otimes \cdots | a_1 \otimes a_2 \otimes \cdots \rangle = a_1(q_1)(pa_2)(q_2) \dots .$$

To preserve duality we define

$$d(p_1 \otimes q_2 \otimes \cdots) = dp_1 \otimes q_2 \otimes \cdots - p_1 \otimes (dq_2) \otimes \cdots + \cdots ,$$
$$dq_1 = q_1 \otimes q_1, \quad dp_2 = p_2 \otimes q_2 + q_2 \otimes p_2.$$

Let subscript zero refer to the origin, then $q_0 = 0$ and $p_0 \otimes p_0$ is closed. The 2-cohomology is carried by $p_0 \otimes p_0'$. In particular,

$$\langle \partial_u \otimes \partial_v | u \otimes v \rangle = 1.$$

Conclusion. The existence of the antisymmetric part of H^2 on the 2-cone is clearly a reflection of the fact that the dimension of tangent space is discontinuous at the singularity, but I cannot say that the existence of the symmetric part has been clarified in any deep sense. I hope to improve this type of analysis and to generalize it for algebraic varieties in general.

7 Homology of cones

Consider the algebra $\mathbb{C}[x_1, \dots, x_n]/g$, where $g = g(x)$ is a second order, homogeneous polynomial,

$$g = g^{ij} x_i x_j .$$

The proof of the following is similar to that of Theorem 4.1; what is used is the fact that the relations are second order, homogeneous.

Theorem 7.1. *Every chain is cohomologous to a linear chain, and no linear chain is exact, so H_k can be identified with the space of closed, linear chains.*

Proposition 7.2. *Let L_k denote the space of alternating, linear k-chains, then $H_1 = L_1$ and for $k \geq 2$, $H_k = \oplus_{p \leq k/2} L_{k-2p}$.*

Examples. For small k the coefficients of representatives of the homology are

$$k = 2 : \alpha^{ij} + \alpha g^{ij},$$
$$k = 3 : \alpha^{ijk} + \alpha^i g^{jk} - \alpha^j g^{ik} + \alpha^k g^{ij},$$
$$k = 4 : \alpha^{ijkl} + \alpha^{kl} g^{ij} + \alpha^{il} g^{jk} - \alpha^{jl} g^{ik} + \alpha^{ij} g^{kl} - \alpha^{kj} g^{li} + \alpha^{ki} g^{lj}$$
$$+ \alpha (g^{il} g^{jk} - g^{ik} g^{jl} + g^{ij} g^{kl}),$$

where the α's are complex, alternating.

8 Cohomology of cones

Exactly as in the case of the 2-cone, one proves that $H^*(\mathrm{Hom}(\mathcal{A}^{\otimes n}, \mathcal{A}))$ is the same as for the restricted complex based on linear, closed chains. For example:

Restricted one-cochains are closed (never exact) if

$$dC(x_i \wedge x_j) = 0 = g^{ij} dC(x_i \otimes x_j),$$

which reduces to $g^{ij} x_i C(x_j) = 0$.

Restricted 2-cochains are defined for arguments $x_i \wedge x_j$ and $g^{ij} x_i \otimes x_j$. A calculation shows that a 2-cochain is closed if

$$g^{ij} x_i C(x_j \wedge x_k) = 0,$$

and exact if it is symmetric and valued in $\mathcal{A}$.

Restricted 3-cochains are defined on $x_i \wedge x_j \wedge x_k$ and on

$$\bar{\alpha} := (\alpha^i g^{jk} - \alpha^j g^{ik} + \alpha^k g^{ij})(x_i \otimes x_j \otimes x_k).$$

They are closed if

$$g^{ij} x_i C(x_j \wedge x_k \wedge x_l) = 0$$

and

$$(g^{il} g^{jk} - g^{jl} g^{ik} + g^{kl} g^{ij}) C(x_j \otimes x_k \otimes x_l) = 0,$$

and exact if they vanish on $x_i \wedge x_j \wedge x_k$ and, for some $Q^i \in \mathcal{A}$, $C(\bar{\alpha}) = \sum_i x_i Q^i$.

Acknowledgements. I thank Daniel Sternheimer and Guiseppe Dito for useful suggestions, Murray Gerstenhaber for information about Harrison cohomology, Didier Arnal for a useful remark and Georges Pinczon for leading me to cones. I thank especially Daniel Sternheimer and Guiseppe Dito for improving the manuscript.

References

[B] Barr, M., Harrison Homology, Hochschild Homology, and Triples, *J. Algebra* 8 (1968), 314–323.

[C] Connes, A., Noncommutative differential geometry, *Inst. Hautes Études Sci. Publ. Math.* 62 (1985), 257–360.

[D] Dito, G., Flato, M., Sternheimer, D., and Takhtajan, L., Deformation Quantization and Nambu Mechanics, *Comm. Math. Phys.* 183 (1997), 1–22.

[G] Gerstenhaber, M., and Schack, S. D., Algebraic cohomology and deformation theory, in *Deformation Theory of Algebras and Structures and Applications* (M. Hazewinkel and M. Gerstenhaber, eds.), NATO ASI Ser. C 247, Kluwer Academic Publishers, Dordrecht 1988, 11–264.

[H] Harrison, D. K., Commutative algebras and cohomology. *Trans. Amer. Math. Soc.* 104 (1962), 191–204.

[HKR] Hochschild, G., Kostant, B., and Rosenberg, A., Differential forms on regular affine algebras, *Trans. Amer. Math. Soc.* 102 (1962), 383–408.

[K] Kontsevich, M., Deformation quantization of Poisson varieties I, `q-alg/9709040`; Operads and motives in deformation quantization, *Lett. Math. Phys.* 48 (1999), 35–72; Deformation quantization of algebraic varieties, *Lett. Math. Phys.* 56 (3) (2001), in press, `math.AG/0106006`.

[N1] Nadaud, F., Generalized Deformations and Hochschild Cohomology, *Lett. Math. Phys.* 58 (2001), 41–55; Generalized deformations, Koszul resolutions, Moyal products, *Rev. Math. Phys.* 10 (1998), 685–704.

[N2] Nadaud, F., On continuous and differentiable Hochschild cohomology, *Lett. Math. Phys.* 47 (1999), 83–95.

[P] Pinczon, G., Noncommutative deformation theory, *Lett. Math. Phys.* 41 (1997), 101–117.

[T] Tamarkin, D., Another proof of M. Kontsevich formality theorem, `math.QA/9803025` v.4; Tamarkin, D. and Tsygan, B., Noncommutative differential calculus, homotopy BV algebras and formality conjectures. *Methods Funct. Anal. Topology* 6 (2000), 85–100, `math.KT/0002116`.

Related semi-classical and Toeplitz algebras

Louis Boutet de Monvel

Institut de Mathématiques de Jussieu
Université Pierre et Marie Curie
4, place Jussieu, 75252 Paris Cedex 05, France
e-mail: boutet@math.jussieu.fr

Abstract. We investigate the relation between Toeplitz algebras and semi-classical algebras (star-products) arising for example in geometric prequantization.

2000 Mathematics Subject Classification: 16S32, 16S80, 32V05, 35S05, 53C05, 58J40

1 Introduction

Prequantization of symplectic manifold, in particular of coadjoint orbits of a Lie algebra, was introduced by B. Kostant to describe many significant representations of Lie groups. The operators used in this description are nicely explained in terms of semiclassical analysis or in terms of Toeplitz operators. In this paper we investigate the relation between the semiclassical and Toeplitz algebras which enter in this description.

Semi-classical algebras are those which appear in deformation quantization: a star product on a symplectic manifold X, as defined in [1], is an associative product $B = \sum h^n B_n$ on the set of formal power series $f = \sum h^n f_n \in C^\infty(X)((h))$, where for each n, B_n is a bidifferential operator, the leading term is the usual product: $B_0(f, g) = fg$, the leading term of the commutator law is the Poisson bracket of X: $B_1(f, g) - B_1(g, f) = \{f, g\}$, and the unit is 1 ($B_n(f, 1) = B_n(1, f) = 0$ if $n > 0$).

Toeplitz operators were introduced in [6, 8]. Modulo smoothing operators they produce a symbolic calculus associated with a contact manifold (or a symplectic cone, see below), locally identical to the calculus of pseudo-differential operators, and very similar to that of semi-classical algebras except that there is no longer a central h (the "small deformation parameter" h should be thought of as the inverse of the size of a large frequency).

Such algebras typically arise in the theory of prequantization [26]. For this the data is

- a compact complex manifold X (necessarily projective if the subsequent conditions are fulfilled)

- a complex line bundle Σ over X equipped with a hermitian metric a, strictly pseudoconvex outside of the zero section (this means that $\omega_\Sigma = i\partial\bar{\partial}a$ is a Kähler form outside of the zero section, in particular it is symplectic).

The algebra of polynomial functions on Σ is $\bigoplus \Gamma(X, L^{\otimes n})$, L the dual bundle of Σ. Σ is a symplectic manifold, so as X (the pull-back of the symplectic form ω_X is $i\partial\bar{\partial}\operatorname{Log} a$). The circle group $U(1)$ acts on the whole situation. We denote θ the infinitesimal generator; it is elementary that it is the hamiltonian field $\theta = H_a$.

In this situation we have an algebra of Toeplitz operators acting on the space of boundary values of holomorphic functions on the unit sphere $Y = \{a = 1\}$ of Σ (they are the operators of the form $T_P : f \mapsto SP(f)$ with P a pseudo-differential operator on Y, S the Szegö projector, i.e. the orthogonal projector onto the set of boundary values of holomorphic functions (CR functions) in $L^2(Y)$). The Toeplitz algebra $\mathcal{A}$ referred to in the title is the algebra of Toeplitz operators mod. smoothing operators; this admits a local definition and is the set of sections of a sheaf of algebras locally isomorphic to the algebra of pseudo-differential operators.

The associated semi-classical algebra (coming out essentially from the construction of [26]) is the sub-algebra $\mathcal{B} \subset \mathcal{A}$ of invariant operators. Its center is the algebra generated by θ, which is invertible (elliptic of degree 1). It has a natural structure of semi-classical algebra, setting $h = \theta^{-1}$.

In this paper we investigate generally how a Toeplitz with a circle group action and a semi-classical algebra can be similarly related. The answer can be roughly summarized as follows: a semi-classical algebra can be embedded in a Toeplitz algebra iff its Fedosov curvature is constant (cf. §5 – the Fedosov curvature is a 2-cohomology class $r = h^{-1}\omega_X + \sum_{k\geq 0} h^k r_k$: except for the leading term $h^{-1}\omega_k$ it must not depend on the deformation parameter h at all); a Toeplitz algebra contains a semi-classical algebra iff its Fedosov curvature is (up to an "exotic" exponent) the pull-back of a 2-form on X. The result is described precisely in §6.

The paper is organized as follows: the definitions and classification results for star-algebras are recalled, in §2–3. In §4 we have grouped some definitions and facts concerning subprincipal symbols and involutions, which are immediately pertinent to the classification and, although easy, do not seem to appear elsewhere. In [19] Fedosov introduced a very elegant description of semi-classical algebras, where the classifying invariant is the cohomology class of a formal closed 2-form on X: $R = \frac{1}{h}\omega_X + \sum h^n r_n$, the curvature of a "Fedosov connection". Fedosov's method can be adapted with trivial modifications to classify Toeplitz algebras, and in the situation above it is interesting to compare the two invariants. This is done in §5. The relation between Toeplitz and semi-classical algebras is investigated in §6, which also contains a related result about compact group actions: any action of a compact group on the symplectic basis X resp. Σ lifts to the semi-classical algebra $\mathcal{B}$ over X resp. Toeplitz $\mathcal{A}$ over Σ; the lifting is unique up to isomorphism.

2 Notations

2.1 Conic manifolds

In this paper we only consider star-products on real manifolds. Star products live on cones; we first recall the notations of [2, 4].

Definition 2.1. A *real cone* is a C^∞ principal bundle Σ with group $\mathbb{R}_+^\times$. The basis is $B\Sigma = \Sigma/\mathbb{R}_+^\times$.

Example 2.2. If D_ω is the long line, $\Sigma = T^*D_\omega - \{\text{the zero section}\}$ is a real non trivial cone. However in this article we will always assume that manifolds are paracompact: a real cone Σ over a paracompact basis $B\Sigma$ is always trivial, isomorphic to $B\Sigma \times \mathbb{R}_+^\times$, but the product structure is not part of the data.

Definition 2.3. (i) We denote $\mathcal{O}(m)$ the sheaf on $B\Sigma$ of homogeneous functions of degree m of Σ, $\mathcal{O} = \bigoplus \mathcal{O}(n)$.

(ii) We denote $\widehat{\mathcal{O}}$ the sheaf on $B\Sigma$ of formal symbols:

$$f \in \widehat{\mathcal{O}} \text{ if } f = \sum_{m \le m_0} f_m \quad \text{with } f_m \in \mathcal{O}(m),\ m \text{ an integer}, m \to -\infty \tag{2.1}$$

(f is an "asymptotic expansion" for $\xi \to \infty$ on Σ).

(iii) For any integer $k \ge 1$ we denote $\widehat{\mathcal{D}}_k$ the sheaf (on $B\Sigma$) of formal k-differential operators: $P(f_1, \dots, f_k) = \sum_{m \le m_0} P_m(f_1, \dots, f_k)$ with P_m a k-linear differential operator homogeneous of degree m with respect to homotheties, m an integer, $m \to -\infty$. For $k = 1$ we just write $\widehat{\mathcal{D}}$.

Locally we may choose homogeneous local coordinates x_j on Σ. Then $P_m(f_1, \dots, f_k)$ is a sum of homogeneous monomials

$$a_\alpha(x)\, \partial^{\alpha_1} f_1 \dots \partial^{\alpha_n} f_n,$$

i.e. a_α is homogeneous of degree $m + \sum \alpha_k \deg(x_k)$. Two cases will be useful:

1) $x_1, \dots, x_{n-1}$ are homogeneous of degree 0 and are local coordinates on the basis $B\Sigma$, x_n is homogeneous of degree 1 or -1;

2) the x_j are all of degree $\frac{1}{2}$. There is no restriction on the order of P_m.

Below "degree" will always refer to the degree with respect to homotheties; thus if $P \in \widehat{\mathcal{D}}_k$ each term P_m of degree m is of finite order, although the resulting infinite sum P may be of infinite order.

We will denote

$$\widehat{\mathcal{D}}^\times \subset \widehat{\mathcal{D}} \tag{2.2}$$

the sheaf of invertible formal differential operators: $P = \sum P_k \in \widehat{\mathcal{D}}^\times$ is invertible iff its leading term $\sigma(P) = P_{m_0}$ is invertible, i.e. P_{m_0} is of order 0, the multiplication by a nonvanishing function homogeneous of degree m_0.

Definition 2.4. We denote by $\widehat{\mathcal{D}}_0^\times$ the subsheaf of those invertible P such that $P(1) = 1$, i.e. P is of degree 0, $P_0 = 1$, $P_m(1) = 0$ if $m < 0$.

In some instances it will be convenient to use the complex extension

$$\Sigma_c = \Sigma \times_{R_+^\times} \mathbb{C}^\times \tag{2.3}$$

to which homogeneous functions of integral degree $f \in \mathcal{O}(n)$ or formal series $f \in \widehat{\mathcal{O}}$ obviously extend.

2.2 Star algebras

Definition 2.5. A *star product* on $B\Sigma$ is a unitary associative algebra law on $\widehat{\mathcal{O}}$

$$B = \sum B_m \in \mathcal{D}_2 : \widehat{\mathcal{O}} \otimes \widehat{\mathcal{O}} \to \widehat{\mathcal{O}} \tag{2.4}$$

for which the unit is $1 \in \widehat{\mathcal{O}}$, $B_0 = 1$ (i.e. $B_0(f, g) = fg$). B is a deformation of the usual product.

A star-algebra on $B\Sigma$ is a sheaf of unitary associative algebras on $B\Sigma$, locally isomorphic to $\widehat{\mathcal{O}}$ equipped with a star-product, where the structural patching sheaf of groups is $\widehat{\mathcal{D}}_0^\times$.

Thus locally, in some frame, $f * g = \sum B_k(f, g)$ with B_k a bidifferential operator homogeneous of degree $-k \to -\infty$, $B_0 = 1$.

Since any $P \in \widehat{\mathcal{D}}_0^\times$ respects the natural filtration of $\mathcal{O}$ and the leading term ($P = 1$ mod. terms of lower degree), a star-algebra $\mathcal{A}$ is equipped with a natural filtration and a canonical "symbol isomorphism" $\sigma : \operatorname{gr} \mathcal{A} \to \mathcal{O}$. This conic framework for star products was described in [2, 4]. The distinction between star-products and star-algebras is not really essential on real manifolds, where a star-algebra is always isomorphic to $\widehat{\mathcal{O}}$ equipped with a star-product because a "total symbol" i.e. a global isomorphism $P : \mathcal{A} \xrightarrow{\sim} \widehat{\mathcal{O}}$ with $P \in \widehat{\mathcal{D}}$ locally always exists (locally, by definition, and on a real manifolds these can be patched using a partition of unity); as mentioned in [2] this is no longer true over holomorphic manifolds: however, the most usual star-algebras such as the algebra of pseudo-differential or semi-classical pseudo differential operators on a manifold are not equipped with a canonical total symbol, so for functorial manipulations they are star-algebras rather than star-products.

In this paper we will be concerned by the two main cases: Toeplitz algebras and semi-classical algebras, and the relation between them.

Let us recall that a deformation quantization or star product on a symplectic manifold X is a unitary associative product law $f * g = \sum h^n B_n(f, g)$ on the space of formal series $f = \sum h^n f_n \in C^\infty(X)[[h]]$, where each B_n is a bilinear differential operator on X, $B_0(f, g) = fg$, the unit is 1 (B_n has no "constant term" if $n > 0$), and the leading term in the commutator law is given by the Poisson bracket of X: $B_1(f, g) - B_1(g, f) = \{f, g\}$. Existence of such star-products was proved in [16].

We will refer to these as "semi-classical" star-products or star-algebras. The corresponding cone is $X \times \mathbb{R}_+$ equipped with the Poisson bracket hc_X, c_X the Poisson bracket of X; the canonical coordinate of $\mathbb{R}_+$ (homogeneous of degree 1, corresponding to a large frequency) is $\frac{1}{h}$; Σ is odd-dimensional. The formal parameter h ("Planck constant"), or its inverse $a = h^{-1}$, is central and part of the data, i.e. for these we only consider automorphisms which preserve h.

Toeplitz algebras (or Toeplitz star-products) are those associated to a symplectic cone Σ, e.g. the algebra of pseudo-differential operators (mod. smoothing operators) on a manifold X: $\Sigma = T^*X - \{\text{its zero section}\}$, equipped with its canonical symplectic structure, or the algebra of Toeplitz operators (mod. smoothing operators) of [8]; Σ is even dimensional, there is no longer a central formal "Planck constant" h, and the center is reduced to constants. The existence of a canonical Toeplitz algebra on a symplectic (even-dimensional) cone was proved in [8].

2.3 Models

The Moyal–Weyl star product gives typical examples of star products: let V be a graded vector space and $c = \sum c_{ij}\partial_i\partial_j \in V \otimes V$ a 2-tensor of degree ≤ -1. Then

$$f * g = \exp \frac{1}{2} c(\partial_\xi, \partial_\eta)\, f(\xi)g(\eta)\, |_{\eta=\xi} \tag{2.5}$$

is a star-product on V. The corresponding Poisson bracket is the top-order part (of degree -1) of the antisymmetrization of c:

$$\{f, g\}(\xi) = \frac{1}{2}\big(c_{-1}(\partial_\xi, \partial_\eta) - c_{-1}(\partial_\eta, \partial_\xi)\big)\, f(\xi)g(\eta)\, |_{\eta=\xi}. \tag{2.6}$$

Toeplitz algebras are locally isomorphic to any such model for which the Poisson bracket is symplectic. The most usual is the differential model: V has a set of "conjugate" coordinates: x_k of degree 0, ξ_k of degree 1, the Poisson bracket is

$$\{f, g\} = \sum \partial_{\xi_k} f \partial_{x_k} g \; - \; \partial_{x_k} f \; \partial_{\xi_k} g$$

and the star product is given by the Leibniz rule:

$$f * g = \sum \frac{1}{\alpha!} \partial_\xi^\alpha f \; \partial_x^\alpha g.$$

We will rather use the more symmetric Weyl calculus: all coordinates are homogeneous of degree $\frac{1}{2}$, c is an antisymmetric bidifferential operator with constant coefficients so it is homogeneous of degree -1, and the star-product is given by formula (2.5) above. Here, and on several occasions below, we use the ξ_j coordinate functions (or other functions of fractional degree) as intermediate tools, but the elements of $\widehat{\mathcal{O}}$ are sum of monomials of integral degree.

For semi-classical star products the canonical local model is the following: X is a vector space (coordinates x_k of degree 0) equipped with a symplectic Poisson bracket

$c = \sum c_{jk}\partial_j \wedge \partial_k$. For the star-product there is an extra central formal variable h:

$$f * h(\xi) = \exp \frac{h}{2} c(\partial_\xi, \partial_\eta) f(\xi) g(\eta)|_{\eta=\xi} \tag{2.7}$$

(h is part of the data and we only consider isomorphisms preserving it).

It is known (and easy) that all Toeplitz algebras (resp. semi-classical star-algebras) are locally isomorphic. Global existence of a star-algebra over a symplectic manifold X, resp. symplectic cone Σ a star-algebra was proved in ([16, 8]).

3 Automorphisms and classification

If $\mathcal{A}$ is a star-algebra, an automorphism U of $\mathcal{A}$ is by definition a formal differential operator $U \in \widehat{\mathcal{D}}$ which preserves the star-product and symbols. Thus $U = 1$ mod. operators of negative degree so $\operatorname{Log} U$ is well defined; it is a derivation of degree ≤ -1. Also the fractional powers $U^s (s \in \mathbb{C})$ are well defined. For a semi-classical algebra $\mathcal{B}$ we require U to preserve h; the set of such automorphisms is denoted $\operatorname{Aut}_h \mathcal{B}$.

An inner automorphisms $U = \operatorname{Ad} P$, i.e. $Uf = PfP^{-1}$ with P of degree 0; inner automorphisms define a sheaf $\operatorname{Int} \mathcal{A} \sim \mathcal{A}_0^\times$ mod. its center, whose sections are locally inner automorphisms. (Inner automorphisms $\operatorname{Ad} P$ with P of integral degree $\neq 0$ also exist, but for these $(\operatorname{Ad} P)^s$ is not, even locally, an inner automorphism). If $U = \operatorname{Ad} P$ is an inner automorphism, $D = \operatorname{Log} U$ is locally an inner derivation $D = \operatorname{ad} p$ with $p = \operatorname{Log} P$, P of degree 0.

We recall some facts from[2].

3.1 Semi-classical algebras

Lemma 3.1. *Automorphisms of semi-classical algebras are locally inner automorphisms.*

If $\mathcal{B}$ is a semi-classical algebra the sheaf $\operatorname{Aut} \mathcal{B}$ of automorphisms is isomorphic to the sheaf of invertible sections of degree 0, mod. its center: $\mathcal{B}_0^\times / \mathbb{C}[[h]]^\times$.

We will rather use the sheaf $\widetilde{\mathcal{B}}_0^\times$ whose sections are pairs φ, f with $f \in \mathcal{B}_0$ invertible, $\varphi \in \mathcal{O}(0)$; $e^\varphi = \sigma(f)$ (equivalently: the sheaf of logarithms); thus we have an exact sequence

$$0 \to \mathbb{C}[[h]] \to \widetilde{\mathcal{B}}_0^\times \to \operatorname{Aut} \mathcal{B} \to 0. \tag{3.1}$$

Although $\widetilde{\mathcal{B}}_0^\times$ is not commutative, it has partitions of unity (it is a "soft" sheaf) so the canonical map $H^1(X, \operatorname{Aut} \mathcal{B}) \to H^2(X, \mathbb{C}[[h]])$ is an isomorphism:

Proposition 3.2. *The set h-$\operatorname{Alg}(X)$ of equivalence classes of semi-classical algebras over X is in 1-1 correspondence with $H^1(X, \operatorname{Aut} \mathcal{B}) \simeq H^2(X, \mathbb{C}[[h]])$.*

(More accurately there is a free transitive action of $H^2(X, \mathbb{C}[[h]])$.)

The automorphisms of a semi-classical algebra $\mathcal{B}$ preserve h (by definition). For further use in §5, it will be useful to note the following fact about automorphisms of $\mathcal{B}$ as a star-algebra: such an automorphism induces an automorphism of the center $\mathbb{C}((h))$, and its restriction is thus completely determined by $U(h) = \sum_1^\infty a_k h^k$ with $a_1 \neq 0$.

Lemma 3.3. *The restriction map* $\operatorname{Aut} \mathcal{B} \to \operatorname{Aut} \mathbb{C}[[h]]$ *is onto* (*there is an exact sequence* $0 \to \operatorname{Aut}_h \mathcal{B} \to \operatorname{Aut} \mathcal{B} \to \operatorname{Aut} \mathbb{C}[[h]] \to 0$).

Equivalently we have, for derivations of degree 0, an exact sequence:

$$0 \to \operatorname{Der}_h \mathcal{B} \to \operatorname{Der} \mathcal{B} \to \operatorname{Der} \mathbb{C}[[h]].$$

This is immediate for the local model, and hence also for global derivations since the sheaves Der of derivations of $\mathcal{B}$ are "soft" (they have partitions of unity, and $H^1(X, \operatorname{Der}_h(\mathcal{B})) = 0$).

3.2 Toeplitz algebras

Lemma 3.4. *Any automorphism of a Toeplitz algebra* $\mathcal{A}$ *is locally of the form* $U = (\operatorname{Ad} P)^s \operatorname{Ad} Q$ *where* P *is elliptic of degree* 1, Q *elliptic of degree* 0.

Definition 3.5. We denote $\underline{\omega}$ the sheaf (on $B\Sigma$) of differential 1-forms on Σ which are closed, homogeneous of degree 0. This is canonically isomorphic, via the Poisson bracket c_Σ, to the sheaf of symplectic vector fields on Σ, homogeneous of degree -1.

We denote $\mathcal{A}_-^\times \subset \mathcal{A}^\times$ the set of invertible elements of symbol 1.

If U is an automorphism, $D = \operatorname{Log} U$ is a well defined derivation of degree ≤ -1, whose leading term $\delta = \sigma(U)$ is a symplectic derivation, homogeneous of degree -1, and defines a section $\sigma(U) \in \underline{\omega}$ (its "symbol"). If $U = (\operatorname{Ad} P)^s \operatorname{Ad} Q$ we have $\sigma(U) = s\frac{dp}{p} + \frac{dq}{q}$ (with p, q the symbols of P, Q).

Then $\sigma(U) = 0$ means that there exists $Q \in \mathcal{A}_-^\times$ such that $U = \operatorname{Ad} Q$. Q is unique since the center of a Toeplitz algebra is $\mathbb{C}$.

Thus we have an exact sequence of sheafs

$$0 \to \mathcal{A}_-^\times \to \operatorname{Aut} \mathcal{A} \to \underline{\omega} \to 0. \tag{3.2}$$

Here again $\mathcal{A}_-^\times$, although not commutative, is soft so the canonical map

$$H^1(B\Sigma, \operatorname{Aut} \mathcal{A}) \to H^1(B\Sigma, \underline{\omega})$$

is one to one:

Proposition 3.6. *The set* T-Alg$(B\Sigma)$ *of equivalence classes of Toeplitz algebras over* $B\Sigma$ *is isomorphic to* $H^1(B\Sigma, \operatorname{Aut} \mathcal{A}) \simeq H^1(B\Sigma, \underline{\omega})$.

The classifying isomorphisms above depend on the choice of an initial star-algebra as base-point. In Section 4 we recall the analysis of Fedosov, which provides a canonical base-point.

3.3 The exponent character

Let Σ be a symplectic cone and $\mathcal{A}$ a Toeplitz algebra on Σ. Denote $Is_{\mathcal{A}}$ the groupoïd of isomorphisms of $\mathcal{A}$, over the groupoïd Sp_Σ of germs of homogeneous symplectic maps of Σ: an element of $\mathrm{Is}_{x' \leftarrow x}$ is an isomorphism $\mathcal{A}_x \to \mathcal{A}'_x$ over a germ of homogeneous symplectic map $\Sigma_x \to \Sigma_{x'}$. Obviously its Lie algebra is the sheaf $\mathcal{G}$ of derivations of degree 0 of $\mathcal{A}$, over the Lie algebra of homogeneous symplectic vector fields. This contains the algebra $\mathcal{G}_1$ of locally inner derivations, which is invariant hence locally integrable.

Choosing a base point $x \in \Sigma$ we get a homomorphism $\delta_{\mathcal{A}} : \pi_1(\Sigma, x) \to \mathrm{Der}\,\mathcal{A}_x / \mathrm{Int}\,\mathcal{A}_x = \mathbb{C}$, which to a loop γ assigns the exponent of U_1 if $U_t \in Is_{\gamma(t) \leftarrow x}$ is any smooth family of germs of isomorphisms such that $U^{-1}\frac{dU}{dt}$ is an inner automorphism, $U_0 = \mathrm{Id}$. (If $\mathcal{A}$ is involutive (see below in §3), we may choose $U(t)$ so that it respects the involution, which implies $\delta_{\mathcal{A}} = 0$, and in particular that the exponent of U_1 vanishes; so in general $\delta_{\mathcal{A}} \in H^1(B\Sigma, \mathbb{C})$ is the exponent part of the Fedosov class of $\mathcal{A}$ – see below §4.)

Proposition 3.7. *Let D be the infinitesimal generator of an action of the circle group $U(1)$ on $\mathcal{A}$ (with symbol H_a, the generator of the corresponding action on Σ). Then the exponent of D is $2\pi\delta_{\mathcal{A}}(\gamma)$ if γ is any orbit.*

In particular the exponent vanishes if the orbit homology class in $H_1(\Sigma, \mathbb{C})$ vanishes, e.g. if there is a fixed point (Σ connected), or if the circle action on the "sphere" $\{a = 1\}$ defines a principal bundle with compact basis and symplectic curvature (in the second case the orbit defines a torsion element of $\pi_1(B\Sigma)$ so its homology class in $H_1(B\Sigma, \mathbb{C})$ vanishes).

4 Subprincipal symbol and involutions

4.1 Subprincipal symbol

Let $\mathcal{A}$ be a Toeplitz resp. semi-classical algebra.

Definition 4.1. A subprincipal symbol on $\mathcal{A}$ is a differential operator of degree -1 (for all m, $a \in \mathcal{A}_m \mapsto \mathrm{sub}_m(a) \in \mathcal{O}(m-1)$) such that

(i) if $a \in \mathcal{A}_m$, $\mathrm{sub}_m(a) = \sigma_{m-1}(a)$ if $\sigma_m(a) = 0$; $\sigma_m(a) = \mathrm{sub}_m(a) = 0$ imply $a \in \mathcal{A}_{m-2}$,

(ii) if a, b are of degree m, m', we have

$$\mathrm{sub}(a * b) = \mathrm{sub}(a)\sigma(b) + \sigma(a)\,\mathrm{sub}(b) + \frac{1}{2}\{\sigma(a), \sigma(b)\} \in \mathcal{O}(m + m' - 1),$$

(iii) if a, b are of degree m, m', we have

$$\mathrm{sub}_{m+m'-2}[a, b] = \{\mathrm{sub}(a), \sigma(b)\} + \{\sigma(a), \mathrm{sub}(b)\} \in \mathcal{O}(m + m' - 2).$$

For example on the standard Weyl algebra (Toeplitz or semi-classical), the canonical subprincipal symbol of $a = \sum a_{m-k}$ is the second term a_{m-1}. If $a(x, D) = a_m(x, D) + a_{m-1}(x, D) + \cdots$ is a pseudodifferential operator on $\mathbb{R}^n$ ($D = \frac{1}{i}\frac{\partial}{\partial x}$) its subprincipal symbol is $a_{m-1} - \frac{1}{2i}\sum \frac{\partial^2 a_m}{\partial x_k \partial \xi_k}$.

Proposition 4.2. 1) *An automorphism U of $\mathcal{A}$ preserves a principal symbol iff it is of degree ≤ -2.*

2) *Two subprincipal symbols are conjugate.*

1) is obvious. Let us prove 2). Let $\mathcal{A}$ be a Toeplitz or semi-classical algebra equipped with a subprincipal symbol sub. In view of (i), (ii) any other subprincipal symbol is of the form $\mathrm{sub}'\,a = \mathrm{sub}\,a + \delta(a_m)$ with δ a differential operator, homogeneous of degree -1. (ii) implies $\delta(ab) = \delta(a_m)b_m + a_m\delta(b_m)$ i.e. δ is a derivation. Then (iii) implies that $\delta(\{a, b\} = \{\delta a, b\} + \{a, \delta b\}$ i.e. δ is symplectic. Then (3.2) δ is the leading term of a star derivation D and we have $\mathrm{sub}'(a) = \mathrm{sub}(e^D a)$. Thus two subprincipal symbols are locally conjugate – in fact globally because the sheaf of automorphisms which preserve a given subprincipal symbol is "soft".

Using partitions of unity it is easy to see that a map satisfying (i), (ii) exists on any semi-classical or Toeplitz algebra (on a real paracompact manifold). But such a map is not sufficient for global computations and does not deserve the name "subprincipal symbol". Subprincipal symbols, satisfying (iii), do not exist on all semi-classical or Toeplitz algebras; in particular they do not always exist on the canonical Toeplitz algebra constructed in [8] (cf. §3.3 below).

4.2 Involutions

An anti-involution on an algebra $\mathcal{A}$ is a linear operator J such that $J^2 = \mathrm{Id}$, $J(ab) = J(b)J(a)$.

On a star-algebra $\mathcal{A}$ we only consider anti-involutions which preserve symbols of degree 0. Since $J([a, b)] = -[Ja, Jb]$ this implies

$$\mathrm{gr}\, J(\sum f_n) = \sum (-1)^n f_n$$

the sublying geometric map is $\xi \mapsto -\xi$ on the cone $\Sigma_c = \Sigma \times_{R_+^\times} \mathbb{C}^\times$; the induced map on the basis $B\Sigma$ is still the identity map. For an anti-involution on a semi-classical algebra we further require $J(h) = -h$.

For example, on the model semi-classical algebra, there is a canonical involution: $a(x,h) \mapsto a(x,-h)$. On the algebra of differential operators on $\mathbb{R}^n$ there is a canonical involution $P(x,\partial_x) = \sum a_\alpha(x)\partial^\alpha \mapsto {}^t P = \sum(-1)^\alpha \partial^\alpha a_\alpha(x)$.

Anticipating on §4, let us note that on the Weyl model for Toeplitz algebras $\widehat{W}$ described below, the canonical involution is $a(x,\xi) \mapsto a(ix,i\xi)$ when using canonical local coordinates x_k, ξ_k of degree $\frac{1}{2}$ (recall that a is sum of functions of integral degree, i.e. of even degree with respect to x, ξ so the i factor does not really appear – otherwise the corresponding geometric map should be interpreted as a map on the double cover $\sqrt{(T\Sigma)_c}$).

Proposition 4.3. (i) *Any two anti-involutions on $\mathcal{A}$ are (globally) conjugate.*

(ii) *If $\mathcal{A}$ has an anti-automorphism, it also has an anti-involution.*

(iii) *An automorphism U which preserves an anti-involution J is of the form* Ad P *with $P \in \mathcal{A}_-^\times$, $JP = P^{-1}$; in particular its symbol is* 0.

Proof. (i) if J, J' are two anti-involutions, $U = J'J$ is an automorphism and $J' = UJ = JU^{-1}$, so $J' = U^{1/2}JU^{-1/2}$.

(ii) If A is an anti-automorphism, A^2 is an automorphism which commutes with A, so $J = (A^2)^{-1/2}A$ is an anti-involution.

(iii) Let J be an anti-involution. If U is an automorphism we set $U^J = JUJ$.

In the Toeplitz case, we obviously have $\sigma(U^J) = -\sigma(U)$. In particular if $U^J = U$, we have $\sigma(U) = 0$, so $U = \operatorname{Ad} P$ for some $P \in \mathcal{A}_-^\times$.

Now if P is elliptic we have $J(PJfP^{-1}) = (JP)^{-1}fJP$ i.e. $(\operatorname{Ad} P)^J = \operatorname{Ad}(JP^{-1})$, so $\operatorname{Ad} P$ commutes with J iff $JP = cP^{-1}$, c central. If $P \in \mathcal{A}_-^\times$ we have necessarily $c = 1$.

In the semi-classical case U is locally an inner automorphism $\operatorname{Ad} P$ with P of degree 0. The condition $U^J = U$ implies that $c = PJP$ is central i.e. an invertible power series $c(h)$; this is even because $Jc = JP\,P = c$. Replacing P by $P' = c^{-\frac{1}{2}}P$, we get $U = \operatorname{Ad} P'$ with again $P' \in \mathcal{A}_-^\times$, $JP' = P'^{-1}$.

A Toeplitz or semi-classical algebra with involution has a subprincipal symbol such that $\operatorname{sub}_m(a) = \frac{1}{2}\sigma_{m-1}(a - (-1)^m Ja)$. In fact our models have both an involution and a subprincipal symbol, and the group of automorphisms which preserve the involution is contained in the group of automorphisms which preserve sub.

On a symplectic cone there is (up to isomorphism) exactly one Toeplitz algebra possessing an anti-involution. As we will see below this is the "base-point" in Fedosov's construction, but it is not necessarily the "canonical" algebra constructed in [8] (§3.3).

For semi-classical algebras the situation is slightly more complicated. We will see below that the base-point $\mathcal{A}_0$ constructed by Fedosov is involutive. Then we can see that the set of equivalence classes of semi-classical algebras is isomorphic to $H^1(X, \operatorname{Aut}^J(\mathcal{A}_0)) \simeq H^2(X, \mathbb{C}^{\text{odd}}[[h]])$.

4.3 Example

If Σ is a real symplectic cone, we constructed in [8] a canonical algebra of operators acting on a suitable subspace $H \subset L^2(B\Sigma)$, whose symbol star-algebra $\mathcal{A}_0$ we have already referred to above.

For example let $X = B\Sigma$ be the boundary of a strictly pseudo-convex complex domain Ω, H is the space of boundary values of holomorphic functions on Ω, and $\mathcal{A}_0$ is a star completion of the algebra of differential operators on Ω. Now the algebra which obviously has an anti-involution (hence also a subprincipal symbol) is the algebra $\mathcal{A}_{\frac{1}{2}}$ of twisted differential operators acting on half-forms (half forms do not always exist, but the twisted algebra always does: it is constructed by twisting $\mathcal{A}_0$ by the cocycle $\mathrm{Ad}(\mathrm{Jacobian})^{\frac{1}{2}}$). The symbol of this twistor is an element of $H^1(X, \underline{\omega}_0) \subset H^1(X, \underline{\omega})$, where $\underline{\omega}_0 \subset \underline{\omega}$ is the sheaf of 1-forms on X (contained in the sheaf of closed 1-forms homogeneous of degree 0 on Σ). Its image in $H^2(X, \mathbb{C})$ is $\frac{1}{2}$ ch Ω (the Chern class of the sheaf of holomorphic differential forms on Ω).

In this case the "canonical" Toeplitz algebra has a subprincipal symbol iff this Chern class vanishes (ch $\Omega = 0$ on X).

Here is a more specific example: let Ω be the unit ball in a complex line bundle L over a complex manifold Y (since Ω is strictly pseudo-convex, Y is projective and L' is ample). Then the condition ch $\Omega_\Omega = 0$ on $X = \partial\Omega$ means that ch L is proportional to ch L on Y. This is always true if $\dim Y = 1$, but no longer necessarily true if $\dim Y \geq 2$ e.g. if Y is the abelian variety $\mathbb{C}^2/\mathbb{Z}^4$.

The Toeplitz algebra $\mathcal{A}_0$ constructed in [8] is not always involutive, but it acts on a Hilbert space – or more accurately on the scale of Hilbert spaces $\mathcal{O}^s$ of boundary values of holomorphic functions which lie in the Sobolev space H^s (more generally the image of H^s by the Toeplitz projector π_Σ).

It is natural to ask which Toeplitz algebras act similarily on a Hilbert space. We will require that this action be microlocally isomorphic to the action above (or microlocally isomorphic to the action of pseudo-differential operators on the scale of Sobolev spaces – in fact it necessarily is):

Let $\mathcal{A}$ be a Toeplitz algebra on Σ acting on a scale of Hilbert spaces as indicated: there is a covering Y_i of the basis Y and over each Y_i an isomorphism $U_i : H \to H_0$ respecting the scale. Then $U_{ij} = U_i U_j^{-1}$ is a Toeplitz operator of degree 0 (it must respect the scale); the difference cocycle for $\mathcal{A}_0$ is $u_{ij} = \mathrm{Ad}\, U_{ij}$; its "exponent" vanishes (there is no exotic component), and its image in $H^2(Y, \mathbb{C}^\times)$ also must vanish. We may conclude:

Proposition 4.4. *$\mathcal{A}$ acts on a scale of Hilbert spaces as above iff the Chern class $c(u) \in H^2(Y, \mathbb{C})$ of the difference cocycle $u = (u_{ij})$ defined by comparison with $\mathcal{A}_0$ is integral.*

Remark 4.5. This does not mean that the cohomology class of the Fedosov curvature defined below is integral, but that the difference with the class of $\mathcal{A}_0$ is integral; this is half-integral and usually not better.

Remark 4.6. If $\mathcal{A}$ is exotic, it may still happen that $c(u)$ is integral, i.e. the cocycle u_{ij} is a "coboundary" of the form $U_i U_j^{-1}$. Then $\mathcal{A}$ will act on a distribution space, locally isomorphic to $\bigcup H^s$. It does not preserve the scale, but it will still preserve a "Sobolev space" of variable order such as used by Višik and Eskin [31].

5 Fedosov connections

In [19], Fedosov gave a very elegant and geometric description for the existence and classification of semi-classical algebras: a semi-classical star algebra $\mathcal{B}$ over a real symplectic manifold X can be embedded as the subalgebra of flat sections of a suitable "Fedosov connection" ad ∇ in a "universal" star algebra $\widehat{W}$ over the tangent bundle TX (the definitions are recalled below); the curvature of ∇ is a closed central 2-form with leading term $h^{-1}\omega_X$, whose cohomology class determines $\mathcal{B}$ up to isomorphism; in particular this construction singles out a base-point (the algebra whose curvature is exactly $h^{-1}\omega_X$). This description can easily be adapted to Toeplitz algebras (more generally to Poisson cones of constant rank [4]). We will need this adapted description, and recall here how it works. In fact we give a direct proof of the embedding theorem, with complements which will be useful in §5.

5.1 Valuations and relative tangent algebra $\widehat{W}$

Let Σ be a Poisson cone. On the tangent bundle $T\Sigma$ each fiber $T_x\Sigma$ inherits of a Poisson bracket c_x with constant coefficients.

We will denote W the sheaf generated by homogeneous polynomials on $T\Sigma$: if we choose homogeneous local coordinates $x = (x_1, \dots, x_n)$ on Σ and $\xi = (\xi_1, \dots, \xi_n)$ denote the corresponding tangent coordinates, a section of W is locally a finite sum of homogeneous monomials $a_\alpha(x)\xi^\alpha$. The degree of a homogeneous monomial is

$$\deg a_\alpha \, \xi^\alpha = \deg a_\alpha + \sum \alpha_k \deg \xi_k. \tag{5.1}$$

The degree valuation is

$$p(f) = \inf(-\deg f_\alpha) \quad \text{if } f \text{ is sum of homogeneous monomials } f_\alpha. \tag{5.2}$$

We define the weight (or order) of a homogeneous monomial $a_\alpha(x)\xi^\alpha$ as

$$w(a_\alpha(x)\xi^\alpha) = -\deg a_\alpha + \frac{|\alpha|}{2}. \tag{5.3}$$

The order valuation is

$$w(f) = \inf w(f_\alpha) \quad \text{if } f \text{ is sum of homogeneous monomials } f_\alpha. \tag{5.4}$$

Thus W is equipped with two valuations $p \le w$.

The Poisson brackets c_x define fiberwise a Weyl product on W:

$$f * g\,(x,\xi) = \exp\frac{1}{2}c_x(\partial_\xi,\partial_\eta)\, f(x,\xi)\, g(x,\eta)\,|_{\eta=\xi}. \tag{5.5}$$

Since we will be dealing simultaneously with several product laws, we will adopt the following notations, to avoid confusions if need be:

$*$ or $*_W$, $*_A$ for the star-product, $\times$ (or no sign) for the usual product. (5.6)

Both p and w are valuations for this product, i.e.

$$p(f*g) \geq p(f) + p(g), \quad w(f*g) \geq w(f) + w(g). \tag{5.7}$$

The graded algebra $\mathrm{gr}_p\, W = \bigoplus_k (W_{p\geq k}/W_{p>k})$ is commutative (its product is the usual product since $p([f,g]) \geq p(f) + p(g) + 1$). The graded algebra $\mathrm{gr}_w\, W$ is not commutative; in fact the star-product $*$ is homogeneous for w, i.e. if f, g are sums of monomials of pure weight k resp. k', then $f * g$ and $[f,g]$ are sums of monomials of pure weight $k + k'$; the whole situation is analogous to the situation for the symbolic calculus of Hermite operators (cf. [7, 23]).

Definition 5.1. We denote $\widehat{W}_\Sigma$ (or $\widehat{W}$ if there is no confusion) the completion of W for the valuation w.

Note that $\widehat{W}$ is a set of formal *functions* on $T\Sigma$, equipped with a star-product.

$\widehat{W}$ is obviously functorial, i.e. if $u : \Sigma \to \Sigma'$ is a morphism of Poisson cones (a smooth homogeneous map $\Sigma \to \Sigma'$ compatible with the Poisson brackets, i.e. $u_* c_\Sigma = c_{\Sigma'}$ or $\{f\circ u, g\circ u\} = \{f,g\}\circ u$ for any smooth functions f, g on Σ') then $u^* : \widehat{W}_{\Sigma'} \to \widehat{W}_\Sigma$ ($f \mapsto f\circ Tu$) is a homomorphism of algebras, both for the star-product and the ordinary product.

In the semi-classical case (X a symplectic manifold), the relevant cone is $\Gamma = X \times \mathbb{R}_+^\times$ equipped with the Poisson bracket hc_X with $h = a^{-1}$, a the canonical coordinate of $\mathbb{R}_+^\times$. We have $T\Gamma = TX \times T\mathbb{R}_+^\times$; the center of $\widehat{W}_\Gamma$ is the commutative algebra, completion of $\mathbb{C}[a,\alpha]$ for the order valuation w, where (a,α) are the canonical coordinates of $T\mathbb{R}_+^\times$ ($a = h^{-1}$, α the corresponding tangent coordinate; $w(a) = -1$, $w(\alpha) = -\frac{1}{2}$).

For semi-classical algebras on X, we will rather use, as Fedosov, the completion $\widehat{W}_h = \widehat{W}_h(X)$ for the valuation w of the sheaf on X of algebras, whose sections are polynomials $f = \sum f_{k,\alpha}(x)h^k\xi^\alpha$; (the valuations p, w are defined as above, ($p(h) = 1$, functions on X are homogeneous of degree 0 ($p = 0$)). $\widehat{W}_h$ is canonically a quotient of the map $\widehat{W}_\Gamma$: $a \mapsto h^{-1}$, $\alpha \mapsto 0$.

5.2 Automorphisms and derivations of $\widehat{W}$

Let Σ be a symplectic cone (resp. X a symplectic manifold). In this section we examine derivations and automorphisms of $\widehat{W}_\Sigma$ (resp. $\widehat{W}_h$). We only consider automorphisms U such that $w(U - \mathrm{Id}) > 0$.

Any derivation or automorphism preserves the center of $\widehat{W}$ (i.e. $\widehat{\mathcal{O}}_\Sigma$, resp. $\widehat{\mathcal{O}}_X((h))$). We will say that D resp. U is an $\mathcal{O}$-derivation (resp. automorphism) if it fixes the center (i.e. D or $U - \mathrm{Id}$ vanishes on the center).

Proposition 5.2. *All $\mathcal{O}$-derivations, resp. automorphisms are inner derivations (resp. automorphisms). More precisely, if D is an $\mathcal{O}$-derivation, resp. U an $\mathcal{O}$-automorphism, there exists a unique section d resp. $u \in \widehat{W}$ such that $d = 0$, resp. $u = 1$ for $\xi = 0$, and $D = \operatorname{ad} d$ resp. $U = \operatorname{Ad} u$ (i.e. $Df = [d, f]$, $Uf = u * f * u^{-1}$).*

Proof. This is obvious on the local model (Σ or X a graded vector space), where all $\widehat{\mathcal{O}}$-derivations (resp. automorphisms) are inner, because $\operatorname{ad} d = \operatorname{ad}(d - d(x, 0))$ and $\operatorname{Ad} U = \operatorname{Ad} \frac{U}{U(x,0)}$. The uniqueness property just reflects the fact that the center is $\widehat{\mathcal{O}}_S$ resp. $\mathcal{O}_X((h))$. The global statement follows (uniqueness ensures that objects constructed on a covering patch together).

Note that if U is an $\widehat{\mathcal{O}}$-automorphism ($w(U - \mathrm{Id}) > 0$), $D = \operatorname{Log} U$ is well defined, $w(D) > 0$, so $D = \operatorname{ad} d$ with some d such that $w(d) > 0$ and $U = \operatorname{Ad} u$ with $u = e^d$; however $d(x, 0) = 0$ is not equivalent to $e^d(x, 0) = 1$ (the set of sections which vanish for $\xi = 0$ is not an ideal for $*$).

An immediate consequence of this proposition is that the sheaf (on $Y = B\Sigma$ or X) of $\widehat{\mathcal{O}}$-derivations or $\widehat{\mathcal{O}}$-automorphisms is "soft" (has partitions of unity).

5.3 Embeddings

Let $\mathcal{A}$ be a Toeplitz algebra over Σ (resp. $\mathcal{B}$ a semi-classical algebra over X). We will say that a homomorphism of algebras $\mathcal{A} \to \widehat{W}$ (resp. $\mathcal{B} \to \widehat{W}_h$) is a good embedding if it respects w ($w(u) \geq 0$ i.e. $w(uf) \geq p(f)$) and locally, in any set of homogeneous coordinates x_j we have $u(x_j) = x_j + \xi_j + r_j$ with $w(r_j) \geq p(x_j) + 1$ (in the semi-classical case we require $u(h) = h$). The second condition does not depend on the choice of homogeneous coordinates.

Theorem 5.3. 1) *For any Toeplitz algebra $\mathcal{A}$ (resp. semi-classical algebra $\mathcal{B}$) there exists a good embedding $\mathcal{A} \to \widehat{W}_\Sigma$ (resp. $\mathcal{B} \to \widehat{W}_h$).*

2) *If u_1, u_2 are two good embeddings there exists a unique $\widehat{\mathcal{O}}$-automorphism U such that $u_2 = U u_1$ (two good embeddings are conjugate).*

Proof. Locally the theorem is immediate: if (x_j) is a set of local homogeneous symplectic coordinates ($\{x_i, x_j\} = c_{ij} =$ constant), a model good embedding is $f(x) \mapsto uf = f(x + \xi)$. As mentioned it is often convenient to choose the x_j, ξ_j homogeneous of degree $\frac{1}{2}$; we may then use them only as intermediate tools since the monomials of WH must be homogeneous of integral degree.

If v is another good embedding, $v(x_j) = x_j + \eta_j$, we have $[\eta_i, \eta_j] = [\xi_i, \xi_j] = c_{ij}$ because the x_j are central in $\widehat{W}$, so by the universal property of the Weyl algebra there exists a unique algebra homomorphism U such that $U\xi_j = \eta_j$ ($Uf = f$ if f is central). Since $w(\eta_j - \xi_j) > w(\xi_j)$, U respects the valuation w and its unique continuous extension to $\widehat{W}$ is bijective.

This unique conjugacy statement is obviously also global.

In the general case there exists an open covering Σ_i of Σ (resp. X_i of X) and for each i a good embedding $u_i : \mathcal{A} \to \widehat{W}$ over Σ_i (resp. ...). For each pair (i, j) there is a unique $\widehat{\mathcal{O}}$-automorphism U_{ij} over $\Sigma_i \cap \Sigma_j$ such that $u_i = U_{ij}u_j$, and this is a cocycle ($U_{ij}U_{jk} = U_{ik}$ because $U_{ij}U_{jk}u_k = U_{ik}u_k = u_i$). Since the sheaf of $\widehat{\mathcal{O}}$-automorphisms (or derivations) is soft, (U_{ij}) is a coboundary: $U_iU_{ij} = U_j$ for a suitable family (U_i) so that the U_iu_i patch together to produce a global good embedding.

Remark 5.4. Another way to state this result is the following: the completion of $\mathcal{O} \otimes \mathcal{A}$ is isomorphic to $\widehat{W}$; the valuation used to define the completion is the unique valuation such that $w(f \otimes g) = p(f) + p(g)$, $w(1 \otimes f - f \otimes 1) = p(f) + \frac{1}{2}$ if $1 \otimes f - f \otimes 1 \neq 0$ i.e. f is not a constant.

Remark 5.5. In the above construction we did not require that the embedding u preserve the homogeneity valuation p. It is easy to show that there also exists a good embedding u which preserves p (see §4.5 below). When this is the case we have $\sigma(uf) = \sigma(f) \circ \bar{u}$ where $\bar{u}$ is a smooth homogeneous map $T\Sigma \to \Sigma$ compatible with the Poisson brackets, and tangent to the map $(x, \xi) \mapsto x + \xi$ along the zero-section of $T\Sigma$. It is still true that two such embeddings u_1, u_2 are conjugate: $u_2 = Uu_1$ with U an $\mathcal{O}$-automorphism ($w(U - \mathrm{Id}) > 0$), but U cannot preserve the homogeneity valuation p if the geometric maps $\bar{u}_1, \bar{u}_2$ are not equal.

5.4 Vector fields with coefficients in $\widehat{W}$

To describe the other derivations or automorphisms, it is convenient to choose a homogeneous symplectic connection ∇^s on Σ (resp. X). We can choose it homogeneous of degree 0 (with respect to homotheties) and torsionless, but any symplectic connection with coefficients in $\widehat{\mathcal{O}}_0 \otimes sp$ (degree ≤ 0) will do. ∇^s extends canonically to $\widehat{W}$, and if V is a vector field on Σ (homogeneous or with coefficients in $\widehat{\mathcal{O}}$), ∇^s_V is a derivation, both for the star-product and the usual product, (it respects both valuations p, w if V is homogeneous of degree 0).

We identify the Lie algebra of sections of $sp(T\Sigma)$ with the sub-Lie algebra W_2 of $\widehat{W}_\Sigma$ of homogeneous polynomials of degree 2 with respect to ξ: $\gamma = \sum \gamma_{jk}\xi_j\xi_k$ (acting by ad : $\mathrm{ad}\,\gamma f = [\gamma, f]$). Thus locally, if we choose homogeneous symplectic coordinates x_j ($\{x_i, x_j\} = c_{ij} = \mathrm{cst}$), we have

$$\nabla^s f(x, \xi) = \sum dx_j \left(\frac{\partial f}{\partial x_j} + \sum [\gamma_{ijk}\xi_j\xi_k, f] \right) \tag{5.8}$$

with $\gamma = \sum dx_i\, \gamma_{ijk}(x)\xi_j\xi_k$ homogeneous of degree 1 ($\mathrm{ad}\,\gamma$ of degree 0).

If D is a derivation of $\widehat{W}_\Sigma$, its restriction to the center is a derivation of $\widehat{\mathcal{O}}_\Sigma$ i.e. a vector field V on Σ with coefficients in $\widehat{\mathcal{O}}$, so we have $D = \nabla^s_V + \operatorname{ad} d$ for some sections $d \in \widehat{W}$.

It is useful to introduce, as Fedosov, "derivations with coefficients in $\widehat{W}$". These are well defined because the symplectic Lie algebra sp is identified with the subalgebra $W_2 \subset \widehat{W}$. They form a Lie algebra Vect_W which is an extension of the Lie algebra $\mathrm{Der}_{\widehat{W}}$ of derivations of $\widehat{W}$. Its sections can be written $\nabla^s_V + d$, where the rule for changes of coordinates for ∇^s_V is the rule for vectors with coefficients in the Lie algebra $sp(T\Sigma) = W_2 \subset \widehat{W}$. The bracket is

$$[\nabla^s_{V_1} + d_1\,,\ \nabla^s_{V_2} + d_2] = [\nabla^s_{V_1}, \nabla^s_{V_2}] + \nabla^s_{V_1}(d_2) - \nabla^s_{V_2}(d_1) + [d_1, d_2] \tag{5.9}$$

where the first bracket is given by the rule for connections with coefficients in $sp = W_2$:

$$[\nabla^s_{V_1}, \nabla^s_{V_2}] - \nabla^s_{[V_1, V_2]} = R^s(V_1 \wedge V_2) \in W_2. \tag{5.10}$$

If V is a derivation with coefficients in $\widehat{W}$ we denote $\operatorname{ad} V$ the derivation it defines. We have $\operatorname{ad} V = 0$ if and only if V is central (of the form $\nabla^s_0 + f$, $f \in \widehat{\mathcal{O}}$), and any derivation $D \in \operatorname{Der} \widehat{W}$ has a unique representative $D = \operatorname{ad}(\nabla^s_V + f)$ with $f(x, 0) = 0$.

5.5 Fedosov connections

Let $\widehat{\Omega}_\Sigma$ (resp $\widehat{\Omega}_X$) denote the exterior algebra of differential forms on Σ (resp. X) with coefficients in $\widehat{\mathcal{O}}$ (resp $\mathcal{O}((h))$).

As for derivations we define connections with coefficients in $\widehat{W}$: such a connection is of the form $\nabla = \nabla^s + \gamma$ with γ a 1-form with coefficients in $\widehat{W}$. It acts on forms with coefficients in $\widehat{W}$ by

$$\operatorname{ad} \nabla f = \nabla^s(f) + [\gamma, f].$$

The curvature is $R = \nabla^2$, identified with a 2-form with coefficients in $\widehat{W}$. We have $(\operatorname{ad} \nabla)^2 = \operatorname{ad} R$ so $\operatorname{ad} \nabla$ is flat (integrable) if and only if the curvature R is central, i.e. a 2-form with coefficients in $\widehat{\mathcal{O}}$ (independent of ξ). We will then say, as Fedosov, that ∇ is abelian.

Definition 5.6. 1) Let (x_j) be local coordinates on Σ, (ξ_j) the corresponding tangent coordinates. We denote $\xi_j^\vee$ the dual linear fiber coordinates on T_Σ, such that $[\xi_i^\vee, \xi_j] = \delta_{ij}$.

2) We denote τ the 1-form with coefficients in $\widehat{W}$, dual to the canonical form of Σ with coefficients in $T\Sigma$ (resp. TX):

$$\tau = \sum dx_j \xi_j^\vee \quad (\text{resp. } \tau = \frac{1}{h} \sum dx_j \xi_j^\vee). \tag{5.11}$$

τ is the unique 1-form with linear coefficients (in ξ) such that for any differential form f with coefficients in $\widehat{W}$:

$$[\tau, f] = \sum dx_j \frac{\partial f}{\partial \xi_j}. \tag{5.12}$$

We have $p(\tau) = -1$, $w(\tau) = -\frac{1}{2}$ (τ is homogeneous of degree 1 and its coefficients vanish of order 1 on the zero section). Also $\tau^2 = \frac{1}{2}[\tau, \tau]$ is the symplectic form:

$$\tau^2 = \omega_\Sigma \quad (\text{resp. } h^{-1}\omega_X). \tag{5.13}$$

Definition 5.7. A Fedosov connection is a connection ∇ with coefficients in $\widehat{W}$ of the form

$$\nabla = \nabla^s - \tau + \gamma \quad \text{with } w(\gamma) \geq 0.$$

Theorem 5.8. *Let $\mathcal{A}$ be a Toeplitz algebra on Σ (resp. $\mathcal{B}$ a semi-classical algebra on X). Then*

1) *For any good embedding $u : \mathcal{A} \to \widehat{W}$ (resp. ...) there exists a unique abelian Fedosov connection ∇ such that $\nabla u = 0$. This is unique up to a central form.*
2) *Conversely if ∇ is an abelian Fedosov connection,* $\ker \nabla \subset \widehat{W}$ *is a Toeplitz (resp. semi-classical) algebra.*

1) Locally, let us choose symplectic coordinates (e.g. homogeneous of degree $\frac{1}{2}$). The standard local embedding $u_0 : f(x) \mapsto u_0 f = f(x+\xi)$ is killed by $\nabla^0 = d_x - \tau$, and obviously this is the only Fedosov connection which kills it, with coefficients vanishing on the zero-section $\{\xi = 0\}$. If u is another good embedding, it is conjugate to u_0: $u = \operatorname{Ad} U^{-1} u_0$ so it is killed by the Fedosov connection $\nabla = \nabla_0 + \gamma$ with $\gamma = U^{-1}\nabla_0(U)$; we have $w(\gamma) > 0$ since $w(U - \mathrm{Id}) > 0$. Note that γ may not vanish for $\xi = 0$, but we can replace it by $\gamma - \gamma(x, 0)$.

2) The converse is immediate: if ∇ is an abelian Fedosov connection, the map $\tilde{f} \in \ker \operatorname{ad} \nabla \mapsto \tilde{f}(x, 0)$ is clearly one to one (for each $f \in \widehat{\mathcal{O}}$ there exists a unique $\tilde{f} \in \widehat{W}$ such $\nabla \tilde{f} = 0$, $\tilde{f}(x, 0) = f(x)$ because ∇ is "transversal" to the zero-section; further since ∇ is a Fedosov connection, we have locally $w(\tilde{f} - f(x+\xi) \geq w(f+1))$.

(The proofs for the semi-classical case are the same).

5.6 Fedosov curvature

Fedosov connections provide a one to one correspondence between Toeplitz or semi-classical algebras and 2-forms on Σ (resp. X):

Theorem 5.9 (Fedosov). 1) *Any closed 2-form $R = \omega_\Sigma + r$ on Σ with coefficients in $\mathcal{O}$ (resp. $\omega_X + r$)...) is the curvature of an abelian Fedosov connection.*

2) *Two abelian connections have the same curvature if and only if they are conjugate. They define isomorphic algebras if and only if their curvatures differ by an exact form.*

Proof. 1) For the sake of completeness, we repeat the proof of Fedosov's, by successive approximations, improving the weight w of the error: if ∇ is a Fedosov connection, the leading term (term of lowest weight w) of ad ∇ is

$$\operatorname{ad}\tau : f \mapsto \sum dx_j \frac{\partial f}{\partial \xi_j} \quad (\text{of weight } -\frac{1}{2}).$$

If $\nabla^2 = R + \rho_N$ with $w(\rho_N) = \frac{N}{2} \geq 0$, one looks for a 1-form γ_{N+1} such that $w(\nabla + \gamma_{N+1})^2 - R) \geq \frac{N+1}{2}$, i.e.

$$[\tau\,,\,\gamma_{N+1}] = \rho_N + (w \geq N). \tag{5.14}$$

Let I_ξ denote the interior product: $I_\xi f = \xi_j \partial_{x_j} \lrcorner f$ (the usual product, not the star-product of $\widehat{W}$). Then if $f = \sum dx_{i_1} \dots dx_{ik} f_\alpha(x)\, \xi^\alpha$ we have

$$I_\xi \operatorname{ad}\tau + \operatorname{ad}\tau I_\xi f = L_\xi f = \sum (|\alpha| + k) dx_{i_1} \dots dx_{ik} f_\alpha\, \xi^\alpha.$$

Since we have $w([\tau, \rho_N]) > \frac{N-1}{2}$, equation 5.14 has a solution (in fact it has a unique solution of pure weight $\frac{N+1}{2}$ killed by I_ξ).

By successive approximations we get an exact solution – in fact there is a unique and canonical solution γ such that $I_\xi \gamma = 0$ (which implies $[I_\xi, \nabla] = 0$).

Remark 5.10. The denomination "canonical" is abusive, because it depends in fact on the choice of a torsionless homogeneous symplectic connection ∇^s on Σ to begin with. In ambiguous cases we will refer to "the canonical Fedosov connection associated with ∇^s".

2) Obviously if $\nabla_1 = U \nabla U^{-1}$ and $R = \nabla^2$ is central, we have $\nabla_1^2 = U R U^{-1} = R$. The converse (i.e. the existence of U such that $\nabla_1 = U \nabla U^{-1}$ if ∇ and ∇_1 have the same central curvature) is proved by successive approximation as above.

Note that ker ad ∇ = ker ad ∇_1 means that $\nabla_1 = \nabla + \gamma$ with γ central, $w(\gamma) \geq 0$; this implies $R_1 = R + d\gamma$, hence the last assertion of 2).

Remark 5.11. The canonical solution is in fact of weight $p(\operatorname{ad}\nabla) \geq 0\,(p(\nabla) \geq -1)$, so that the corresponding embedding u is associated to a geometrical map $\underline{u} : T\Sigma \to \Sigma$, preserving Poisson brackets and tangent to the map $(x, \xi) \mapsto x + \xi$.

Remark 5.12. If $\mathcal{A}$ is a Toeplitz or semi-classical algebra, an embedding u defines a total symbol $\sigma_U(f) = uf(x, 0)$ which is an isomorphism $\mathcal{A} \to \widehat{\mathcal{O}}$ (locally in $\widehat{\mathcal{D}}_0^\times$), and a star-product B_u $(\sigma_u(f *_A g) = B_u(\sigma_u f, \sigma_u g))$. If we replace u by Uu, U an automorphism of $\widehat{W}$, σ_u is replaced by $P_{U,u}\sigma_u$ for some well defined asymptotic operator $P_{U,u} \in \widehat{\mathcal{D}}_0^\times$ $(P_{UV,u} = P_{U,Vu} P_{V,u})$, and B_u is replaced by $P_{U,u} B_u$ $((PB)(f, g) = P\,B(P^{-1}f, P^{-1}g))$.

The set of transition operators $P_{U,u}$ is a strict subset of $\widehat{\mathcal{D}}_0^\times$; for example we have $P_{U,u} = 1$ if U is even, i.e. $U = 1 + \sum u_\alpha(x)\xi^\alpha$, $\alpha > 0$ even). Likewise the set of all total symbols σ_u is a strict subset of the set $\widehat{\mathcal{D}}_0^\times \sigma_u$ of all possible total symbols equivalent to σ_u, and the set of all corresponding star-products B_u is a strict subset of the set of all star-products equivalent to a given one; for instance one always has $B_u(f, g) = fg + \frac{1}{2}\{g, g\} + \cdots$. I will not describe further here this special class of star-products. In fact I do not know if it has been distinguished before.

5.7 Base-point

Fedosov's construction gives a canonical base-point in the set of star-algebras, viz. the algebra corresponding to a connection with "trivial curvature" $R = \omega_\Sigma$ (resp. $R = h^{-1}\omega_X$).

Note that the Weyl algebra $\widehat{W}$ has a canonical involution (and a sub-principal symbol, fiberwise). The algebra $\mathcal{A}_\nabla$ is a sub-involutive algebra of $\widehat{W}$ if ∇ respects the involution. If this is the case the curvature R also respects the involution, i.e. it is odd (an odd power series $h^{-1}\omega_X + \sum h^{2k+1}\omega_k$ in the semi-classical case, $R \sim 0$ in the Toeplitz case). Conversely if R is odd, Fedosov's construction obviously yields a connection ∇ which respects the involution. Thus the base-point is involutive; in the Toeplitz case, it is the unique involutive algebra; in the semi-classical case we have seen that there are many other nontrivial involutive algebras.

For further use let us also note the following result: let G_h denote the group of symbol preserving automorphisms of $\mathbb{C}((h))$ (i.e. automorphisms U of $\mathbb{C}((h))$ with its standard commutative algebra structure, such that $w(U(h) - h) > 0$). G_h obviously acts on the set of semi-classical algebras on X (replacing h by $U(h)$).

Proposition 5.13. *If X is compact, the action of G_h on h-*Alg(X) *is free.*

On any symplectic manifold X, a symbol-preserving isomorphism $\tilde{U}$ preserves the center, hence induces an automorphism $U \in G_h$. Recall that the local model for semi-classical algebras is the algebra $\mathcal{B}_0$ of (jets along $\{\xi = 0\}$ of) pseudodifferential operators on $\mathbb{R}^n \times \mathbb{R}$ which commute with $h^{-1} = \frac{\partial}{\partial t}$ ($X = T^*\mathbb{R}^n$, $\xi = h\frac{\partial}{\partial x}$). Clearly any $U \in G_h$ lifts to Aut $\mathcal{B}_0$ (e.g. $\tilde{U}$ such that $\tilde{U}(x) = x$, $\tilde{U}(\frac{\partial}{\partial x}) = \frac{\partial}{\partial x}$, $\tilde{U}(t)$ such that $[U(\frac{\partial}{\partial t}), \tilde{U}(t)] = 1$).

The proposition states that if X is compact, an isomorphism which preserves symbols, but not necessarily h, in fact fixes h.

Proof. If $U \in G_h$, it acts on $\widehat{W}_h$. If R is the Fedosov curvature of a semi-classical algebra $\mathcal{B}$, the curvature of $U(\mathcal{B})$ is $U(R)$. Now the leading term of R is $h^{-1}\omega_X$. If $U \neq 0$ the leading term of $U - 1$ is $ch^k\frac{\partial}{\partial h}$ for some integer $k \geq 2$ and constant $c \neq 0$, so that the leading term of $U(R) - R$ is $-ch^{k-2}\omega_X$; this is $\neq 0$ if X is compact (ω_X is not a coboundary).

Therefore if $U \in G_h$ and $U\mathcal{B}$ is isomorphic to $\mathcal{B}$, then $U(h) = h$ i.e. $U =$ Id. Equivalently if $\tilde{U}$ is a symbol-preserving isomorphism of $\mathcal{B}$, it fixes the center ($U(h) = h$).

6 Related Toeplitz and semi-classical algebras

6.1 Homomorphisms between star algebras

If $\mathcal{A}$, $\mathcal{A}'$ are two star-algebras over cones Σ, Σ' there is an obvious notion of homomorphism $U : \mathcal{A} \to \mathcal{A}'$ (preserving the filtrations); the symbol map is $f \mapsto u^*f = f \circ u$ where u is a smooth homogeneous map $\Sigma' \to \Sigma$ preserving the Poisson brackets, i.e. $u^*\{f, g\} = \{u^*f, u^*g\}$.

For instance the algebra of semi-classical pseudo-differential $P(x, h\partial, h)$ on a manifold V is isomorphic to the algebra of germs of pseudo-differential operators $P(x, t, \partial_x, \partial_t)$ on $X \times \mathbb{R}$ which commute with ∂_t (i.e. do not depend on t), near the line-bundle $\{\xi = 0\}$: the map takes h to ∂_t^{-1}. It is immediate that any "1-codimensional" embedding of a semi-classical algebra to a Toeplitz is locally isomorphic to the embedding above.

Using this example it is easy to embed a semi-classical algebra $\mathcal{A}_X$ over a symplectic manifold X in a Toeplitz algebra $\mathcal{A}_\Sigma$ where Σ is a disjoint union of pieces as above. However without more information on the projection $B\Sigma \to X$ this only gives very poor information: to reconstruct $\mathcal{A}_X$ from $\mathcal{A}_\Sigma$ one would need to know how various components patch together, which requires further non trivial information.

Definition 6.1. We will say that a semi-classical algebra $\mathcal{B}$ on X is *related* to $\mathcal{A}$ if there exists an injective homomorphism $\mathcal{B} \to \mathcal{A}$, where the corresponding projection $Y \to X$ is a principal circle bundle (this definition will be slightly refined below).

A typical example is as in the introduction: Σ is a holomorphic line bundle over a projective manifold X, Y is the unit sphere bundle for a strictly pseudo-convex hermitian norm a.[1]

Let $u : \mathcal{B} \to \mathcal{A}$ be a relating homomorphism, $\bar{u} : U \to X$ (or $\Sigma \to \Gamma = X \times \mathbb{R}_+$) the corresponding map. The symbol of the operator $A \in \mathcal{A}$ corresponding to h^{-1} is $a = h^{-1}$. The integral curves of the hamiltonian field H_a are the fibers, which are closed circles, and since a is homogeneous of degree 1 its period is locally constant, constant if X is connected.

[1]The definition of homotheties must be modified to make $\omega_\Sigma = \frac{i}{2}\partial\bar{\partial}a$ homogeneous of degree 1, i.e. $t.z = \sqrt{t}z$ for $t > 0$.

Proposition 6.2. *If $\mathcal{A}$ and $\mathcal{B}$ are related, the symplectic cone Σ is equipped with a homogeneous symplectic action of the circle group $U(1)$.*

Replacing h by $h\times$constant, we may as well suppose (and will do so in the rest of this paragraph) that the infinitesimal generator is H_a, with period 2π.

We determine below which pairs $\mathcal{A}$, $\mathcal{B}$ of Toeplitz and semi-classical algebras can thus be related, but first investigate how the circle group or more generally a compact group acts on a Toeplitz (or semi-classical) algebra.

6.2 Action of a compact group

Let $\mathcal{A}$ be a Toeplitz algebra on Σ (resp. an h-algebra on X) as above.

Theorem 6.3. *Let G be a compact group acting on Σ by symplectic homogeneous isomorphisms. We suppose $g^*\mathcal{A} \sim \mathcal{A}$ for all $g \in G$, i.e. $g^*R_A - R_A$ is exact, with R_A the Fedosov curvature of $\mathcal{A}$ (this is always true if G is connected). Then:*

1. *The action of G lifts to $\mathcal{A}$.*
2. *Any two liftings are conjugate through an automorphism of $\mathcal{A}$.*

This follows from the fact that $\operatorname{Aut}\mathcal{A}$ is a complete filtered group, and $\operatorname{gr}\mathcal{A}$ is a G-vector space, so since G is compact any continuous cocycle is a coboundary. The first assertion is also seen using Fedosov connections:

1. If G acts on Σ (resp. X), its action extends functorially to $\widehat{W}$ resp. $\widehat{W}_h$. By hypothesis the cohomology class of $\mathcal{A}$ is invariant, so it has an invariant representative R. Then Fedosov's construction yields an invariant connection ∇ (we first choose ∇^s invariant so the starting point $\nabla^s - \tau$ is invariant), and G acts on $\ker\nabla \sim \mathcal{A}$.

2. Likewise if $U_0(g)$, denoted below $f \mapsto gf \in \mathcal{A}$, and $U_1(g) = U_g$ are two liftings, we set

$$u_g = U_g\, g^{-1} \in \operatorname{Aut}\mathcal{A} \quad \text{so that we have} \quad u_{gh} = u_g\, g u_h g^{-1}.$$

This means that $\sigma(u_g)$ is a 1-G-cocycle with coefficients in $\operatorname{gr}\operatorname{Aut}\mathcal{A}$ (i.e. we have $\sigma(g\cdot u_h) - \sigma(u_{gh}) + \sigma(u_g) = 0$), so it is of the form $\sigma(u_g) = g.\sigma(v) - \sigma(v)$ (the cohomology vanishes in positive degree since G is compact). By successive approximations we get $v \in \operatorname{Aut}\mathcal{A}$ such that $u_g = v -^{-1} gvg^{-1}$, i.e. $U_g = v^{-1}gV$.

6.3 Circle action

We suppose now that G is the circle group $G = U(1)$. The action of G on Σ has an infinitesimal generator θ, which is a symplectic vector field homogeneous of degree 0, and a generating function a which is a homogeneous function of degree 1:

$$\theta = H_a, \quad \text{with } a = I_\theta(\lambda_\Sigma) \tag{6.1}$$

where $I_\theta = \sum \theta_j L_{\frac{\partial}{\partial x_j}}$ denotes the interior product by θ, λ_Σ the Liouville form of Σ.

We will also denote

$$L_\theta \text{ the vector field with coefficients in } \widehat{W} \text{ lifting } \theta. \tag{6.2}$$

Let $\mathcal{A}$ be a Toeplitz algebra on Σ. As noted above (theorem 6.3) the circle group action lifts to $\mathcal{A}$. The infinitesimal generator D of this action is well defined; it is unique up to conjugation by an automorphism of $\mathcal{A}$,[2] and there exists an equivariant embedding $u : \mathcal{A} \to \widehat{W}$, e.g. we choose an invariant representative of the curvature R: the corresponding canonical Fedosov connection ∇ is then invariant:

$$[L_\theta, \nabla] = 0. \tag{6.3}$$

We denote $\widetilde{\mathcal{A}} \subset \widehat{W}$ the image of $\mathcal{A}$, and set

$$\nabla_\theta = [I_\theta, \nabla], \qquad R_\theta = [I_\theta, R] \tag{6.4}$$

here and everywhere else $[,]$ is the superbracket: $[f, g] = (-1)^{\bar{f}\bar{g}}(f * g - g * f)$, where $\bar{f}, \bar{g}$ denotes the degree of f resp. g as differential forms. ∇_θ is a vector field with coefficients in $\widehat{W}$, and we have

$$L_\theta = \nabla_\theta + \alpha \quad (\alpha \in \widehat{W}). \tag{6.5}$$

We have $\alpha(x, 0) = 0$ if ∇ "vanishes" on the zero section, in particular if ∇ is the canonical connection associated to R.

We have $R_\theta = [I_\theta, \nabla^2] = [[I_\theta, \nabla], \nabla]$. Since $[I_\theta, \nabla] = \nabla_\theta = L_\theta - b$ and ∇ is invariant, i.e. $[L_\theta, \nabla] = 0$, we have

$$R_\theta = [L_\theta - \alpha, \nabla] = \nabla(\alpha). \tag{6.6}$$

Lemma 6.4. *Notations being as above, the infinitesimal generator D of the action of $U(1)$ on $\mathcal{A}$ is an inner derivation if and only if R_θ is exact (as a form on Σ with coefficients in $\widehat{\mathcal{O}}$).*

Proof. ad L_θ coincides with ad α on $\widetilde{\mathcal{A}}$. In other words the $\widehat{\mathcal{O}}$-derivation of $\widehat{W}$ corresponding to D is ad α. If D is an inner derivation there exists $b \in \widetilde{\mathcal{A}}$ ($\nabla b = 0$) such that $\alpha - b$ is central, so $R_\theta = \nabla\alpha = d(\alpha - b)$ is exact.

[2] Here is an alternate proof of this: we first choose a derivation D_0 of degree 0 in $\mathcal{A}$ such that $\sigma(D_0) = \partial_\theta$, for instance $D_0 = \operatorname{ad} A^0$ where $A^0 \in \mathcal{A}$ is any element with symbol a. Then for $t \in \mathbb{R}$, $\exp tD_0$ is a well defined group of isomorphisms above $\exp t\theta$. In particular $e^{2\pi D_0}$ is an automorphism of $\mathcal{A}$ (over $\exp 2\pi\theta = \mathrm{Id}$): it is of the form $\exp 2\pi\delta$ with δ a derivation of degree -1 which commutes with D_0. Then $D = D_0 - \delta$ is the infinitesimal generator of an action of $U(1)$, lifting the action on Σ ($U_t = \exp tD$).

If $D_1 = D + \delta_1$ with δ_1 a derivation of degree -1), we see by successive approximations that D_1 is conjugate to $D + \delta_2$ where δ_2 commutes with D (if $U = 1 + u$ with u of degree $-N$, $U^{-1}D_1U = [D, u] + v$ with v of degree $-N - 1$, and if the Fourier series of u is $u = \sum u_k$, we have $[D, u] = \sum iku_k$). Then if $\exp 2\pi D_1 = \mathrm{Id}$ we have $\exp 2\pi\delta_2 = \mathrm{Id}$ i.e. $\delta/2 = 0$ since δ_2 is of degree < 0.

6.4 Elliptic circle action

Definition 6.5. An action on Σ of the circle group $U(1)$ is elliptic if its generating function a is > 0 (there is an obviously symmetric case $a < 0$).

From now on we suppose that Σ is equipped with a free elliptic action of $U(1)$, with generating function $a > 0$ ($\theta = H_a, a \in \mathcal{O}_\Sigma(1)$). We will use a as privileged radial coordinate.

The basis of Σ is identified with the"unit sphere"

$$Y = \{a = 1\} \subset \Sigma \sim \Sigma/\mathbb{R}_+^\times. \tag{6.7}$$

Y is a principal $U(1)$-bundle, with basis X and connection form λ_Y:

$$X = Y/U(1)\,, \quad \lambda_Y = \lambda_\Sigma|Y, \tag{6.8}$$

where as above λ_Σ denotes the Liouville form λ_Σ of Σ; we will also denote $\lambda_Y = \frac{\lambda_\Sigma}{a}$ the pull-back to Σ.

The basis $X = Y/U(1)$ is a symplectic manifold with symplectic form

$$\omega_X \quad \text{such that } p^*(\omega_X) = d\lambda_Y. \tag{6.9}$$

The $U(1)$ action lifts canonically to $\widehat{W}_\Sigma$, and as above L_θ denotes the vector field with coefficients in $\widehat{W}_\Sigma$ defined by the infinitesimal generator θ.

Let $\Gamma = \Sigma/U(1)$. We have $T\Gamma = T\Sigma/TU(1)$, where the tangent generator is (obviously) the $\sum \theta_j \frac{\partial}{\partial \xi_j}$; it corresponds to a vector field τ_θ with coefficients in $\widehat{W}$:

$$\tau_\theta = I_\theta \cdot \tau, \tag{6.10}$$

where $I_\theta = \sum \theta_j I(\frac{\partial}{\partial x_j})$ is the extension to $\Omega \otimes \widehat{W}$ of the interior product by θ, and τ is the canonical 1-form of $\widehat{W}$ (def. 5.6).

Lemma 6.6. 1) *$\widehat{W}_\Gamma$ is identified with the subalgebra of $\widehat{W}_\Sigma$ of sections invariant by $TU(1)$, i.e. killed both by L_θ and* ad τ_θ.

2) *The Weyl algebra $\widehat{W}_h(X)$ is identified with the quotient $\widehat{W}_\Gamma$ / ($\tau_\theta = 0$)*

(If $U(1)$ acts by translations $(x_1, \dots, x_n) \mapsto (x_1 + u, x_2, \dots, x_n)$, which is always true in a suitable set of local coordinates, then $TU(1)$ acts by $(x, \xi) \mapsto (x_1 + u, \dots, x_n, \xi_1 + v, \dots, \xi_n)$).

Let now $\mathcal{A}$ be a Toeplitz algebra on Σ, equipped with an extension of the action of $U(1)$ (we have seen this exists, and is unique up to conjugation).

Up to equivalence, $\mathcal{A}$ is determined by the cohomology class of its Fedosov curvature R. We may choose R invariant, so as the Fedosov connection (if we start with an invariant torsionless symplectic connection, the canonical construction of §4.6 produces an invariant connection). There is a corresponding equivariant embedding, which we will adjust further suitably below.

The curvature is $R = \omega_\Sigma + r$, with r closed, of weight $w(r) \geq 0$. As any closed 2-form on Σ, r is cohomologous to an invariant form homogeneous of degree 0, so we may suppose:

$$r = \mu_X + \lambda_Y \nu_X + (\gamma_X + c\lambda_Y)\frac{da}{a} \tag{6.11}$$

where μ_X, ν_X, γ_X are the pull-backs of 2 and 1-forms on X.

Since $dr = 0$, c is a constant, and $d\gamma_X + c\omega_X = 0$, $d\nu_X = 0$, $d\mu_X + \omega_X \nu_X = 0$.

We necessarily have $c = 0$ if ω_Σ is not exact (on any component of X); this is always the case if X is compact.

Since $I_\theta \omega_\Sigma = -da$ we have, with the notations above:

$$R_\theta = I_\theta R = -da + \nu_X + c\frac{da}{a}. \tag{6.12}$$

We have set $L_\theta = \nabla_\theta + \alpha$. Since $\nabla_\theta = [I_\theta, \nabla$ and $[L\theta, \nabla] = 0$ we get $R_\theta = [I_\theta, \nabla^2] = [\nabla_\theta, \nabla] = -[\alpha, \nabla$, i.e.

$$R_\theta = \nabla(\alpha). \tag{6.13}$$

Let us further note that the leading term of α is $\tau_\theta = I_\theta(\tau)$.

Lemma 6.7. *There exists an invertible section U of $\widehat{W}$ such that $U\alpha U^{-1} = \tau_\theta$.*

This is immediate by successive approximations as above: if $w(\alpha_n) \geq \frac{n}{2}$ there exists u such that $w(u) \geq \frac{n+1}{2}$, $[\tau_\theta, u] = \sum \theta_j \frac{\partial}{\partial \xi_j} = \alpha_n$ (there is a unique solution such that $u(, \xi) = 0$ if ξ belongs to the hyperplane orthogonal to the Liouville form λ, which is transversal to ad τ_θ). Then $(1+u)^{-1}\tau_\theta(1+u) = \tau_\theta + \alpha_n + (w \geq \frac{n+1}{2})$.

Let $\mathcal{B} = \mathcal{A}^{U(1)} \subset \mathcal{A}$ be the invariant subalgebra. The image $\widetilde{\mathcal{B}} \subset \widetilde{\mathcal{A}}$ is

$$\widetilde{\mathcal{B}} = \ker L_\theta \cap \widetilde{\mathcal{A}} = \ker \nabla \cap \ker \operatorname{ad} \alpha. \tag{6.14}$$

It is also contained in $\widehat{W}_\Gamma$ if $\alpha = \tau_\theta$.

Lemma 6.8. *$\mathcal{B}$ possesses a non-trivial central element iff the component ν_X of the Fedosov curvature of $\mathcal{A}$ vanishes. Then $\mathcal{B}$ possesses a structure of related semi-classical algebra.*

Proof. Let $B \neq 0$ be a central element of $\mathcal{B}$, of degree $k \neq 0$. The symbol of B is $c\,a^k$ for some constant $c \neq 0$, and replacing B by $(\frac{B}{c})^{\frac{1}{k}} \in \mathcal{B}$ we may suppose $\sigma(B) = a = h^{-1}$. Then the infinitesimal generator is necessarily of the form

$$D = \operatorname{ad}(B'), \quad \text{with} \quad B' = B + c \operatorname{Log} B + \sum_1^\infty c_k B^{-k}.$$

With the notations above, $\alpha' = \alpha - \tilde{B}'$ is central, and $T_\theta = \nabla\alpha = d\alpha'$. The only term of B' not in $\mathcal{B}$ is $c \operatorname{Log} B$ with differential $c\frac{da}{a}$+exact, so $R_\theta - c\frac{da}{a}$ is exact, i.e. $\nu_X \sim 0$ and c is the coefficient of $\frac{da}{a}$ which appears in (6.11), (6.12).

6.5 Related Toeplitz and semi-classical algebras

We may now finish the analysis of related Toeplitz and semi-classical algebras. We will suppose the basis X compact (or more generally that ω_X is not exact, on each component of X). We leave as an exercise the case where ω_X is exact so the constant c above can be $\neq 0$.

We refine Definition 6.1 as follows:

Definition 6.9. We will say that a relating homomorphism $\mathcal{B} \to \mathcal{A}$ is good (or that we have a good relation) if the operator A corresponding to h^{-1} is an infinitesimal generator of $U(1)$

Theorem 6.10. *We suppose that X is compact.*

1) *Let $\mathcal{A}$ be a Toeplitz algebra over Σ with elliptic action of $U(1)$ as above, and Fedosov curvature R_A, and let $\mathcal{B}$ be the invariant subalgebra. Then $\mathcal{B}$ possesses a structure of semi-classical algebra if and only if $\nu_X = I_\theta R \sim 0$. In this case the infinitesimal generator of the action of $U(1)$ on $\mathcal{A}$ is an inner derivation* ad A *and there exists a unique semiclassical structure on $\mathcal{B}$ such that $h = f(A) = A^{-1} + \cdots$ for any formal series $f(T) = T^{-1} + \sum_{k>1} f_k T^{-k}$. It is well-related to $\mathcal{A}$ iff $h^{-1} = A + c$, $c \in \mathbb{C}$ a constant.*

2) *Let $\mathcal{B}$ be a semiclassical algebra over $X = Y/U(1)$ as above. Then $\mathcal{B}$ has a related Toeplitz algebra if and only if its Fedosov curvature R_B is constant mod. ω_X, i.e. of the form $R_B = h^{-1}\omega_X +$ constant $+ \varphi(h)\omega_X$. It is well-related $\mathcal{A}$ iff $R_B = h^{-1}\omega_X + \mu$, μ a constant 2-form on X (independent of h).*

The Fedosov curvature of $\mathcal{A}$ is then $R_A = p^* R_B + \frac{da}{a} p^*\gamma_X$ for some $\gamma_X \in H^1(X)$. In particular there is a unique (up to isomorphism) "non exotic" such $\mathcal{A}$ ($\gamma_X = 0$).

Proof. Let $\mathcal{A}$ be a Toeplitz algebra on Σ, and choose R, ∇ as above, so $L_\theta = \nabla_t h + \alpha$, $\alpha = \tau_\theta$. Since X is compact we have $c = 0$ anyway. If $\mathcal{A}$ contains a semi-classical algebra, we have $\nu_X \sim 0$ so we may as well suppose $\nu_X = 0$, $R = \omega_\Sigma + \mu_X + \gamma_X\frac{da}{a}$.

This being so we have $R_\theta = \nabla\alpha = -da$ so $\nabla(a + \alpha) = 0$. The infinitesimal generator is ad A where A is the element with total symbol a ($\tilde{A} = a + \alpha$). Moreover with this choice $\tilde{A}$ is a polynomial of degree 1 with respect to ξ so $\tilde{A} * f = \tilde{A}f + \frac{1}{2}\{\tilde{A}, f\}$. It follows in particular that $\tilde{A} * f = \tilde{A}f$ (or $a * f = af$ in $\mathcal{A}$ for the star product and total symbol defined by this choice of embedding) if A commutes with f, and we get a well related semiclassical algebra by setting $h = A^{-1}$.

If V is a vector field on X, we denote $\tilde{V}$ the unique vector field on Σ which projects on V, such that $I_\theta\tilde{V} = I_\rho\tilde{V} = 0$, with ρ the generator of homotheties (i.e.

$\tilde{V}$ is orthogonal to the conic rays and fiber circles and projects on V – in particular it is homogeneous of degree 0 and rotation invariant). With our choice of embedding ($\tilde{D} = \operatorname{ad}\alpha, \alpha = \tau_\theta$), $\nabla_{\tilde{V}}$ obviously preserves $\Omega_\Gamma = \ker L_\theta \cap \ker \operatorname{ad}\alpha$, and kills a and α so goes down to Ω_h. The Fedosov connection ∇^B of $\mathcal{B}$ is then obtained by restriction:

$$\nabla^B_V = \text{ the derivation of } \Omega_h \text{ defined by } \nabla_{\tilde{V}}|_{\Omega_h}. \tag{6.15}$$

The curvature R^B is the induced by R: $R^B = \mu_X$. Note that it does not depend on γ_X, and changing γ_X, which is equivalent to twisting $\mathcal{A}$ by a cocycle $A^{s_{ij}}$, (s_{ij}) a cocycle on X with coefficients in $\mathbb{C}$, does not change $\mathcal{B}$, the commutator of A. Among the Toeplitz algebras related to $\mathcal{B}$ the only one which is not "exotic" has curvature μ_X.

Any other related semi-classical structure is obtained by replacing h by $\varphi(h)$ for some formal series $\varphi(h) = h + \sum_2^\infty \varphi_k h^k$; as seen in Proposition 5.13, the corresponding semi-classical algebras are pairwise non-isomorphic. Among them, are well-related to $\mathcal{A}$ those for which $\varphi(h) = \frac{h}{1+\gamma h}$ $(\varphi(h)^{-1} = h^{-1} + \gamma)$.

Remark 6.11. If ν_X is not cohomologous to 0, the invariant subalgebra $\widetilde{\mathcal{B}}$ is not a semiclassical algebra, because its center is not big enough. However its lifting to the universal cover $\widetilde{X}$ is; $\widetilde{\mathcal{B}}$ itself is obtained by gluing together local models of semi-classical algebras with isomorphisms which preserve symbols, but do not fix h $(h \mapsto U(h) = h + \sum_{k\geq 2} u_k h^k)$.

Example 6.12. In the canonical model we have $\Sigma = \mathbb{C}^n - \{0\}$, $a = |z|^2$. The symplectic form is $\omega_\Sigma = i\partial\bar{\partial}a = i\sum dz_j d\bar{z}j$ (twice the usual one – the coordinates $z_j, \bar{z}_j$ are homogeneous of degree $\frac{1}{2}$ so a is of degree 1); we have $\{\bar{z}_p, z_q\} = i\delta_{pq}$ (the Poisson bracket is $c = i\sum \partial_{\bar{z}_j} \wedge \partial_{z_j}$. The infinitesimal generator of homotheties is $\rho = \frac{1}{2}\sum z_j \frac{\partial}{\partial \bar{z}_j} + z_j \frac{\partial}{\partial \bar{z}_j}$. Rotations are the usual ones: $z \mapsto e^{it}z$, ith infinitesimal generator $\theta = i\sum z_j \frac{\partial}{\partial \bar{z}_j} - z_j \frac{\partial}{\partial \bar{z}_j} = H_a$.

$\mathcal{A}$ is the Weyl algebra, with product $f * g = \exp \frac{1}{2}c(\partial_x, \partial_y) f(x)g(y)|_{y=x}$. The action of $U(1)$ lifts to $\mathcal{A}$, with generator $\operatorname{ad} a$. The canonical embedding $\mathcal{A} \to \widehat{W}$ is $f \mapsto f(x+\xi)$, the corresponding connection is $\nabla = d - \tau$, $\tau = i(dz_j\bar{\zeta}_j - d\bar{z}_j\zeta_j)$.

The invariant subalgebra $\mathcal{B}$ is a semiclassical algebra over $X = P_{n-1}(\mathbb{C})$ with $h = a^{-1}$.

Note that in this case, with the notations above, we have $a + \alpha = |z + \zeta|^2 \neq \tau_\theta$ (we still have to modify ∇ to get the connection used above).

References

[1] Bayen, F., Flato, M., Frønsdal, C., Lichnerowicz, A., Sternheimer, D., Deformation theory and quantization I, II, *Ann. Phys* 111 (1977), 61–131.

[2] Boutet de Monvel, L., Complex star algebras, *Math. Phys. Anal. Geom.* 2 (1999), 113–139.

[3] Boutet de Monvel, L., Symplectic cones and Toeplitz operators, *Contemp. Math.* 205 (1997) 15–24.

[4] Boutet de Monvel, L., Star products on conic Poisson manifolds of constant rank, *Mat. Fiz. Anal. Geom.* 2 (1995), no. 2, 143–151.

[5] Boutet de Monvel, L., Opérateurs pseudodifférentiels à bicaractéristiques périodiques, *Séminaire Bony-Meyer-Sjöstrand*, Ecole Polytechnique, 1984, exposé n°20.

[6] Boutet de Monvel, L., On the index of Toeplitz operators of several complex variables, *Invent. Math.* 50 (1978/79), no. 3, 249–272.

[7] Boutet de Monvel, L., Hypoelliptic operators with double characteristics and related pseudodifferential operators. *Comm. Pure Appl. Math.* 27 (1974), 585–639.

[8] Boutet de Monvel, L., Guillemin, V., *The Spectral Theory of Toeplitz Operators*, Ann. of Math. Stud. 99, Princeton University Press, Princeton, NJ, 1981.

[9] Boutet de Monvel, L., Sjöstrand, J., Sur la singularité des noyaux de Bergman et de Szegö, *Astérisque* 34–35, Soc. Math. France, Paris 1976.

[10] Bressler, P., Nest, R., Tsygan, B. A Riemann-Roch type formula for the microlocal Euler class, *Internat. Math. Res. Notices* 20 (1997), 1033–1044.

[11] Cahen, M., Gutt, S., Rawnsley, J., Quantization of Kähler manifolds II, *Trans. Amer. Math. Soc.* 337 (1993), 73–98.

[12] Cahen, M., Gutt, S., Rawnsley, J., Quantization of Kähler manifolds. I. Geometric interpretation of Berezin's quantization. *J. Geom. Phys.* 7 (1990), no. 1, 45–62.

[13] Charles, L., Quantification des observables sur les variétés Kähleriennes, Thèse Université Paris-Dauphine, Dec. 2000.

[14] De Wilde, M., Van Hauten, F., Derivations and automorphisms of the Lie algebra structure of the Weyl algebra, *Comm. Algebra* 22 (1994), no. 13, 5231–5248.

[15] De Wilde, M., Lecomte, P., Existence of star-products revisited, *Note Mat.* 10 (1990), suppl. 1, 205–216 (1992).

[16] De Wilde, M., Lecomte, P. B. A., Existence of star-products and of formal deformations of the Poisson Lie algebras of arbitrary symplectic manifolds, *Lett. Math. Phys.* 7 (1983), no. 6, 487–496.

[17] Fedosov, B. V., *Deformation quantization and index theory*, Math. Top. 9. Akademie Verlag, Berlin 1996.

[18] Fedosov, B. V., The index theorem for deformation quantization, in *Boundary value problems, Schrödinger operators, deformation quantization*, Math. Top. 8, Akademie Verlag, Berlin 1995, 206–318 .

[19] Fedosov, B. V., A simple geometrical construction of deformation quantization. *J. Differential Geom.* 40 (1994), no. 2, 213–238.

[20] Guillemin, V., Okikiolu, K., Subprincipal terms in Szegö estimates, *Math. Res. Lett.* 4 (1997), no. 1, 173–179.

[21] Guillemin, V., Star products on compact pre-quantizable symplectic manifolds, *Lett. Math. Phys.* 35 (1995), no. 1, 85–89.

[22] Guillemin, V., Sternberg, S., *Geometrical asymptotics*, Amer. Math. Soc. Surveys 14, Amer. Math. Soc., Providence, RI, 1977.

[23] Guillemin, V., Symplectic spinors and partial differential equations, in *Géométrie symplectique et physique mathématique*, Coll. Internat. CNRS, Aix-en-Provence 1974, Éditions C.N.R.S., Paris 1975, 217–252.

[24] Kashiwara, M., Quantization of Contact Manifolds, *Publ. Res. Inst. Math. Sci.* 32 (1996), 1–7.

[25] Kontsevich, M., Deformation Quantization of Poisson Manifolds, preprint, I.H.E.S., aeXiv: q-alg/9709040, 1997.

[26] Kostant, B., Quantization and unitary representations, in *Lectures in Modern Analysis and Applications*, Lecture Notes in Math. 170, Springer-Verlag, Berlin 1970, 87–208.

[27] Lichnerowicz, A., Equivalence of deformations of twisted products on a symplectic manifold, *Lett. Math. Phys.* 3 (1979), 495–502.

[28] Moyal, J. E., Quantum mechanics as a statistical theory. *Proc. Cambridge Philos. Soc.* 45, (1949), 99–124.

[29] Nest, R., Tsygan, B., Formal deformations of symplectic manifolds with boundary. *J. Reine Angew. Math.* 481 (1996), 27–54.

[30] Vey, J., Déformation du crochet de Poisson sur une variété symplectique, *Comment. Math. Helv.* 50 (1975), 421–454.

[31] Višik, M. I., Eskin, G. I., Convolution equations of variable order, *Trudy Moskov. Mat. Obšč.* 16 (1967), 25–50.

[32] Weinstein, A., Deformation quantization, *Séminaire Bourbaki* no. 789, Juin 1994.

[33] Weyl, H., *The theory of groups and quantum mechanics*, Dover, New York 1931.

Fedosov connections on jet bundles and deformation quantization

Alberto S. Cattaneo, Giovanni Felder**, and Lorenzo Tomassini***

Department of Mathematics, Harvard University, 1 Oxford St., Cambridge, MA 02138, U.S.A.
and
Institut für Mathematik, Universität Zürich
Winterthurerstrasse 190, 8057 Zürich, Switzerland
email: `asc@math.unizh.ch`

D-MATH, ETH-Zentrum
8092 Zürich, Switzerland
email: `felder@math.ethz.ch`

D-MATH, ETH-Zentrum
8092 Zürich, Switzerland
email: `lorenzo@math.ethz.ch`

Abstract. We review our construction of star-products on Poisson manifolds and discuss some examples. In particular, we work out the relation with Fedosov's original construction in the symplectic case.

1 Introduction

In [8] Kontsevich solved the problem of describing associative deformations of the algebra of functions on a manifold. In particular, he gave an explicit formula for the associative deformations ("star-products" [2]) $f \star g = fg + \epsilon B_1(f, g) + \epsilon^2 B_2(f, g) + \cdots$ of the product of functions on an open subset M of $\mathbb{R}^d$, with bidifferential operators B_j and so that $f \star 1 = 1 \star f = f$. Here ϵ is a formal parameter ($\epsilon = i\hbar/2$ in the notation of physics). He also described a (non-explicit) way to obtain star-products on arbitrary manifolds, as a corollary of his formality theorem, see [8] and, for more details, Appendix A.3 of [9].

Here we review a more explicit version [5] of the construction of star-products on arbitrary manifolds. For a more physics-oriented review and the relation to quantum field theory, see [4]. Our approach is based on the notion of Fedosov connections on jet bundles. A special case of this notion was introduced by Fedosov [7] to construct star-products such that the Poisson bracket defined by $f \star g - g \star f = 2\epsilon\{f, g\} + O(\epsilon^2)$ comes from a symplectic structure.

*Work partially supported by SNF Grant No. 20-63821.00
**Work partially supported by SNF Grant No. 21-65213.01

In general, one is given a Poisson manifold (M, α), which is a manifold M with a bivector field $\alpha \in \Gamma(M, \wedge^2 TM)$ so that the bracket $\{f, g\} = \alpha(df \otimes dg)$ obeys the Jacobi identity and asks for star-products as above so that $\frac{1}{2}(B_1(f, g) - B_1(g, f)) = \{f, g\}$.

To construct such products using Kontsevich's local formula, we first describe the local data given by the Poisson structure. Let E_0 be the bundle of (infinite) jets of smooth functions on M. The fiber over $x \in M$ consists of jets of functions at x, namely equivalence classes of smooth functions defined on some open neighborhood of x, where two functions are considered to be equivalent iff they have the same Taylor expansion at x (with respect to any choice of local coordinates around x). Similarly, one considers jets of multivector fields and multidifferential operators. To any function $f \in C^\infty(M)$ there corresponds a section of E_0, the jet of f: its value at x is the jet of f at x. The bundle E_0 comes with a canonical flat connection $D_0 : \Gamma(M, E_0) \mapsto \Omega^1(M, E_0)$, which is the unique connection so that its horizontal sections are precisely the jets of globally defined smooth functions. Moreover the jets of α define a Poisson bracket on each fiber and we have the Leibniz rules

$$D_0(fg) = D_0(f)g + fD_0(g), \quad D_0\{f, g\} = \{D_0(f), g\} + \{f, D_0(g)\},$$

for any sections $f, g \in \Gamma(M, E_0)$. Thus E_0 is a bundle of Poisson algebras with flat connection. The Leibniz rules imply that we have an isomorphism of Poisson algebras

$$\iota : C^\infty(M) \to H^0(E_0, D_0) = \mathrm{Ker}(D_0),$$

from the algebra of smooth functions onto the algebra of horizontal sections.

The idea is now to "quantize" (E_0, D_0): we construct a bundle $E = E_0 \otimes \mathbb{R}[[\epsilon]]$ of associative $\mathbb{R}[[\epsilon]]$-algebras using Kontsevich's local formula in each fiber. Then we deform the connection D_0 to a flat connection $\bar{D} = D_0 + O(\epsilon)$ on E obeying the Leibniz rule

$$\bar{D}(f \star g) = \bar{D}(f) \star g + f \star \bar{D}(g), \quad f, g \in \Gamma(M, E).$$

Thus the product induces a product on the space of horizontal sections $H^0(E, \bar{D})$. Finally we construct a *quantization map*, an isomorphism of $\mathbb{R}[[\epsilon]]$-modules $\rho : H^0(E_0, D_0)[[\epsilon]] \to H^0(E, \bar{D})$ given by differential operators and such that $\rho(1) = 1$. Then $f \star_M g = \rho^{-1}(\rho(f) \star \rho(g))$ is a star-product on M.

2 Formality theorem for $\mathbb{R}^d$

Let $\mathcal{A}_d = \mathbb{R}[[y^1, \dots, y^d]]$ be the algebra of formal power series in d indeterminates. A (formal) Poisson bracket on $\mathcal{A}_d$ is a Lie algebra structure on $\mathcal{A}_d$ of the form

$$\{f, g\} = \sum_{i,j=1}^{d} \alpha^{ij} \frac{\partial f}{\partial y^i} \frac{\partial g}{\partial y^j}, \quad f, g \in \mathcal{A}_d,$$

for some $\alpha^{ij} \in \mathbb{R}[[y^1, \dots, y^d]]$. M. Kontsevich constructed in [8] an $\mathbb{R}[[\epsilon]]$-bilinear associative product $\star_\alpha$ on $\mathcal{A}_d[[\epsilon]]$ which is a deformation of the product on $\mathcal{A}_d$: for $f, g \in \mathcal{A}_d$, the product has the form

$$f \star_\alpha g = fg + \sum_{n=1}^{\infty} \frac{\epsilon^n}{n!} U_n(\alpha, \dots, \alpha)(f \otimes g),$$

where $U_n(\alpha, \dots, \alpha) \in \mathrm{Hom}_{\mathbb{R}}(\mathcal{A}_d \otimes \mathcal{A}_d, \mathcal{A}_d)$ is a certain bidifferential operator whose coefficients are homogeneous polynomials of degree n in the partial derivatives of the coordinates α^{ij} of the Poisson bivector field α. For example, $U_1(\alpha)(f \otimes g) = \sum_{i,j=1}^{d} \alpha^{ij} \partial_i f \partial_j g = \{f, g\}$.

More generally, $U_n(\alpha_1, \dots, \alpha_n)$ is defined not only for Poisson bivector fields but also for multivector fields (skew-symmetric contravariant tensor fields) α_i. If α_i has rank m_i, $i = 1, \dots, n$, then $U_n(\alpha_1, \dots, \alpha_n) \in \mathrm{Hom}_{\mathbb{R}}(\mathcal{A}_d^{\otimes m}, \mathcal{A}_d)$, where $m = \sum m_i - 2n + 2$, is a multidifferential operator. The coefficients of these multidifferential operators are polynomials in the partial derivatives of the coordinates of $\alpha_1, \dots, \alpha_n$, linear in each α_i. They are given in terms of integrals of differential forms over configuration spaces of n-points in the upper half plane, see [8], and can be interpreted as Feynman amplitudes for a topological string theory [3].

The associativity of the product is a special case of Kontsevich's *formality theorem*, which is a sequence of quadratic relations for the operators U_n. For our purpose we need a further set of special cases, namely all cases where α_i is either a given Poisson bivector field, a vector field or a function. Leaving functions apart for the time being, the relations may be expressed as follows: introduce the generating series

$$P(\alpha) = \sum_{n=0}^{\infty} \frac{\epsilon^n}{n!} U_n(\alpha, \dots, \alpha) \in \mathrm{Hom}_{\mathbb{R}}(\mathcal{A}_d \otimes \mathcal{A}_d, \mathcal{A}_d), \tag{2.1}$$

$$A(\xi, \alpha) = \sum_{n=0}^{\infty} \frac{\epsilon^n}{n!} U_{n+1}(\xi, \alpha, \dots, \alpha) \in \mathrm{Hom}_{\mathbb{R}}(\mathcal{A}_d, \mathcal{A}_d), \tag{2.2}$$

$$F(\xi, \eta, \alpha) = \sum_{n=0}^{\infty} \frac{\epsilon^n}{n!} U_{n+2}(\xi, \eta, \alpha, \dots, \alpha) \in \mathcal{A}_d. \tag{2.3}$$

Let $W_d = \oplus_1^d \mathcal{A}_d \partial/\partial y^i$ be the Lie algebra of formal vector fields on $\mathbb{R}^d$. This Lie algebra acts on $\mathcal{A}_d$ and thus acts on the Hochschild cochain complex $\mathrm{Hoch}^{\cdot}(\mathcal{A}_d, \mathcal{A}_d) = \oplus_{n=0}^{\infty} \mathrm{Hom}_{\mathbb{R}}(\mathcal{A}_d^{\otimes n}, \mathcal{A}_d)$. Let

$$C^{\cdot}_{\mathrm{Lie}}(W_d, \mathrm{Hoch}^{\cdot}(\mathcal{A}_d, \mathcal{A}_d)) = \mathrm{Hom}_{\mathbb{R}}(\wedge^{\cdot} W_d, \mathrm{Hoch}^{\cdot}(\mathcal{A}_d, \mathcal{A}_d))$$

be the corresponding Lie algebra complex, with differential δ. The maps $\xi_1 \wedge \cdots \wedge \xi_k \mapsto U_{n+k}(\xi_1, \dots, \xi_k, \alpha, \dots, \alpha)$, for $k = 0, 1, 2$ can be then considered as elements of this complex. The formality theorem for these objects may be written as

(i) $P(\alpha) \circ (P(\alpha) \otimes \mathrm{Id}) = P(\alpha) \circ (\mathrm{Id} \otimes P(\alpha))$.

(ii) $P(\alpha) \circ (A(\xi, \alpha) \otimes \mathrm{Id} + \mathrm{Id} \otimes A(\xi, \alpha)) - A(\xi, \alpha) \circ P(\alpha) = \delta P(\xi, \alpha)$.

(iii) $P(\alpha) \circ (F(\xi, \eta, \alpha) \otimes \mathrm{Id} - \mathrm{Id} \otimes F(\xi, \eta, \alpha)) - A(\xi, \alpha) \circ A(\eta, \alpha) + A(\eta, \alpha) \circ A(\xi, \alpha)$
$= \delta A(\xi, \eta, \alpha)$.

(iv) $-A(\xi, \alpha) \circ F(\eta, \zeta, \alpha) - A(\eta, \alpha) \circ F(\zeta, \xi, \alpha) - A(\zeta, \alpha) \circ F(\xi, \eta, \alpha)$
$= \delta F(\xi, \eta, \zeta, \alpha)$.

The first equation is the associativity of the product. The second equation says that changing coordinates leads to an equivalent product. Indeed $\delta P(\xi, \alpha) = \frac{d}{dt} P(\phi_{t*}\alpha)|_{t=0}$ is the infinitesimal variation of the product under a coordinate transformation given by the flow ϕ_t of ξ. Before explaining the meaning of the remaining equations, which will appear in Proposition 5.1 below, we need to discuss the action of the Lie algebra $gl_d \subset W_d$ of linear vector fields and the lowest order terms of P, A, F. The action of gl_d on P, A, F is described by two properties:

A. The operators $U_n(\alpha_1, \dots, \alpha_n)$ are gl_d equivariant, in the sense that

$$g^* \circ U_n(g_*\alpha_1, \dots, g_*\alpha_n) = U(\alpha_1, \dots, \alpha_n) \circ (g^*)^{\otimes m},$$

where $g^* f(y) = f(gy)$ and $g_*\alpha_i(y) = (g \otimes \cdots \otimes g)\alpha_i(g^{-1}y)$, if $g \in \mathrm{GL}(d, \mathbb{R})$, $f \in \mathcal{A}_d$, and α_i is a formal multivector field. Therefore P, A, F are gl_d-equivariant.

B. For any $\xi \in gl_d$, $\eta \in W_d$, $A(\xi, \alpha) = \xi$, viewed as a first order differential operator and $F(\xi, \eta, \alpha) = 0$. This property follows from the vanishing of certain integrals, see property P5 in Sect. 7 of [8].

The lowest order terms in ϵ are also the result of explicit calculation of simplest Feynman integrals:

C. $P(\alpha)(f \otimes g) = fg + \epsilon\alpha(df \otimes fg) + O(\epsilon^2)$.

D. $A(\xi, \alpha) = \xi + O(\epsilon)$.

E. $F(\xi, \eta) = O(\epsilon)$.

It is worth noticing that the equations (i)–(iv) may be written succinctly as a Maurer–Cartan equation $\delta S + \frac{1}{2}[S, S]_G = 0$ for $S = P + A + F$, where $[\,\cdot\,,\cdot\,]_G$ is the Gerstenhaber bracket on the Hochschild complex.

3 Formal geometry

In order to apply Kontsevich's formula to the jet bundle E_0 we need to introduce coordinates. This is done using ideas of formal geometry, see [8]: let M^{coor} be the manifold of jets of coordinates systems on M. A point in M^{coor} is an infinite jet at zero of local diffeomorphisms $U \subset \mathbb{R}^d \to M$ defined on some open neighborhood U of $0 \in \mathbb{R}^d$. Two such maps define the same infinite jet if and only if their Taylor expansions at zero (for any choice of local coordinates on M) coincide. We have a projection $\pi : M^{\text{coor}} \to M$ sending φ to $\varphi(0)$.

The group $\mathrm{GL}(d,\mathbb{R})$ of linear diffeomorphisms acts on M^{coor} and we set $M^{\text{aff}} = M^{\text{coor}}/\mathrm{GL}(d,\mathbb{R})$. Define $\tilde{E}_0 := M^{\text{coor}} \times_{\mathrm{GL}(d,\mathbb{R})} \mathbb{R}[[y^1,\dots,y^d]]$. Moreover the Lie algebra W_d of formal vector fields acts on M^{coor} by infinitesimal coordinate transformations. The action is given by an isomorphism from W_d to the tangent space at any point of M^{coor}. The inverse map defines the *Maurer–Cartan form* ω_{MC}, a W_d-valued one-form on M^{coor}. The fact that this one-form comes from a Lie algebra action implies that it obeys the Maurer–Cartan equation $d\omega_{\text{MC}} + \frac{1}{2}[\omega_{\text{MC}}, \omega_{\text{MC}}] = 0$. As a consequence, $\tilde{D}_0 = d + \omega_{\text{MC}}$ is a flat connection on the trivial bundle $M^{\text{coor}} \times \mathcal{A}_d$ over M^{coor} and it is easy to check that it descends to $\tilde{E}_0$.

We will need the fact that the fibers of the bundle $M^{\text{aff}} \to M$ are contractible so that there exist sections $\varphi^{\text{aff}} : M \to M^{\text{aff}}$. For example, the exponential map of a torsion free connection on the tangent bundle defines such a section.

Lemma 3.1. *Let $\varphi^{\text{aff}} : x \mapsto [\varphi_x]$ be a section of M^{aff}. The map sending a point (x, f) of the jet bundle E_0 (f is the jet of a function at x) to the class of $(\varphi_x$, Taylor expansion of $f \circ \varphi_x)$ is an isomorphism of vector bundles with flat connection from (E_0, D_0) to $(\varphi^{\text{aff}*}\tilde{E}_0, \varphi^{\text{aff}*}\tilde{D}_0)$.*

From now on, we fix a section φ^{aff} and identify (E_0, D_0) with $(\varphi^{\text{aff}*}\tilde{E}_0, \varphi^{\text{aff}*}\tilde{D}_0)$. Here is an explicit local description of E_0 and D_0. On any open contractible subset $U \subset M$ we may choose a section φ of $M^{\text{coor}} \to M$ such that $\pi \circ \varphi = \varphi^{\text{aff}}$ where $\pi : M^{\text{coor}} \to M^{\text{aff}}$ is the canonical projection. We call such a section a *local lift* of φ^{aff}. Local lifts on U differ by a $\mathrm{GL}(d,\mathbb{R})$-gauge transformation $\varphi \mapsto \varphi \circ g$, $g : U \to \mathrm{GL}(d,\mathbb{R})$. A local lift induces a local trivialization of $E_0|_U \simeq U \times \mathcal{A}_d$, so that the isomorphism $C^\infty(M) \to H^0(E_0, D_0)$ sends f to the Taylor expansion at zero of $f \circ \varphi$. One way to choose a local lift is induced by a choice of local coordinates: let $x^1, \dots, x^d : U \to \mathbb{R}$ be coordinates on $U \subset M$. Then there is a unique local lift $x \mapsto \varphi_x$ so that $\varphi_x^j = x^j \circ \varphi_x$ has the form

$$\varphi_x^j(y) = x^j + y^j + \text{higher order terms in } y.$$

The flat connection D_0 on a local section $x \mapsto f_x(y) \in \mathbb{R}[[y^1,\dots,y^d]]$ is then

$$D_0 f_x(y) = \sum_{j=1}^{d} dx^j \left(\frac{\partial f_x(y)}{\partial x^j} - \sum_{k,l=1}^{d} T_j^l(x;y) \frac{\partial \varphi_x^k(y)}{\partial x^l} \frac{\partial f_x(y)}{\partial y^k} \right),$$

where $T(x;y)$ is the matrix inverse to $(\partial \varphi_x^j(y)/\partial y^k)$. Indeed, it follows from the chain rule that Taylor expansions of globally defined functions are D_0-closed sections. Conversely, observe that $\sum_l T_i^l(x;y)\frac{\partial \varphi_x^k}{\partial x^l}$ is a formal power series in y beginning with δ_i^k and whose coefficients are smooth in x. By this property it follows immediately that the coefficients of a section σ of E_0 satisfying $D_0\sigma = 0$ are determined by the zeroth coefficient $\sigma^0(x)$. If we set $\tilde{\sigma} = \sigma^0 \circ \varphi$, we have $D_0(\sigma - \tilde{\sigma}) = 0$ and $(\sigma - \tilde{\sigma})|_{y=0} = 0$; but this implies $\sigma = \tilde{\sigma}$. This shows that a section of E_0 is the Taylor expansion of a globally defined function if and only if it is D_0-closed.

4 A deformation of the canonical connection

Let $E = E_0[[\epsilon]]$ be the bundle of jets of $\mathbb{R}[[\epsilon]]$-valued functions. By definition, sections of E are formal power series in ϵ whose coefficients are sections of E_0. We fix as above a section φ^{aff} of M^{aff}. Then E, with fiberwise Kontsevich product associated to the Taylor expansion of α, is a bundle of $\mathbb{R}[[\epsilon]]$-algebras: choose a local lift $x \mapsto \varphi_x$ of φ^{aff} to a section of M^{coor}. This induces a local trivialization of E_0 so that local sections are given by $\mathcal{A}_d[[\epsilon]]$-valued functions $x \mapsto f_x(y)$. The product of local sections f, g is then the section

$$x \mapsto (f \star g)_x = P(\alpha_x)(f_x \otimes g_x), \quad \alpha_x = \text{Taylor expansion of } (\varphi_x^{-1})_* \alpha.$$

It does not depend of the choice of local lift since P is $\mathrm{GL}(d, \mathbb{R})$-equivariant, see Property A in Sect. 2.

We now use A of equation (2.2) to construct a deformation $D = D_0 + \epsilon D_1 + \cdots$ of the canonical connection D_0. Let $x \in M$ and $\xi \in T_x M$. Again we choose a local lift φ of φ^{aff} and set

$$\hat{\xi}_x = \varphi^* \omega_{\text{MC}}(\xi) \in W_d.$$

The connection is then defined as

$$D f_x = d f_x + A_x^M f_x,$$

where $A_x^M(\xi) = A(\hat{\xi}_x, \alpha_x)$, $\xi \in T_x M$. By properties A and B, this formula defines a connection on E which is independent of the choice of local lift. By property D, it is a deformation of D_0.

Equation (ii) in the formality theorem now implies the crucial statement:

Proposition 4.1. *For any $f, g \in \Gamma(M, E)$, we have the Leibniz rule*

$$D(f \star g) = Df \star g + f \star Dg.$$

However D is not flat, so that the algebra of horizontal sections with respect to D will not be isomorphic to $C^\infty(M)[[\epsilon]]$ as an $\mathbb{R}[[\epsilon]]$-module. We are going to discuss this difficulty in the next section.

5 Flattening the connection D

The connection D can be extended to a (graded) derivation

$$D : \Omega^{\cdot}(M, E) \to \Omega^{\cdot+1}(M, E)$$

of the algebra of differential forms on M with values in E (the product on this algebra is defined by the star-product in the fibers and the wedge product on differential forms). Its curvature D^2 is then an $\mathrm{End}(E)$ valued two-form. Let $F^M \in \Omega^2(M, E)$ be the two-form $x \mapsto F_x^M$ with

$$F_x^M(\xi, \eta) = F(\hat{\xi}_x, \hat{\eta}_x, \alpha_x), \quad \xi, \eta \in T_x M.$$

By properties A and B, F^M is independent of the local lift φ_x needed to define $\hat{\xi}_x$, $\hat{\eta}_x$ and α_x.

Then the formality identities (iii), (iv) translate into the following statements.

Proposition 5.1. *For any $f \in \Gamma(M, E)$,*

$$D^2 f = F^M \star f - f \star F^M, \tag{5.1}$$

$$DF^M = 0. \tag{5.2}$$

The two-form F^M is called the *Weyl curvature* of D. In general, a connection on a bundle of associative algebras with the above properties, i.e. to be a derivation whose curvature is an inner derivation such that its Weyl curvature satisfies the Bianchi identity (5.2), is called a *Fedosov connection*.

We want now to modify D so that it becomes flat still remaining a derivation. The first observation is that

$$\bar{D} := D + [\gamma, \cdot]_\star$$

is still a derivation for any $\gamma \in \Omega^1(M, E)$. The star-commutator is defined by $[a, b]_\star = a \star b - b \star a$. Moreover, $\bar{D}$ turns out to be again a Fedosov connection with Weyl curvature $\bar{F}^M = F^M + D\gamma + \gamma \star \gamma$. If we are able to find γ so that $\bar{F}^M = 0$,

$$F^M + D\gamma + \gamma \star \gamma = 0, \tag{5.3}$$

then $\bar{D}$-closed sections will form a nontrivial subalgebra of $\Gamma(M, E)$.

The one-form γ can be found as follows. Since F^M starts at order ϵ, see property E, we may write $F = \epsilon F_1 + \epsilon^2 F_2 + \cdots$. The correction γ does not need to have a term of order zero, so we write $\gamma = \epsilon\gamma_1 + \epsilon^2\gamma_2 + \cdots$. The equation $\bar{F} = 0$ at order ϵ reads

$$F_1 + D_0\gamma_1 = 0,$$

while the Bianchi identity imply at this order $D_0 F_1 = 0$. Hence, $D_0\gamma_1$ is equal to a D_0-closed expression. But the D_0-cohomology is trivial in degree 2, so it is possible to find a γ_1 that solves the equation. At higher order in ϵ, one proves by induction that one always has an equation of the form $D_0\gamma_k$ equal to a known D_0-closed form depending on the lower order coefficients of γ and F^M, so a solution exists by the same argument.

Thus we have a flat connection $\bar{D}$ on the bundle of algebras E, obeying the Leibniz rule. Let $H^0(E, \bar{D})$ be the $\mathbb{R}[[\epsilon]]$-algebra of horizontal sections. We are left to show the existence of a quantization map, an $\mathbb{R}$-linear homomorphism

$$\rho : C^\infty(M) \simeq H^0(E_0, D_0) \to H^0(E, \bar{D}),$$

inducing an isomorphism $C^\infty(M)[[\epsilon]] \to H^0(E, \bar{D})$ of $\mathbb{R}[[\epsilon]]$-modules. Then the pull-back of the product defines a star-product on $C^\infty(M)[[\epsilon]]$. To construct ρ one looks for a bundle map, also denoted by ρ, from E_0 to E of the form $\rho = \mathrm{Id} + \epsilon\rho_1 + \cdots$

so that

$$\rho \circ D_0 = \bar{D} \circ \rho, \tag{5.4}$$

and that ρ_i are given by differential operators. Again, there are no cohomological obstructions to find such a ρ: the ρ_i can be found recursively by solving equations of the form $D_0(\rho_i) := \rho_i \circ D_0 - D_0 \circ \rho_i =$ known. The known term on the right is D_0-closed by the induction hypothesis, and ρ_i can be found as a differential operator, as the D_0-cohomology of the subbundle of $\mathrm{End}(E_0)$ given by differential operators is trivial in degree 1. Furthermore ρ is unique if we require that $\rho(f)_x(y=0) = f_x(y=0)$, for any section f. The result is then:

Theorem 5.2. *Let (M, α) be a Poisson manifold. Fix a section $\varphi^{\mathrm{aff}} : M \to M^{\mathrm{aff}}$. Let $\star$, D, F^M be the corresponding product on $\Gamma(M, E)$, the connection on E and its curvature, respectively. Let $\gamma \in \Omega^1(M, E)$, and $\rho \in \Gamma(M, \mathrm{Hom}(E_0, E))$ be solutions of* (5.3), (5.4), *respectively. Then $\bar{D} = D + [\gamma, \cdot]_\star$ is a flat connection on E obeying the Leibniz rule $\bar{D}(f \star g) = \bar{D}(f) \star g + f \star \bar{D}(g)$ and ρ induces an isomorphism of $\mathbb{R}[[\epsilon]]$-modules $C^\infty(M)[[\epsilon]] \to H^0(E, \bar{D})$. The pull-back of the product to $C^\infty(M)[[\epsilon]]$ is a star-product with first order term α.*

Thus the construction of a star-product for Poisson manifolds requires solving an equation for γ and an equation for ρ. We want to show that these equations can be solved in an explicit way. The equation for $\gamma = \sum \epsilon^j \gamma_j$ can be solved recursively: at the jth step of the recursion, one has an equation of the form

$$D_0 \gamma_j = \beta_j. \tag{5.5}$$

where $\beta_j \in \Omega^2(M, E_0)$ obeys $D_0 \beta_j = 0$ and is known. The point is that γ_j may be constructed by a purely local calculation: namely, let us introduce local coordinates and a local trivialization of E_0 as in Sect. 3.

It is useful to define the total degree of a form on M taking values in sections of E_0 as the sum of the form degree and the degree in y. Then we write

$$D_0 = -\delta + D_0',$$

where

$$\delta := \sum_{i=1}^{d} dx^i \frac{\partial}{\partial y^i}$$

is the zero-degree part and D_0' has positive degree. It follows immediately that $\delta^2 = 0$. We can define a dual operator to δ on E_0-valued differential forms:

$$\delta^* := \sum_{i=1}^{d} y^i \iota_{\frac{\partial}{\partial x^i}}, \tag{5.6}$$

where ι denotes inner multiplication. It is easy to verify that $(\delta\delta^* + \delta^*\delta)\rho = k\rho$ for every form ρ of total degree k. Thus, if we restrict to δ-closed forms of positive total degree k, we may then invert δ by $\delta^{-1}\rho = \frac{1}{k}\delta^*\rho$. This inverse yields the unique form σ such that $\delta\sigma = \rho$ and $\delta^*\sigma = 0$. This proves that the δ-cohomology is concentrated in degree zero, i.e., functions on M (independent of y).

Then the solution of eq. (5.5) obeying $\delta^*\gamma_i = 0$ is

$$\gamma_i = -\sum_{n=0}^{\infty}(\delta^{-1}D_0')^n\delta^{-1}\beta_i.$$

The infinite sum converges in the sense of formal power series since the n-th term is of degree at least n in y.

The equation for ρ can be solved similarly.

6 Casimir and central functions

We describe variants of our constructions which are relevant for the treatment of the center and for the comparison with Fedosov's construction in the symplectic case.

6.1 Central closed two-forms

Let $\omega = \omega_0 + \epsilon\omega_1 + \cdots \in \Omega^2(M, E)$ be such that $D\omega = 0$ and $[\omega, f]_\star = 0$ $\forall f \in \Gamma(M, E)$. Then we may construct a more general flat connection $\bar{D}$ by replacing (5.3) by

$$F^M + D\gamma + \gamma \star \gamma = \omega. \tag{6.1}$$

Indeed, the Bianchi identity holds also for $F^M - \omega$ and $\bar{D}^2 = 0$ since ω is central. We get thus a family of products parametrized by D-closed central two forms ω.

6.2 The second construction of a quantization map

The second variation concerns the quantization map. It is possible to make it compatible with the center in the following sense: let $Z_0(M) = \{f \in C^\infty(M) \mid \{f, \cdot\} = 0\}$ be the algebra of Casimir functions. There exists a quantization map $\rho : H^0(E, D_0) \to H_0(E, \bar{D})$ whose restriction to $Z_0(M)$ induces an algebra isomorphism from $Z_0(M)[[\epsilon]]$ onto the center of $H^0(E, \bar{D})$. It is constructed using the two remaining special cases of the formality theorem involving the Poisson vector field, vector fields and functions, see [5].

This quantization map can be used to construct a linear map from the space of D_0-closed Casimir two-forms $B = \{\omega \in \Omega^2(M, E_0) \mid D_0\omega = 0, \{\omega, \cdot\} = 0\}$ onto the space of $\bar{D}$-closed central two-forms. Thus we may parametrize our products by two-forms in $B[[\epsilon]]$.

6.3 Dependence on choices

If we choose the homotopy δ^{-1} to determine canonical choices of γ and ρ, our construction of star-products depends on a Poisson bivector field α, a section φ^{aff} and a central closed 2-form in $B[[\epsilon]]$. It is possible to show that different choices of φ^{aff} lead to equivalent products (two star-products $\star$, $\star'$ are called equivalent if there is a series $\psi = \mathrm{Id}+\epsilon\psi_1+\epsilon^2\psi_2+\cdots$ so that $\psi(f\star g) = \psi(f)\star'\psi(g), \forall f, g \in C^\infty(M)$). Also, if we replace $\omega \in B[[\epsilon]]$ by $\omega + D_0\beta$ for a Casimir one-form $\beta \in \Omega^1(M, E_0)$, we obtain an equivalent star-product. Still in general we do not get all star-products up to equivalence in this way. To obtain all equivalence classes of star-products, we have to take α to be a power series in ϵ. Then [8] equivalence classes of star-products are in one-to-one correspondence with $\mathbb{R}[[\epsilon]]$-valued Poisson vector fields α modulo formal paths in the group of diffeomorphisms generated by flows of vector fields $\epsilon\xi_1+\epsilon^2\xi_2+\cdots$. This correspondence is realized by taking $\omega = 0$ in our construction. If we take another ω, we get thus a product corresponding to $\omega = 0$ for a different α. This defines a map from $B[[\epsilon]]$ modulo D_0-exact central two-forms to the equivalence classes of $\mathbb{R}[[\epsilon]]$-valued Poisson bivector fields. It would be interesting to describe this map. For infinitesimal ω and to leading order in ϵ this map has the following description. Let $C^{\cdot,\cdot} = \Omega^\cdot(M, \wedge^\cdot \mathrm{Der}(E_0))$ be the double complex of differential forms with values in the jets of formal multivector fields. The differentials are D_0 and the differential δ_α of Poisson cohomology (the Schouten–Nijenhuis bracket with α) on the fibers. Then ω defines a class in $H^2(H^0(C, \delta_\alpha), D_0)$. As the D_0-cohomology is trivial except in degree zero, we have a map

$$j : H^2(H^0(C^{\cdot,\cdot}, \delta_\alpha), D_0) \to H^0(H^2(C^{\cdot,\cdot}, \delta_\alpha), D_0) = H^2_\alpha(M),$$

given by $\omega \mapsto \delta_\alpha D_0^{-1}\delta_\alpha D_0^{-1}\omega$. Thus j sends the class of ω to an element of the second Poisson cohomology group, which consists of equivalence classes of infinitesimal variations of the Poisson bivector field α.

7 Examples, related results

7.1 Open subsets of $\mathbb{R}^d$

Let us check that our construction gives back the original Kontsevich formula for open subsets of $\mathbb{R}^d$. In this case we may take φ^{aff} to be given by

$$\varphi_x^j(y) = x^j + y^j.$$

Then $D_0 = \sum_{j=1}^d dx^j\left(\frac{\partial}{\partial x^j} - \frac{\partial}{\partial y^j}\right)$. Since $A(\xi, \alpha) = \xi$ for constant ξ, we have $D = D_0$ and $F = 0$. We may then choose $\gamma = 0$ and $\rho = \mathrm{Id}$ and the result is Kontsevich's formula $f \star g = \sum_{n=0}^\infty \frac{\epsilon^n}{n!} U_n(\alpha, \ldots, \alpha)(f \otimes g)$.

7.2 Leaves of Poisson foliations [1]

If N is a foliated Poisson manifold so that the Poisson bivector field is tangent to the leaves, then each leaf is a Poisson submanifold. An *adapted chart* with domain $V \subset M$ is a chart $\psi : V \to \mathbb{R}^d$ so that the leaves are given by equations $\psi^j(x) = \text{const}$, $j = m+1, \ldots, d$. A star-product is called *tangential* if on the domain V of any adapted chart and for any $f \in C^\infty(M)$ so that $f|_V$ is constant on each leaf, one has $f \star g|_V = fg|_V$, $\forall g \in C^\infty(M)$. In [1] it is shown that if φ^{aff} is chosen to be *adapted* to the foliation, in the sense that for all $x \in M$, there is a representative φ_x of $\varphi^{\text{aff}}(x)$ so that φ_x^{-1} is an adapted chart, then $\star$, with ρ from the "second construction", is tangential. Using this property, one can restrict star-products to the leaves of a foliated Poisson manifold.

7.3 The case of a symplectic manifold

Suppose (M, α) is a Poisson manifold whose Poisson bracket comes from a symplectic form Ω.

A *Darboux section* of M^{aff} is a section $x \mapsto [\varphi_x]$ of M^{aff} such that, for all $x \in M$, $\varphi_x^*\Omega$ is a constant two-form. Symplectic manifolds always have Darboux sections. For example, a torsion free symplectic connection gives rise to a Darboux section (but not through the exponential map), see [7], Section 2.5.

In Darboux coordinates the Kontsevich product reduces to the Moyal product, and also the other objects A, F may be described explicitly. As a result, we have explicit formulae in terms of a local lift of φ^{aff} with the property that $\alpha_x = (\varphi_x^{-1})_*\alpha$ is a constant bivector field α_0. In this case, $\varphi_x^*\omega_{MC}$ is a 1-form on M with values in the Hamiltonian formal vector fields. The corresponding hamiltonian functions h_x, normalized by $h_x(y=0) = 0$, define an E_0-valued one-form $x \mapsto h_x$. Then we have

(i) $(f \star g)_x(y) = f_x \star g_x(y) = \exp\left(\epsilon \sum \alpha_0^{ij} \frac{\partial}{\partial y_1^i} \frac{\partial}{\partial y_2^j}\right) f_x(y_1) g_x(y_2)|_{y_1=y_2=y}$.

(ii) $Df_x = d_x f_x + \frac{1}{2\epsilon}[h_x, f_x]_\star$.

(iii) $F_x^M(\xi, \eta) = \frac{1}{4\epsilon^2}\left([\langle h_x, \xi\rangle, \langle h_x, \eta\rangle]_\star - 2\epsilon\{\langle h_x, \xi\rangle, \langle h_x, \eta\rangle\}\right)$.

Moreover, in the symplectic case, the space B of closed Casimir two-forms consists of closed two-forms in $\Omega^2(M)[[\epsilon]] \subset \Omega^2(M, E)$. Using these formulae the comparison with Fedosov's original construction [7] becomes clear. The latter starts with the definition of the Weyl bundle: let $F(M)$ be the principal $\mathrm{GL}(d, \mathbb{R})$-bundle of frames (bases of tangent spaces). The Weyl bundle is the associated vector bundle $W = F(M) \times_{\mathrm{GL}(d,\mathbb{R})} \mathbb{R}[[y^1, \ldots, y^d]][[\epsilon]]$. The fiber over x is the space of (formal) functions on the tangent space at x, a symplectic vector space. So there is a Moyal product in each fiber, and W is a bundle of algebras. A symplectic connection ∇ on the tangent bundle induces a connection D_{F} on W obeying the Leibniz rule on sections.

Its curvature is $D_F^2 = (2\epsilon)^{-1}[R, \cdot]$, where the W-valued 2-form $R = -\frac{1}{2}\Omega(\nabla^2 y, y)$ is the quadratic form on TM associated with the curvature of ∇. For any closed two-form of the form $\Omega(\epsilon) = -(2\epsilon)^{-1}\Omega + \Omega_0 + \epsilon\Omega_1 + \cdots \in \Omega^2(M)[[\epsilon]]$, Fedosov shows that the equation $D_F\gamma_F + \gamma_F \star \gamma_F + R = \Omega(\epsilon)$ has a solution γ_F of the form $(2\epsilon)^{-1}\Omega_{ij}y^i dx^j + \gamma_{F1} + \epsilon\gamma_{F2} + \cdots$, obeying the normalization condition $\gamma_F|_{y=0} = 0$. It follows that connection $\bar{D}_F = D_F + [\gamma_F, \cdot]$ is flat.

Let ∇ be the symplectic connection on TM whose Christoffel symbols are obtained from the 2-jet of the Darboux section via

$$\varphi_x^j(y) = x^j + y^j - \frac{1}{2}\sum_{k,l}\Gamma_{kl}^j(x)y^k y^l + \cdots .$$

We have a bundle isomorphism $E \to W$ sending the jet at x of a function f to the class of $(e, f \circ \varphi_x)$, where φ_x is the unique representative of $\varphi^{\mathrm{aff}}(x)$ that maps the standard frame of $\mathbb{R}^d$ to the frame e. By (i) this is an isomorphism of algebra bundles. The connection $\bar{D}_F$ is then of the form $\bar{D} = D + [\gamma, \cdot]$ with $\gamma_F = (2\epsilon)^{-1}h_x + \gamma$, where γ is a solution of (6.1) with $\omega = \Omega_0 + \epsilon\Omega_1 + \cdots$.

The result is that our star-product constructed using a closed two-form $\omega \in \Omega^2(M)[[\epsilon]]$ is equivalent to Fedosov's star-product associated to the class of $\Omega(\epsilon) = -(2\epsilon)^{-1}\Omega + \omega$. Details will be presented elsewhere [6].

References

[1] D. Arnal, N. Dahbene, A. Gammella and M. Masmoudi, Tangential deformation on foliated Poisson manifolds, preprint 2001

[2] F. Bayen, M. Flato, C. Frønsdal, A. Lichnerowicz and D. Sternheimer, Deformation theory and quantization I, II, *Ann. Phys.* 111 (1978), 61–110, 111–151.

[3] A. S. Cattaneo and G. Felder, A path integral approach to the Kontsevich quantization formula, `math.QA/9902090`, *Comm. Math. Phys.* 212 (2000), no. 3, 591–612.

[4] A. S. Cattaneo and G. Felder, On the globalization of Kontsevich's star product and the perturbative Poisson sigma model, in *Proceedings of the International Workshop on Non-commutative Geometry and String Theory*, Keio 2001, to appear.

[5] A. S. Cattaneo, G. Felder and L. Tomassini, From local to global deformation quantization of Poisson manifolds, `math.QA/0012228`, to appear in *Duke Math. J.*

[6] A. S. Cattaneo, G. Felder and L. Tomassini, in preparation.

[7] B. Fedosov, *Deformation quantization and index theory,* Akademie Verlag, Berlin 1996.

[8] M. Kontsevich, Deformation quantization of Poisson manifolds, I, `q-alg/9709040`.

[9] M. Kontsevich, Deformation quantization of algebraic varieties, `math.AG/0106006`.

Quantization of Lie bialgebras via the formality of the operad of little disks

Dimitri Tamarkin

Harvard University
Department of Mathematics
1 Oxford St., Cambridge, MA 02138, U.S.A.
e-mail: `tamarkin@math.harvard.edu`

Abstract. We give a proof of Etingof–Kazhdan theorem on quantization of Lie bialgebras based on the formality of the chain operad of little disks and show that the Grothendieck–Teichmüller group acts non-trivially on the corresponding quantization functors.

2000 Mathematics Subject Classification: 53D55 (81T70, 81-XX, 16E40)

1 Introduction

1.1

In this paper we give a proof of Etingof–Kazhdan theorem on quantization of Lie bialgebras based on the formality of the chain operad of little disks.

Any known construction of such a formality involves multiple zeta values; in particular there is no canonic way to establish such a formality over $\mathbb{Q}$. For example, in the construction from [8] one needs to choose an associator over $\mathbb{Q}$. In [9] it is shown that different associators produce homotopically non-equivalent formalities of chain operad of little disks. Each of these formalities, in turn, produces a certain quantization procedure of Lie bialgebras and we prove that these procedures are not isomorphic. This can be considered as a step in studying the action of Grothendieck–Teichmüller group on quantization functors originated in [5].

1.2 Idea of the construction of quantization

1.2.1 Let $\mathfrak{g}$ be a Lie bialgebra with bracket $[,]$ and cobracket δ. Call $\mathfrak{g}$ *conilpotent* if for any $x \in \mathfrak{g}$ there exists an N such that any N-fold iteration of δ applied to x produces zero.

1.2.2 Let H be a Hopf algebra with product $\cdot$, coproduct Δ, unit 1, and counit ϵ (we do not assume that the antipode exists). Let $\Delta'(x) = \Delta(x) - 1 \otimes x - x \otimes 1$. Call H *conilpotent* if for any x such that $\epsilon(x) = 0$ there exists an N such that any N-fold iteration of Δ' applied to x produces zero.

1.2.3 Note that in any conilpotent H there exists an antipode map and it is uniquely defined.

1.2.4 We are going to construct a functor Q from the category of conilpotent Lie bialgebras to the category of conilpotent Hopf algebras

1.2.5 Let $\mathfrak{g}$ be a conilpotent Lie bialgebra. Let $C^\bullet(\mathfrak{g})$ be its cochain complex with respect to the cobracket. This means that $C^\bullet(\mathfrak{g}) = S(\mathfrak{g}[-1])$ is a free graded commutative algebra equipped with a differential D defined on the space of generators $\mathfrak{g}[-1] \subset S(\mathfrak{g}[-1])$ by the cobracket $\delta : \mathfrak{g}[-1] \to S^2(\mathfrak{g}[-1])$.

$C^\bullet(\mathfrak{g})$ has a structure of Gerstenhaber algebra so that the bracket on $\mathfrak{g}[-1] \subset C^\bullet(\mathfrak{g})$ is defined by the bracket on $\mathfrak{g}$.

1.2.6 Let e_2 be the operad of Gerstenhaber algebras and $\mathbf{ho}e_2$ a free resolution of e_2 Let M be the operad of brace structures described in 6.10-6.10.3 and let $\mathbf{ho}\mathcal{M} \to M$ be a free resolution We have a quasi-isomorphism $\mathbf{ho}\mathcal{M} \to e_2$ ([6],[8]). Therefore, there is a way to construct an M-algebra out of an e_2-algebra. Denote this way by W (it is a functor from the category of Gerstenhaber algebras to the category of M-algebras (= brace algebras). Thus, $W(C^\bullet(\mathfrak{g}))$ is a brace algebra.

1.2.7

Remark 1. It is exactly on this step that the associators or integrals are being used.

Remark 2. One of the steps of the proof of the formality of M in [6] is linking M with the operad of singular chains of the operad of little disks (this step is purely "combinatorial", it does not use transcendental methods). Thus, the formality of M follows from the formality of the operad of singular chains of the operad of little disks.

1.2.8 In 6.10.3 it is explained that there exists a simple way H to construct a conilpotent Hopf algebra out of a brace algebra. Thus $h(\mathfrak{g}) = HWC^\bullet(\mathfrak{g})$ is a differential graded Hopf algebra. We prove that

Proposition 1.1. *The cohomology $H^i(h(\mathfrak{g}))$ vanishes for all $i \neq 0$.*

Set $Q(\mathfrak{g}) = H^0(h(\mathfrak{g}))$. We prove that Q is an equivalence of categories.

1.3 Universal language

Algebraic structures of Lie bialgebra and of Hopf algebra cannot be described in the language of operads. One has to consider a certain generalization of operad called PROP (5).

A PROP P is a symmetric monoidal category with certain properties. A P-*algebra* in a symmetric monoidal category C is the same as a symmetric monoidal map $P \to C$.

If we have a definition of a certain algebraic structure, then the PROP P describing this structure is defined as a unique up to an isomorphism symmetric monoidal category P such that the notion of P-algebra coincides with the definition of our algebraic structure.

1.3.1 Let $\mathcal{LBA}$ be the PROP of Lie bialgebras and $\mathcal{BA}$ be the PROP of Hopf algebras. Thus, a Lie bialgebra in the category of complexes can be described as a symmetric monoidal functor $\mathcal{LBA} \to$ **complexes**. Let $\mathcal{LBA}^\vee$ be the category of all functors (without symmetric monoidal structure) $\mathcal{LBA} \to$ **complexes**. We can introduce a symmetric monoidal structure on $\mathcal{LBA}^\vee$ (see 3.9). For a symmetric monoidal functor $F : \mathcal{LBA} \to$ **complexes**, its symmetric monoidal structure defines a commutative algebra structure on F as an element of symmetric monoidal category $\mathcal{LBA}^\vee$.

Remark Note that a commutative algebra structure on $F \in \mathcal{LBA}^\vee$ corresponds to a *weak* symmetric monoidal structure on F; for a commutative algebra F in $\mathcal{LBA}^\vee$ to be a strong symmetric monoidal functor $\mathcal{LBA} \to$ **complexes**, one has to impose an additional condition on F.

Similar statements are true for the PROP $\mathcal{BA}$.

How to describe conilpotent Lie bialgebras and conilpotent Hopf algebras in this language? They correspond to *conilpotent* functors $\mathcal{LBA} \to$ **complexes** and $\mathcal{BA} \to$ **complexes** with symmetric monoidal structure.

Let S (resp L) be the full subcategory of all functors in $\mathcal{BA}^\vee$ (resp. $\mathcal{LBA}^\vee$) which are quasi-isomorphic to conilpotent functors $\mathcal{BA} \to$ **complexes** (resp. $\mathcal{LBA} \to$ **complexes**) (we do not assume any symmetric monoidal structure on these functors). These categories have symmetric monoidal structures defined by the restriction from $\mathcal{LBA}^\vee$ and $\mathcal{BA}^\vee$. Let $L_0 \subset L$ be the full subcategory consisting of conilpotent functors $\mathcal{LBA} \to$ **vect**, let $S_0 \subset S$ be the similar thing.

The constructions of 1.2.4 can be rewritten in terms of a symmetric monoidal functor $\mathbf{Q} : L \to S$ so that given a conilpotent Lie bialgebra $\mathfrak{g} \in L$, $\mathbf{Q}(\mathfrak{g}) \in S$ is a conilpotent Hopf algebra.

We prove that $\mathbf{Q}$ induces an equivalence of certain 'good' subcategories $L' \subset L$ and $S' \subset S$.

1.3.2 To prove Proposition 1.1, one introduces t-structures on L' and S' and shows that Q is a t-functor. (More precisely, first one takes the categories DL' (resp. DS') with the same objects as in L' (resp. S') but $\hom_{DL'}(X, Y) = H^0 \hom_{L'}(X, Y)$

(similarly for DS') and endows them with triangular structure. See [1] for the theory of triangular categories and t-structures. In this paper we have tried to give all the necessary definitions.)

This implies that $\mathbf{Q}$ induces an isomorphism of the cores of L' and S' which are L_0 and S_0, which implies Proposition 1.1 and the fact that Q is an equivalence of categories.

1.3.3 Let Pe_2 be the PROP generated by the operad of Gerstenhaber algebras and PM be the PROP generated by the operad of brace algebras. Let $Pe_2\vee$ be the symmetric monoidal category of functors $Pe_2 \to$ **complexes** and $PM\vee$ be the same thing for PM. Then the map Q decomposes as a sequence of maps

$$\mathbf{Q} : L \to Pe_2\vee \to PM\vee \to S. \tag{1.1}$$

We define certain 'good' subcategories of each of the categories involved, endow them with t-structures and show that all the arrows are t-functors and symmetric monoidal homotopy equivalences.

Also we construct the inverse functor $DQ : S \to L$ by inverting each of the arrows in (1.1).

1.3.4 In the last part of the paper we investigate how the functors Q and DQ depend on a choice of associator by means of which the formality $\mathbf{hoe}_2 \to M$ is constructed.

We know that the variety of associators is a torsor over the Grothendieck–Teichmüller group. In [9] it is shown that the graded Lie algebra $\mathfrak{grt}$ acts on an appropriate resolution $\mathbf{hoe}_2$ of e_2. Let t be a formal infinitesimal parameter, then any $x \in \mathfrak{grt}$ produces a family of associators A_t^x and a family $F_t^x : P\mathbf{hoe}_2^{\frown} \to P\mathbf{hoe}_2^{\frown}[[t]]$ such that $F_t^x = \mathrm{Id} \pmod t$. Let $f_A : \mathbf{hoe}_2 \to M$ be the formality constructed by means of A. and let $Q_t : L_0 \to S_0[[t]]$ be the family of equivalences constructed using A_t. The functor $Q = Q_0$ produces the constant functor $Q' : L_0 \stackrel{Q}{\to} S_0 \to S_0[[t]]$. We prove that there is no equivalence of Q_t and Q. Without loss of generality we can assume that this equivalence is equal to identity modulo t. This statement easily reduces to non-existence of an isomorphism J_t between the map $T_{F_t^x}$ induced by F_t^x on the core of $Pe_2^{\frown}$ and Id, such that $J_t = \mathrm{Id} \mod t$.

Next, we use the theorem which states that the assumption that the automorphisms of the core induced by F_t^x and Id are isomorphic implies that F_t^x is *deformationally equivalent* to Id. Then we use section 4 to show that this implies that $f_t : \mathbf{hoe}_2 \to \mathbf{hoe}_2[[t]]$ produces zero in the deformational complex of $\mathbf{hoe}_2$ which contradicts to the result of [9].

1.4 Content of the paper

The first four sections of the paper are preparatory. They explain some terminology that is not commonly used and include some simple propositions to be used in the sequel.

The first part contains a review of differential graded symmetric monoidal categories. We assume that in each such a category a class of morphisms called quasi-isomorphisms is chosen so that it behaves nicely with respect to tensor product. We discuss the notion of a *weak* symmetric monoidal functor F between two such categories. Weakness means that the maps $F(X) \otimes F(Y) \to F(X \otimes Y)$ of the symmetric monoidal structure on F are not obliged to be isomorphisms. If, though, all these maps are quasi-isomorphisms, then we call F *essentially strong*.

Then we discuss the notion of homotopy equivalence of two symmetric monoidal categories and of homotopy equivalence of two weak symmetric monoidal functors between two fixed symmetric monoidal categories. We do not use the usual term "weak homotopy equivalence" as the word "weak" is used with symmetric monoidal functors.

After that we discuss the notion of deformational homotopy equivalence, which will be only used to prove nonequivalence of different quantizations.

The second part deals with how to define the class of quasi-isomorphisms in a dg category C. Proposition 3.1 provides us with such a recipe. It has a drawback that the class of quasi-isomorphisms defined in this way is very narrow. For example, let A be a usual ring and let C be the category of dg A-modules. Let M and M' be A-modules and let $p : M' \to M$ be a free resolution. Most likely, p is not a quasi-isomorphism. Although, if A is semi-simple, the definition works fine.

How can we fix this problem? First, let us analyze how the notion of quasi-isomorphism is usually defined (say for C being the category of dg A-modules). We take the forgetful functor F from our category to some semi-simple category **triv** and call an arrow f in our category quasi-isomorphic if such is $F(f)$ according to Definition–Proposition 3.1.

But now our quasi-isomorphisms in C do not satisfy Proposition 3.1, which makes it impossible to work with them. In many cases this situation can be corrected by taking a subcategory $\mathbf{Cof}_F C \subset C$ within which any quasi-isomorphism satisfies Proposition 3.1. An object of $\mathbf{Cof}_F(C)$ is called F-cofibrant. In the example $C = A - \mathrm{mod}$ any complex of projective A-modules bounded from the right hand side is in $\mathbf{Cof}_F(C)$. Also, the total of any bounded complex (see 3.0.9) of F-cofibrant objects is cofibrant. Unfortunately, it may happen that there are not enough cofibrant objects.

After this is done we restrict ourselves to the so-called saturated categories. The key property of them is that they are (informally speaking) closed with respect to taking bounded complexes of its objects. For any dg category C the category $C\vee$ of all functors $C \to$ **complexes** is such and we only study such categories. Any functor $F : D \to C$ induces a functor $F^{-1} : D\vee \to C\vee$. We show that the category $C^\frown := \mathbf{Cof}_{F^{-1}} C\vee$ possesses a variety of good properties.

Next, we study the case when C is symmetric monoidal. We see that in this case $C^\frown$ has a natural symmetric monoidal structure.

After that, we discuss t-structures on such a $C^\frown$. It is shown that under certain restrictions on C we can define them.

Finally, we study the action of deformationally equivalent weak symmetric monoidal functors on certain operads (see section 4). This section is only needed for studying the Grothendieck–Teichmüller group action.

In the fifth part the theory developed above is applied to the case when C is a PROP after which we discuss a notion of generalized maps of PROP's. They formalize the situation when we have a recipe how given a P-algebra we may construct a Q-algebra (as for example in 1.2.5.) One of the variants of such a definition is standard and can be found say in [4]. This variant is not completely satisfactory for us because it is difficult to incorporate the notion of conilpotent algebra in it (for this one has to consider topologic PROP's) Thus, we develop another variant of such a definition and show how to pass from one of them to another. There is a problem with our definition because in general given a P-algebra it produces only a *weak* Q-algebra. Thus we need to impose some restrictions to insure that we always obtain a usual (= strong) Q-algebra. That's how we arrive at the definition of a strong generalized map of PROP's.

After all this preparation we follow the plan announced in 1.3–1.3.4. First we define the rightmost arrow in (1.1) and a quasi-inverse to it and show that both of them are homotopy equivalences and then we do the same for the remaining arrows in (1.1) in order from the right to the left.

We conclude with studying the action of Grothendieck–Teichmüller group on quantizations.

2 Symmetric monoidal categories

Let C, D be dg symmetric monoidal categories such that in each of them a class of morphisms called quasi-isomorphisms is chosen and the tensor product of quasi-isomorphisms is a quasi-isomorphism.

Let $F : C \to D$ be a functor preserving quasi-isomorphisms; such functors are called *exact*. *A weak symmetric monoidal structure* on F is a natural transformation of functors $e_{XY} : F(X) \otimes F(Y) \to F(X \otimes Y)$ which is commutative and associative and is not necessarily an isomorphism. *A strong symmetric monoidal structure* on F is a weak structure in which all e are isomorphisms; *an essentially strong* symmetric monoidal structure is a weak symmetric monoidal structure in which all e_{XY} are quasi-isomorphisms.

2.0.1 F is called a homotopy equivalence if it has an essentially strong symmetric monoidal structure, $F : \hom(X, Y) \to \hom(F(X), F(Y))$ is a quasi-isomorphism for any X, Y and any $Z \in D$ is quasi-isomorphic to some $F(X)$.

2.0.2 Let $F, G : C \to D$ be weak symmetric monoidal functors. A homotopy equivalence $F \to G$ is a collection of a homotopy equivalences $I : C' \to C$ and

$J : D \to D'$ and a natural symmetric monoidal transformation $Q : JFI \to JGI$ such that all Q_X are quasi-isomorphisms.

2.0.3 F, G are called homotopy equivalent if they can be connected by a chain of homotopy equivalences.

2.0.4 Let $F, G : C \to D[[t]]$ be functors such that $F = G \bmod t$; for definition of $D[[t]]$ see 4.1.2. A *deformational* homotopy equivalence between F and G is a collection of homotopy equivalences $I : C' \to C$ and $J : D \to D'$ and a natural transformation $Q : JFI \to JGI$ such that all Q_X are quasi-isomorphisms and $Q = \mathrm{Id} \bmod t$.

2.0.5 Call F, G deformationally homotopy equivalent if they are connected by a chain of deformational homotopy equivalences.

3 dg-categories

3.0.6 Let now S be any dg category. Let $f : X \to Y$ be a morphism in S.

Proposition 3.1. *The following properties are equivalent:*

1) *For any Z, $f^* : \mathrm{hom}(Y, Z) \to \mathrm{hom}(X, Z)$ is a quasi-isomorphism;*

2) *For any Z, $f_* : \mathrm{hom}(Z, X) \to \mathrm{hom}(Z, Y)$ is a quasi-isomorphism;*

3) *There exists a map $g : Y \to X$ such that both $fg - \mathrm{Id}_Y = du$ and $gf - \mathrm{Id}_X = dv$ for some u, v.*

Proof. 1) $\to$ 3). Take $Z = Y$. Then by 1), there exists $g : Z \to X$ such that $f_*g = fg$ is homologous to Id_Y. We thus have $fg - \mathrm{Id}_Y = du$. Consider now $f_*(gf - \mathrm{Id}_X) = fgf - f = d(uf)$. Since f_* is a quasi-isomorphism, $gf - \mathrm{Id}_X$ is homologous to 0.

3) $\to$ 1) Note that both g_*f_* and f_*g_* induce identities on cohomology.

2) $\to$ 3) and 3) $\to$ 2) is proved in the same way as above. □

Any f satisfying one of the conditions of the proposition is called *natural quasi-isomorphism.*

3.0.7 Let $F : C \to D$ be a functor of dg-categories. Call a morphism m in C an F-quasi-isomorphism if $F(m)$ is a natural quasi-isomorphism. For an object X of C denote $h_X(Y) = \mathrm{hom}_C(X, Y)$; $h_X : C \to$ **complexes** is a functor. Call X F-cofibrant if h_X maps F-quasi-isomorphisms to quasi-isomorphisms of complexes.

3.0.8 Let $\mathbf{Cof}_F(C)$ be the full subcategory of F-cofibrant objects in C. In $\mathbf{Cof}_F(C)$ any F-quasi-isomorphism is natural and vice versa in virtue of Proposition 3.1,3.

3.0.9 An N-bounded complex in C is a collection of objects X_n, $n \geq N$ and maps $d_{nm} : X_n \to X_m[m-n+1]$, $n > m$ such that $\sum_m d_{nm}d_{mk} = 0$ for any n, k. We refer to such a complex as $(X_\bullet, d)$ or simply $X_\bullet$. A bounded complex is the same as an N-bounded complex for some N.

For $Y \in C$ set $h_X(Y)_k = \hom(X_k, Y)$. We see that $H = h_X(Y)_\bullet$, $H^{ij} = (h_{X_j}(Y))^i$ is a complex of complexes, i.e. a bicomplex and we can take its total complex; denote it by $h_X(Y)$, $h_X(Y)^N = \prod_i H^{i,N-i}$. $h_X : C \to$ **complexes** is a functor. If h_X is representable, denote by $|X_\bullet|$ the representing object.

Call C *saturated* if $|X_\bullet|$ exists for any non-positive X.

3.0.10 Let Com C be the dg category of bounded complexes in C. One can show that for any saturated C, there exists a functor $||$: Com $C \to C$ such that $|X|$ represents h_X for any X.

3.0.11 Categories C^+ and C^{++}. Let C be an arbitrary dg category. Let C^+ be the category an object X of which is a pair consisting of a family $X_i \in C, i \in I$ indexed by a set I_X. and a family n_i^X, $i \in I_X$ of integers. Set

$$\hom(X, Y) = \prod_{j \in I_X} \oplus_{i \in I_Y} \hom(X_j, Y_i)[n_i^Y - n_j^X].$$

Let C^{++} be the category of bounded complexes in C^+. It is clear that C^{++} is saturated; if C is saturated, then we have a natural functor $\iota : C^{++} \to C$.

Assume C is symmetric monoidal. Then C^+ and C^{++} are also naturally symmetric monoidal and ι has a strong symmetric monoidal structure.

3.1 Triangulated category structure

3.1.1 Let C be saturated. For $X \in C$ and an integer n set $X'_k = 0$ if $k \neq n$; $X'_n = X$; $d_r = 0$ for any r. $X'_\bullet$ is then a bounded complex. Set $X[n] = |X'|$. We have canonic isomorphisms $X[0] \cong X$ and $X[n][m] \cong X[n+m]$.

3.1.2 Let $f : X \to Y$ be a map in C. Set $Z_1 = X$; $Z_0 = Y$; $Z_n = 0$ for $n \neq 0, 1$; set $d_{0,-1} : Z_0 \to Z_{-1}$ to be equal to f. Call Cone $f := |Z|$.

3.1.3 We have a natural sequence of maps $\cdots \to X \xrightarrow{f} Y \to \text{Cone } f \to X[1] \to Y[1] \to \cdots$. Call this sequence *a special distinguished triangle*.

3.1.4 Let DC be the category with the same objects as in C but now $\hom_{DC}(X, Y) = H^0(\hom_C(X, Y))$. A distinguished triangle in DC is any triangle isomorphic to the image in DC of a special distinguished triangle. This endows DC with a structure of triangulated category.

3.2 Compatible morphisms

3.2.1 Let C, D be saturated categories and $F : C \to D$ a functor. Let $(X_\bullet, d)$ be a bounded complex in C. Then $(F(X_\bullet), F(d))$ is a complex in D. We have a natural map $F' : h_X(Y) \to h_{F(X)}(F(Y))$. Let $\mathrm{Id} \in h_X(|X|)$. Then $F'(\mathrm{Id})$ defines a map $F'' : |F(X)| \to F(|X|)$. Call F *compatible* if F'' is an isomorphism for any X.

3.2.2 Assume that C, D are saturated and fix a compatible map of dg categories $F : C \to D$. Assume that F has a left adjoint G.

Lemma 3.2. *G is compatible.*

3.3 Standard bar resolution

The material of this subsection is standard and can be found in e.g. [2].

3.3.1 Denote $S = GF$. The adjointness implies that we have maps of functors $\epsilon : S \to I$ and $\Delta : I \to I \circ I$ such that Δ is associative and ϵ is a counit. In other words, S is a monad. This allows us to construct a simplicial object $\Sigma_\bullet$ in the category of endofunctors of C as follows. Set $\Sigma_i = S^{\circ(i+1)}$. Define the face map $d_i : \Sigma_n \to \Sigma_{n-1}$ by

$$S^{n+1} \cong S^{\circ n-i} \circ \mathbf{S} \circ S^{(i-1)} \xrightarrow{\epsilon} S^{\circ n-i} \circ S^{(i-1)} \to S^n,$$

where ϵ is applied to the bold S (it is the same S as usual S; it is made bold to show where ϵ is applied.) The i-th degeneration $s_i : \Sigma_{n-1} \to \Sigma_n$, $0 \leq i \leq n-1$ is application of δ to the i-th S from the right hand side of $S^{\circ n}$. The map $\epsilon : \Sigma \to \mathrm{Id}$ is an augmentation. Let $\Sigma' := C_\bullet(\Sigma_\bullet)$ be the chain complex of the simplicial object $\Sigma_\bullet$. The augmentation induces a map $\iota : \Sigma' \to \mathrm{Id}$; let Σ'' be the augmented complex.

3.3.2 Let $e : \mathrm{Id}_D \to FG$ be the canonical map Define maps $\gamma_i : F\Sigma''_i \to \Sigma''_{i+1}$ by $\gamma_i : FS^{i+1} \to FGFS^{i+1} \cong \Sigma''_{i+1}$. Let $\gamma = \sum \gamma_i$. It is well known that $d\gamma + \gamma d = \mathrm{Id}$. Therefore, $F\Sigma''(X)$ is contractible for any X and $F(\iota)$ is a universal quasi-isomorphism. Let $\Sigma(X) = |\Sigma'(X)|$. We have the induced map $p : \Sigma(X) \to X$ which is an F-quasi-isomorphism.

3.3.3 $\Sigma(X)$ is cofibrant for any X.

3.3.4 Let $\mathbf{Cof}_F(C)$ be the full subcategory of cofibrant objects of C.

Proposition 3.3. *$\mathbf{Cof}_F$ is saturated.*

Proof. One needs to check that total of a bounded complex of cofibrant objects is cofibrant which is immediate. □

The two notions of F-quasi-isomorphism and natural quasi-isomorphism coincide in $\mathbf{Cof}_F(C)$ and for any object X in C there exists an object $\Sigma(X)$ which is F-quasi-isomorphic to X

3.3.5 Assume a symmetric monoidal structure given on C and D such that F has a weak symmetric structure and preserves $\mathbf{Cof}_F(C)$.

3.4 Symmetric monoidal structure on Σ

Consider the composition

$$X \otimes Y \to FGX \otimes FGY \to F(GX \otimes GY)$$

and assume that the adjoint map $m : G(X \otimes Y) \to G(X \otimes Y)$ is an isomorphism. Then G and hence GF are symmetric monoidal functors and the transformations $FG \to \mathrm{Id}$ $\mathrm{Id} \to GF$ both preserve the tensor structures. Therefore, we have a morphism of simplicial objects in C: $\Sigma_\bullet(X) \otimes \Sigma_\bullet(Y) \to \Sigma_\bullet(X \otimes Y)$. Since the chain functor has a symmetric monoidal structure, we have the induced map $\Sigma(X) \otimes \Sigma(Y) \to \Sigma(X \otimes Y)$ which determines a tensor structure on Σ.

3.5 Categories $C\vee$

Let C be a small dg category. Denote by $C\vee$ the dg category of functors $C \to$ **complexes**. We have the Yoneda inclusion $i : C^{op} \to C\vee$ sending $X \in C$ to $i(X) : C \to$ **complexes**, where $i(X)(Y) = \mathrm{hom}(X, Y)$.

3.5.1 $C\vee$ is saturated.

3.6 Functoriality

Let $F : C \to D$ be a functor of small dg categories. The composition with F defines a functor $F^{-1} : D\vee \to C\vee$ so that $F^{-1}A(X) = A(F(X))$.

Proposition 3.4. *F^{-1} has the left and the right adjoint.*

Proof. Let us construct the left adjoint functor $F_!$. For $X \in C$ and $Y \in D$ denote

$$h(X, Y) = \hom_D(F(X), Y); \tag{3.1}$$

h is a functor $C^{op} \otimes D \to$ **complexes**. For $A \in C\vee$ set $F_!A = A \otimes_C h$. It is clear that $\hom_D(F_!A, B) \cong \hom(A, F^{-1}B)$.

Define also the right adjoint functor F_* by $F_*(A) = \hom_C(h, A)$. □

3.6.1 Let **triv** be a small k-linear category such that the abelian category of functors **triv** $\to$ **vect** is semi-simple. Let C be a small category equipped with a map s : **triv** $\to C$. Call an s-quasi-isomorphism in $C\vee$ simply a quasi-isomorphism.

3.6.2 Denote by $C^\frown = \mathbf{Cof}_s C\vee$ the full subcategory consisting of cofibrant objects. Let $F : C \to D$ be a functor. An Fs-quasi-isomorphism in D will be called quasi-isomorphism.

Proposition 3.5. *The functor F^{-1} maps quasi-isomorphisms to quasi-isomorphisms. The functor $F_!$ maps $C^\frown$ to $D^\frown$ and it maps any quasi-isomorphism in $C^\frown$ to a quasi-isomorphism in $D^\frown$.*

3.7

Let $F : C \to (D^{op})\vee$ be a map of categories. Define $F^\# : D^\frown \to C\vee$ by $F^\#(U)(X) = U \otimes_D F(X)$. One sees that $\Sigma(F^\#)$ is compatible.

3.8 t-structures

3.8.1 Let C be a saturated dg-category. A *t-structure on C* is a pair of full subcategories $\mathcal{D}^{\geq 0}$, $\mathcal{D}^{\leq 0}$ of C satisfying certain properties; to formulate them introduce a notation

3.8.2 Set $\mathcal{D}^{\geq n} \subset C^\frown$ to be the full subcategory of objects U such that $U[n] \in \mathcal{D}^{\geq 0}$. Set $\mathcal{D}^{\leq n} \subset C$ to be the full subcategory of objects U such that $U[n] \in \mathcal{D}^{\leq 0}$.

3.8.3 Axioms of t-structure

1 $\mathcal{D}^{\leq 0} \subset \mathcal{D}^{\leq 1}$;

2 $\mathcal{D}^{\geq 1} \subset \mathcal{D}^{\geq 0}$;

3 For any $X \in \mathcal{D}^{\leq 0}$ and $Y \in \mathcal{D}^{\geq 0}$, $H^0 \hom(X, Y) = 0$;

4 For any X in C there exists a distinguished triangle $\cdots \to A \to X \to B \to \cdots$ with $A \in \mathcal{D}^{\leq -1}$ and $B \in \mathcal{D}^{\geq 0}$.

3.8.4 It follows that in 4. the objects A and B are defined by X uniquely up to a quasi-isomorphism. Furthermore there exist functors $\tau_{\leq n}$ and $\tau_{\geq n}$ so that $\tau_{\leq n}X \cong (\tau_{\leq 0}X[-n])[n]$ and $\tau_{\geq n}X \cong (\tau_{\leq 0}X[n])[-n]$; and there are natural transformations $\tau_{\leq -1} \to \mathrm{Id} \to \tau_{\geq 0}$ so that

$$\to \tau_{\leq -1}X \to X \to \tau_{\geq 0}X \to$$

is a distinguished triangle for any X.

3.8.5 Example. Assume that the spaces of homomorphisms for any two objects in C lie totally in the zeroth degree. Set $\mathcal{D}^{\geq 0} \subset C^{\frown}$ to consist of all functors U such that $H^{<0}U(X) = 0$ and $\mathcal{D}^{\leq 0}$ to include all U such that $H^{>0}U(X) = 0$ for any X.

3.8.6 Example. Assume that C is a dg category such that for any $X, Y \in C$, $\hom_C(X, Y)^i = 0$ for $i < 0$ and $d_0 : C(X, Y)^0 \to C(X, Y)^1$ is equal to 0. Assume that **triv** is the category with the same objects as in C but $\hom_{\mathbf{triv}}(X, Y) = 0$ if $X \neq 0$ and $\hom_{\mathbf{triv}}(X, X) = k\,\mathrm{Id}$ and let $s : \mathbf{triv} \to C$ to be the inclusion. Define a category B with the same set of objects as in C in which $\hom_B(X, Y) = \hom_C(X, Y)^0$. We have a map $p : C \to B$ and $i : B \to C$. Furthermore, s takes values in B so that we have a map $s' : \mathbf{triv} \to B$. Call a map $f \in C\vee$ (resp. $g \in B\vee$) a quasi-isomorphism if such is $s^{-1}f$ (resp. $(s')^{-1}g$).

3.8.7 Let $\mathcal{D}^{\geq 0} \subset C^{\frown}$ be the full subcategory of objects U such that $H^i(U(X)) = 0$, for any $i < 0$ and any X.

3.8.8 Let $\mathcal{D}^{\leq 0} \subset C^{\frown}$ be the full subcategory of objects V such that $H^i((p_!V)(X)) = 0$ for any $i > 0$ and any X.

3.8.9

Proposition 3.6. *Assume that $i_! : \mathcal{B}^\vee \to C\vee$ maps quasi-isomorphisms to quasi-isomorphisms. Then the pair $\mathcal{D}^{\geq 0}$ and $\mathcal{D}^{\leq 0}$ described above is a t-structure.*

Proof. 1, 2 are clear.

3. Let $X' = \Sigma(X)$. It suffices to prove the statement for $\hom(X', Y)$. Set $(F^iY)^j = Y^j$ if $j \geq i$ $(F^iY)^j = 0$ otherwise. Note that

$$\hom(X, Y) \to \mathrm{liminv}\ \hom(X', Y/F^iY)$$

is an isomorphism of complexes and the maps

$$p_i : \hom(X, Y/F^{i+1}Y) \to \hom(X, Y/F^iY)$$

are componentwise surjective. Therefore, it suffices to check the statement for $\hom(X', F^iY/F^{i+1}Y)$, $i > 0$ and for $\hom(X', Y/F^1Y)$.

Note that $F^iY/F^{i+1}Y = p^{-1}Z$ for some Z (because all morphisms from Ker p act trivially), and the statement follows from the adjointness axioms; in the second case, we have a quasi-isomorphism in $C\vee$ $Y/F^1Y \to H^0(Y/F^1Y) = Y'$ and again $Y' = p^{-1}Z'$ for some Z'.

4. We can write that X' is the functor $||$ applied to a complex

$$\cdots \to s_!A_n \xrightarrow{D} \cdots .$$

One can assume without loss of generality that all A_n are complexes with zero differential. Let $U_n = s'_!A_n$. Therefore, $s_!A_n \cong i_!U_n$. Then each U_n has zero differential. In what follows we say that $X \in C\vee$ is in $D^{\leq 0}$ if so is ΣX (or any other cofibrant object quasi-isomorphic to X). We see that if $0 \to U \to V \to W \to 0$ is a sequence of objects from $C\vee$ such that $0 \to U(X) \to V(X) \to W(X) \to 0$ is an exact sequence of complexes for any $X \in C$ and $U, W \in D^{<0}$, then so is V.

For $U \in B\vee$ being a complex with zero differential and $X \in B$ set $U_{<n}(X) = \tau_{<n}U(X)$, where for a complex K we set

$$\begin{aligned} \tau_{<n}K^i &= 0, \quad i \geq n; \\ \tau_{<n}K^i &= K^i, \quad i \leq n-1. \end{aligned}$$

Set

$$V_n = i_!\tau_{<n}U_n + Di_!\tau_{<n+1}U_{n+1} \subset i_!U_n.$$

It is clear that $V_\bullet \subset U_\bullet$ is a subcomplex and that $H^{<0}(U_\bullet/V_\bullet) = 0$. Show that $|\Sigma(V)|$ is in $\mathcal{D}^{<0}$. Set $F^iV_n = V_n$, $n < i$; $F^iV_i = i_!\tau_{<i}U_i$; $F^iV_n = 0$, $n > i$. It suffices to check that $|\Sigma(F^{i+1}V/F^iV)|$ is in $\mathcal{D}^{<0}$. Let $G = F^{i+1}V/F^iV$. Note that

$$\begin{aligned} G_n &= 0, \quad n \neq i, i+1; \\ G_i &= (Di_!\tau_{<i+1}U_{i+1})/(Di_!\tau_{<i+1}U_{i+1} \cup \tau_{<i}U_i); \\ G_{i+1} &= i_!\tau_{<i+1}U_{i+1}. \end{aligned}$$

Note that $D(i_!\tau_{<k}U_{i+1}) \subset i_!\tau_{<k+1}U_i$ (because $\hom_C(X,Y)^{<0} = 0$). Let $K_{i+1} = i_!\tau_{<i}U_{i+1} \subset G_{i+1}$; $K_n = 0$, $n \neq i+1$. The above remark shows that K is a subcomplex of G. Since $K \in \mathcal{D}^{<0}$, it suffices to prove that $G/K \in \mathcal{D}^{<0}$.

Consider the following complex.

$$D' : i_!(\tau_{<i+1}U_{i+1}/\tau_{<i}U_{i+1})[i+1] \to i_!(\tau_{<i+1}U_i/\tau_{<i}U_i)[i],$$

where D' is induced by D. Then G/K is a subcomplex of this complex. We have $\tau_{<i+1}U/\tau_{<i}U = U^i[-i]$ for any complex with zero differential. Therefore, G/K is isomorphic to the total of $D' : i_!U^i_{i+1} \to D'(i_!U^i_{i+1})$ which is quasi-isomorphic to Ker $D'[1]$.

One sees that the map $D' : i_!U^i_{i+1} \to i_!U^i_i$ is induced by a certain map $\delta : U^i_{i+1} \to U^i_{i+1}$. Because of the exactness of $i_!$ on $B'\vee$, this implies that Ker $D'[1] \cong i_!$Ker $\delta[-1]$, hence is in $\mathcal{D}^{\leq 0}$. □

3.8.10 let C, D be as above. A compatible functor $F : D^\frown \to C^\frown$ is called a t-functor if it maps $\mathcal{D}^{\geq 0}$ (resp. $\mathcal{D}^{\leq 0}$) to the corresponding subcategories in $C^\frown$.

3.8.11 Let $F : C \to D$ be a homotopy equivalence of dg-categories and $F(\mathcal{D}^{\geq 0}) \subset \mathcal{D}^{\geq 0}$ Then F is a t-functor.

3.9 Symmetric monoidal structures

3.9.1 Assume C is symmetric monoidal. Define the tensor product on $C\vee$ as follows. Let $X, Y \in C\vee$ and let $m : C \otimes C \to C$ be the tensor product. Set $X \otimes Y = m_!(X \boxtimes Y)$

Proposition 3.7. *Let $f : C \to D$ be a symmetric monoidal functor. Then $f_!$ has a symmetric monoidal structure and f^{-1} has an induced weak symmetric monoidal structure.*

3.9.2 Assume that **triv** is symmetric monoidal and s : **triv** $\to C$ is symmetric monoidal map.

Proposition 3.8. *$C^\frown$ is preserved by the tensor product on $C\vee$.*

Proposition 3.9. *Assume C is as in* 3.8.6. *Then $\mathcal{D}^{\leq n} \otimes \mathcal{D}^{\leq m} \subset \mathcal{D}^{\leq n+m}$.*

3.9.3 For A, B in T_C set $A \otimes_T B = \tau_{\geq 0}(A \otimes B)$. The above result implies that $\otimes_T$ is a tensor product in T_C

3.9.4 Let $F : C^\frown \to D^\frown$ be a symmetric monoidal functor with a weak symmetric structure on it. Then $F_T : T_C \to T_D$ has an induced weak symmetric monoidal structure.

3.9.5 Assume that a (symmetric monoidal) category C is such that $\hom(X, Y)$ are always in degree 0. Then T_C as a (symmetric monoidal) category is equivalent to the category of functors $C \to$ **vect**.

4 Full operads and weak symmetric monoidal functors

4.1 Definitions

4.1.1 Let C be a symmetric monoidal dg category and $V \in C$. For a finite set X denote $\mathcal{O}_V(X) = \hom_C(V^{\otimes X}, V)$. We see that $\mathcal{O}_V$ is a dg operad.

4.1.2 Denote by $C[[t]]$ the category with the same objects as in C but

$$\hom_{C[[t]]}(X, Y) = \hom_C(X, Y)[[t]],$$

where t is a formal parameter. We have natural functors $i : C \to C[[t]]$ and $\bmod t : C[[t]] \to C$.

4.1.3 We will call natural quasi-isomorphisms in C simply "quasi-isomorphisms".

4.1.4 Let $F : C \to D$ be a weak symmetric monoidal functor. We have an induced map $F_. : \mathcal{O}_V \to \mathcal{O}_{F(V)}$. Similarly, for $F_t : C \to D[[t]]$, we have an induced map $F_{t.} : \mathcal{O}_V \to \mathcal{O}_{F_t(V)}[[t]]$.

4.1.5 Let $\mathcal{O}$ be an arbitrary dg-operad and let $X \in \mathcal{O}([1])$ be an invertible element. It defines a map $\mathrm{Ad}_X : \mathcal{O} \to \mathcal{O}$ such that for $f \in \mathcal{O}([n])$ $\mathrm{Ad}_X f = X^{-1} \circ f \circ X^{\otimes n}$.

4.1.6 Let $\mathcal{O}_1, \mathcal{O}_2$ be arbitrary dg operads. Assume a map of operads $F_t : \mathcal{O}_1 \to \mathcal{O}_2[[t]]$ is given. Denote by $F_0 = i(F_t \bmod t) : \mathcal{O}_1 \to \mathcal{O}_2[[t]]$ to be the term of zeroth order in t. F_t is called *deformationally trivial* if there exists a quasi-isomorphism $q : \mathcal{O}_1' \to \mathcal{O}_2$, an $X_t \in \mathcal{O}_{F(V)}([1])[[t]]$, $X_t = \mathrm{Id} \pmod{t}$ and an extension $\gamma : \mathcal{O}_1' \to \mathcal{O}_2[[t, dt]]$ of

$$\mathrm{Ad}_{X_t} F_0 q : \mathcal{O}_1' \to \mathcal{O}_2[[t]],$$

where for an arbitrary operad $\mathcal{O}$ we define the operad $\mathcal{O}[t, dt]$ to be $\mathcal{O} \otimes \mathbf{k}[t, dt]$, where $\mathbf{k}[t, dt]$ is a commutative dg ring freely generated by the variables t, $\deg t = 0$ and dt, $\deg t = 1$, the differential is defined by $dt = dt$.

4.1.7

Proposition 4.1. *Let $q : O_1' \to \mathcal{O}_1$ and $p : \mathcal{O}_2 \to \mathcal{O}_2'$ be quasi-isomorphisms and let $F_t : \mathcal{O}_1 \to \mathcal{O}_2[[t]]$. Then F_t is deformationally trivial if and only if such is pF_tq*

4.2

Let $F_t : C \to C[[t]]$ be a weak symmetric functor. We have a map $F_{t.} : \mathcal{O}_V \to \mathcal{O}_{F(V)}[[t]]$. Assume we have another functor G_t having the same properties as F_t and a natural symmetric monoidal transformation $a : G_t \to F_t$ which is a homotopy equivalence.

Proposition 4.2. *$F_{t.}$ is a homotopically trivial deformation if and only if $G_{t.}$ is such.*

We may assume that C and D are saturated (if not, replace them with C^{++} and D^{++}).

Let $A = F(V)$, $B = G(V)$. We have a quasi-isomorphism $\alpha : A \to B$ in $D[[t]]$ induced by the natural transformation. It follows that the diagram

$$\begin{array}{ccc} A^{\otimes n} & \xrightarrow{F_{t.}} & A \\ \Big\downarrow{\scriptstyle \alpha^{\otimes n}} & & \Big\downarrow{\scriptstyle \alpha} \\ B^{\otimes n} & \xrightarrow{G_{t.}} & B \end{array} \tag{4.1}$$

commutes.

Let $h : P \to Q$ be any map in C. Let s, ds be variables such that degree of s is 0 and degree of ds is 1. Let $Q[s]$ be the direct sum of countably many copies of Q (each copy denoted by $Qs^k, k \geq 0$) and $Q[s]ds$ be the direct sum of countably many copies of $Q[-1]$ (each copy denoted by $Qs^k ds, k \geq 0$.

Set $D_k : Qs^k \to Qs^{k-1}ds$ to be the map equal to k Id; The maps D_k define a map

$$D : Q[s] \to Q[s]ds[1]$$

of degree 0. Set $Q[s, ds] = \operatorname{Cone} D$. Consider the map $e : Q[s, ds] \to Q[ds]$ which is evaluation at $s = 0$. This map splits in C; therefore the colimit of the diagram

$$P \to Q[ds] \leftarrow Q[s, ds]$$

exists. Denote it Ch We have canonic quasi-isomorphic maps $p_1 : Ch \to P$ and $p_2 : Ch \to Q$, p_1 being the canonic projection of the colimit onto P and p_2 being the evaluation at $s = 1$. The following important property holds:

Lemma 4.3. *For any $X \in C$ the maps of complexes $\hom(X, Ch) \xrightarrow{p_{1*}} \hom(X, P)$ and $\hom(X, Ch) \xrightarrow{p_{2*}} \hom(X, Q)$ are componentwise surjective and quasi-isomorphic.*

Let $h : P \to Q$ and $g : P' \to Q'$ be quasi-isomorphisms. We have maps

$$D_1 : \hom(Ch, Cg) \oplus \hom(P, P') \to \hom(Ch, P'),$$

and

$$D_2 : \hom(Ch, Cg) \oplus \hom(Q, Q') \to \hom(Ch, Q')$$

given by $D_i(a \oplus b) = p_{i*}a - p_i^*b$. It follows that D_i are componentwise surjective. Denote $\hom_i(Ch, Cg) = \operatorname{Ker} D_i$. The natural maps $\hom_i(Cf, Cg) \to \hom(Ch, Cg)$, $\hom_1(Ch, Cg) \to \hom(P, P')$, and $\hom_2(Ch, Cg) \to \hom(Q, Q')$ are quasi-isomorphisms.

Also we have a map

$$Ch \otimes Ch' \to C(h \otimes h'). \tag{4.2}$$

which is a quasi-isomorphism.

The commutativity of (4.1) means that we have a commutative diagram

$$
\begin{array}{ccccc}
\hom(V^n, V) & \longrightarrow & \hom_1(C(\alpha)^{\otimes n}, C(\alpha)) & \xrightarrow{\approx} & \hom(A^{\otimes n}, A)[[t]] \\
\downarrow & & \downarrow \approx & & \\
\hom_2(C(\alpha)^{\otimes n}, C(\alpha)) & \xrightarrow{\approx} & \hom(C(\alpha)^{\otimes n}, C(\alpha)) & & \\
\downarrow \approx & & & & \\
\hom(B^{\otimes n}, B)[[t]]) & & & &
\end{array}
$$

Note that each term of this diagram forms an operad and all the arrows are maps of operads. The composition of the two arrows in the upper row is equal to the map $F_{t.}$; the composition of the two arrows in the leftmost column is equal to $G_{t.}$.

Now the statement follows from 4.1. □

4.2.1 Let now $I : C' \to C$ and $J : D \to D'$ be homotopy equivalences of symmetric monoidal categories. Let $V' \in C'$ be an element such that $I(V')$ is quasi-isomorphic to V.

Proposition 4.4. *The deformation $(JF_tI)_. : \mathcal{O}'_V \to \mathcal{O}_{JF_tI(V')}[[t]]$ is trivial if and only if such is $F_{t.}$.*

Proof.

Lemma 4.5. *Let $f : W \to V$ be a quasi-isomorphism in C. and $H_t : C \to D[[t]]$ a weak symmetric monoidal functor. Then the deformation $H^V_{t.} : \mathcal{O}_V \to O_{F_t(V)}[[t]]$ is trivial if and only if $H^W_{t.} : \mathcal{O}_W \to \mathcal{O}_{F_t(W)}[[t]]$ is trivial.*

Proof. First consider the case when f is a retraction, i.e. there exist $g : V \to W$ such that $fg = \mathrm{Id}$. Let $P = gf$, $P^2 = P$. We have a map $q : \mathcal{O}_V \to \mathcal{O}_W$ such that $\phi : V^{\otimes n} \to V$ goes into $g\phi f^{\otimes n} : W^{\otimes n} \to W$.

Similarly, we have a map $q_t : \mathcal{O}_{H_t(V)} \to \mathcal{O}_{H_t(W)}[[t]]$ such that $\phi : H_t(V)^{\otimes n} \to V$ goes into $H_t(g)\phi H_t(f)^{\otimes n}$:.

Sublemma 4.6. *The deformation q_t is trivial*

Proof. First, show that There exists a family $U_t \in \hom(H_t(W), H_t(W))[[t]]$ such that $U_t = \mathrm{Id} \bmod t$ and $H_t(F) = U_t F_0$ for some $F_0 \in \hom(H_t(W), H_t(W))$. Such a U_t can be given by the formula

$$U_t = \mathrm{Id} + H_t(F)H_0(G) - H_0(FG).$$

Now, $\mathrm{Ad}_{U_t} q_t = \mathrm{Id}$, whence the statement. □

The composition $\mathcal{O}_V \to \mathcal{O}_W \to \mathcal{O}_{H_t(W)}$ is equal to $\mathcal{O}_V \to \mathcal{O}_{H_t(V)} \to \mathcal{O}_{H_t(W)}$, whence the statement for f being a retraction.

If f is arbitrary, then there exists $W' \in C$ and quasi-isomorphic retractions $W' \to V$ and $W' \to W$, whence the statement in general case. □

Now our statement reduces to the following one: the deformation $F_{t.} : \mathcal{O}'_V \to \mathcal{O}_{IF_tJV'}$ is trivial if and only if such is $\mathcal{O}_{JV'} \to \mathcal{O}_{F_tJV'}$. But $F_{t.}$ decomposes as $\mathcal{O}_{V'} \to \mathcal{O}_{JV'} \to \mathcal{O}_{F_tJV'} \to \mathcal{O}_{IF_tJV'}$, whence the statement. □

4.2.2 The two previous subsections imply the following result: Let $F_t, G_t : C \to D[[t]]$ be deformationally homotopically equivalent. Then for any $V \in C$ the triviality of the deformation $F_{t.}$ implies the triviality of the deformation $G_{t.}$.

5 PROP's

Let **setf** be the category of finite sets.

5.0.3 Trivial PROP's. Consider the groupoid **iso** of finite sets and their isomorphisms. It has an obvious symmetric monoidal structure in which $X \otimes Y = X \sqcup Y$. Let **triv** be the **k**-span of **iso**. A dg PROP is a symmetric monoidal dg category P with its objects being finite sets equipped with a morphism of symmetric monoidal categories s : **triv** $\to P$ which induces the identity on the classes of objects. It is clear that any PROP is a small category.

5.0.4 Example Let V be an object of an arbitrary symmetric monoidal dg category $\mathcal{A}$. We thus have for any finite set X the corresponding tensor product $\otimes_X V$. For $X = \{1, \dots, n\}$,

$$\otimes_X V \cong V^{\otimes n}.$$

Define a PROP P_V in which

$$\hom_{P_V}(X, Y) = \hom_{\mathcal{A}}(\otimes_X V, \otimes_Y V).$$

the tensor structure and the map s are straightforward. The PROP P_V is universal among all PROP's P equipped with a symmetric monoidal map of categories $f : P \to \mathcal{A}$ such that $f([1]) = V$.

5.0.5 P-algebras In the notation of a previous section, a P-algebra structure on V is a map of PROP's $P \to P_V$.

5.1 Weak P-algebras

A weak P-algebra in C is a weak symmetric monoidal map $P \to C$.

5.1.1 Let $C = Q^{op}\vee$ Then any weak P-algebra in C is defined by a weak symmetric monoidal map $A : P \to Q\vee$ which is the same as a map $A' : P \otimes Q^{op} \to$ **complexes** or the same as a weak symmetric monoidal map $A'' : Q^{op} \to P\vee$.

5.1.2 Let $s^{-1} : P\vee \to \mathbf{triv}\vee$. A is a strong symmetric monoidal map if and only if such is $s^{-1}A''$

5.2 Generalized maps of PROP's

5.2.1 Weak generalized maps A weak generalized map $P \to Q$ is by definition a weak symmetric monoidal map $F : Q\vee \to P\vee$. For any usual map of PROP's $f : P \to Q$, f^{-1} is a weak generalized map.

5.2.2 Strong generalized map Let $s_Q^{-1} : Q\vee \to \mathbf{triv}\vee$ and $s_P^{-1} : P\vee \to \mathbf{triv}\vee$. A weak generalized map $F : Q\vee \to P\vee$ is called *strong* if for any symmetric monoidal category C and any weak symmetric monoidal map $A : C \to Q\vee$ such that $s_Q^{-1}A$ is strong symmetric monoidal so is $s_P^{-1}FA$.

5.2.3 Homotopic versions A weak map of PROP's $P \to Q$ is by definition a weak symmetric monoidal map $F : Q^\frown \to P^\frown$. For any usual map of PROP's $f : P \to Q$, Σf^{-1} is a weak generalized map. A weak map $F : Q\vee \to P\vee$ is called strong if for any symmetric monoidal category C and any weak symmetric monoidal map $A : C \to Q\vee$ such that $s_Q^{-1}A$ is essentially strong symmetric monoidal so is $s_P^{-1}FA$.

5.2.4 Let $F : P \to Q^{op}\vee$ be a symmetric monoidal map. Consider the map $F\# : Q\vee \to P\vee$ defined in 3.7. Define a symmetric monoidal structure on $F\#$. it suffices to construct maps $F\#(U)(X) \otimes F\#(V)(Y) \to F\#(U \otimes V)(X \otimes Y)$ for each X, Y, U, V. We have the following sequence of natural maps

$$F\#(U)(X) \otimes F\#(V)(Y) \cong F(X) \otimes_Q U \otimes F(X) \otimes_Q V \to F(X \otimes Y) \otimes_Q (U \otimes V).$$

5.2.5 If F is an essentially strong symmetric monoidal map, then $F\#$ is essentially strong.

5.2.6 Let $F : P \to Q^{op}\vee$ be a strong symmetric monoidal map. We have a natural extension $F' : P^{op}\vee \to Q^{op}\vee$.

5.2.7

Proposition 5.1. *Assume that $F([1])$ is quasi-isomorphic to $h_{[1]}$ in $Q^{op}\vee$, where $h_{[1]}(X) = \hom_Q([1], X)$, and that the induced map $F_* : P \to P_{f[1]}$ is a quasi-isomorphism. Then $\Sigma(F\#) : Q^\frown :\to P^\frown$ is a quasi-isomorphism of PROP's.*

6 Conilpotent Hopf algebras

6.1 PROP of bialgebras

6.1.1 Let C be a symmetric monoidal category with unit. We say that $X \in C$ has a structure of bialgebra if C has structure of associative algebra with unit and a coassociative coalgebra with counit such that the coproduct map $X \to X \otimes X$ and the counit map $X \to \mathbf{1}$ are morphisms of unital associative algebras.

6.1.2 Say that $m \in C$ has a $\mathcal{BA}$-structure if $X := \mathbf{1} \oplus m$ exists and we have a bialgebra structure on X with the natural inclusion of $\mathbf{1}$ being the unit and the natural projection onto $\mathbf{1}$ being the counit.

6.1.3 There exists a PROP $\mathcal{BA}$ of BA-algebras. $\mathcal{BA}$ is uniquely specified by the condition that $\mathcal{BA}$-structures on $m \in C$ are in 1-1 correspondence with the maps of PROP's $\mathcal{BA} \to P_m$.

6.1.4 A $\mathcal{BA}$-structure on m implies structures of associative algebra without unit and of a coassociative algebra without unit on m. Let **assoc** (resp. **coass**) be the PROP's describing associative algebra (resp. coassociative algebra) structure. We thus have maps of PROP's $i : \mathbf{assoc} \to \mathcal{BA}$ and $j : \mathbf{coass} \to \mathcal{BA}$. The composition defines a map $\phi_Z : \mathbf{assoc}(Y, Z) \otimes_{S_Y} \mathbf{coass}(X, Y) \to \mathcal{BA}(X, Z)$. Therefore, we have a map

$$\phi = \sum \phi_n : \bigoplus_{n=|X|+1}^{\infty} \mathbf{assoc}([n], Z) \otimes_{S_n} \mathbf{coass}(X, [n]) \to \mathcal{BA}(X, Z)$$

Proposition 6.1. *ϕ is an isomorphism*

6.1.5

Corollary 6.2. *The tensor product on $\mathcal{BA}^\vee$ is exact.*

Proof. We see that $i^{-1}(X \otimes Y) \cong i^{-1}(X) \otimes i^{-1}(Y)$. Therefore it suffices to show that the tensor product in **assoc**$\vee$ is exact. Let A, B be finite sets. Call a map $h : A \to B$ in **assoc** simple if it is a composition of several products (no linear compositions are allowed). To specify such a map is the same as to prescribe a surjective map of sets $h' : A \to B$ and a total order on each $f^{-1}a$. Let now $A = A_1 \sqcup A_2$. Call h' irreducible if for any $a \in A$ and any neighboring $x, y \in f^{-1}a$ (i.e. there are no elements in $f^{-1}a$ between x and y), either $x \in A_1$, $y \in A_2$ ore vice versa. Any simple map $h : A_1 \sqcup A_2 \to B$ in **assoc** uniquely splits as a product $h_1 h_2$, where $h_2 : A_1 \sqcup A_2 \to B_1 \sqcup B_2$ maps A_i to B_i and $h_1 : B_1 \sqcup B_2 \to B$ is irreducible.

Let now $X, Y \in \mathbf{assoc}\vee$. We have

$$X \otimes Y([k]) = \oplus X([m]) \otimes X([n]) \otimes_{S_m \otimes S_n} \mathbf{Irred}\ ([m] \sqcup [n], [k]),$$

which is manifestly exact. □

6.2 Conilpotent functors

A functor $F : \mathcal{BA} \to$ **complexes** is called conilpotent if for any $n > 0$ and any $x \in F([n])$ there exists an m such that for any $m' \geq m$

$$j(\mathbf{coass}([n], [m])x = 0.$$

A $\mathcal{BA}$-algebra X in **complexes** is called conilpotent if such is X as a functor $\mathcal{BA} \to$ **complexes**. Denote by S the full subcategory of $\mathcal{BA}\vee$ consisting of conilpotent functors. Denote by $I : S \to \mathcal{BA}\vee$ the obvious inclusion.

6.2.1 Denote by $S' \subset \mathcal{BA}^\frown$ the full subcategory of objects quasi-isomorphic in $\mathcal{BA}^\vee$ to elements from S.

6.3 Study of S'

6.3.1 Let $\mathcal{BA}_n$ be the subcategory of $\mathcal{BA}^\vee$ consisting of objects X such that $X([m]) = 0$ for all $m > n$. We have $\mathcal{BA}_n \subset S$.

Proposition 6.3. *Any object X from S admits an increasing filtration $X_0 \subset X_1 \subset X_2 \subset \cdots = X$ such that $X_n \in \mathcal{BA}_n$.*

Proof. Set X_n to be the subobject generated by all $x \in X([k])$ such that

$$j(\mathbf{coass}([k], [n]))x = 0.$$ □

6.3.2

Proposition 6.4. *Let $\otimes$ be the tensor product in $\mathcal{BA}\vee$. Then*

1. $\mathcal{BA}_n \otimes \mathcal{BA}_m \subset \mathcal{BA}_{n+m}$;
2. $S \otimes S \subset S$ *and*
3. $S' \otimes S' \subset S'$.

Proof. Since the tensor product exact, 3. follows from 2. In virtue of Proposition 6.3, 2. follows from 1. Finally, 1. can be checked directly. □

6.3.3 Denote by $S'_n \subset S'$ the full subcategory of objects quasi-isomorphic to objects from $\mathcal{BA}_n$. We have $S'_n \otimes S'_m \subset S'_{n+m}$.

Proposition 6.5. *Any object X from S' is quasi-isomorphic to an object X' admitting an increasing exhausting filtration $X'_0 \subset X'_1 \subset \cdots$ such that $X'_k \in S'_k$ and all the quotients X'_{k+1}/X'_k are cofibrant.*

Proof. We have that X is quasi-isomorphic to an object $Y \in S$. Let $Y_0 \subset Y_1 \subset \cdots$ be a filtration as in 6.3.1. Set $X' = \Sigma X$ and $X'_n = \Sigma X_n$. □

6.4 Functors between $\mathcal{BA}_n$ for different n

6.4.1 Let C be a category whose objects are finite sets. Let $C_{\leq n}$ be its full subcategory consisting of sets with at most n elements. Let $i_n^C : C_{\leq n} \to C$ be the inclusion.

6.4.2 Let $\mathfrak{I}_n \subset \mathcal{BA}$ be the double-sided deal generated by all $\hom([p],[q])$ with $p, q \geq n$, $p < q$. Let $\mathcal{BA}'_n = \mathcal{BA}/I_n$, $p_n : \mathcal{BA} \to \mathcal{BA}'_n$ be the natural projection and $S_n = (BA'_n)_{\leq n}$. We have: $S_n(p,q) \cong \oplus_{r=2}^n \mathbf{coass}(p,r) \otimes \mathbf{assoc}(r,q)$ We have natural maps $p'_{nm} : \mathcal{BA}'_n \to \mathcal{BA}'_m$; $p_{nm} : S_n \to (\mathcal{BA}'_m)_{\leq n}$; $i_{mn} : S_m \to (\mathcal{BA}'_m)_{\leq n}$, where $m \leq n$ and $i_{mn} = i_m^{(\mathcal{BA}'_m)_{\leq n}}$.

6.4.3 We have $\mathcal{BA}_n \cong S_n\vee$ and the inclusion $I_n : \mathcal{BA}_n \to \mathcal{BA}\vee$ is equivalent to $p_n^{-1}(i_n^{\mathcal{BA}})_! \cong p_n^{-1}(i_n^{\mathcal{BA}})_*$.

6.4.4 The functor $T_n = (i_n^{\mathcal{BA}})^{-1} p_{n!}$ is the left adjoint to I_n.

6.4.5 Similarly, we have maps $I_{mn} : \mathcal{BA}_m \to \mathcal{BA}_n$, $m < n$ given by $I_{mn} = p_{nm}^{-1}(i_{mn})_! = p_{nm}^{-1}(i_{mn})_*$. Therefore, I_{mn} has a left adjoint $T_m = i_{mn}^{-1}(p_{nm})_!$; $T_m : \mathcal{BA}_n \to \mathcal{BA}_m$.

6.4.6 A similar notation can be introduced for the category **coass**. Let $C_n = \mathbf{coass}_{\leq n}$ and $i_n : C_n\vee \to \mathbf{coass}\vee$ be the functor of the extension by zero. The map j defines maps $j_n : C_n \to S_n$. We have the left adjoint functor $T_n : \mathbf{coass}\vee \to C_n\vee$ for the functor i_n. We have a natural quasi-isomorphism $T_n j_! K \to j_{n!} T_n K$ for any $K \in \mathbf{coass}^\frown$.

6.4.7 Define the symmetric monoidal structure on $S_n\vee$ by setting for $U, V \in \mathcal{BA}_n$, $U \otimes V = T_n(\mu_{nn})_!(U \boxtimes V)$ The functor $T_n : \mathcal{BA} \to \mathcal{BA}_n$ has then a symmetric monoidal structure.

6.5 Properties of T_n and I_n

6.5.1

Proposition 6.6. *Let $A \in S_n^{\frown}$. Then the natural map $m_n : T_n \Sigma I_n A \to T_n I_n A \cong I_n A$ is a quasi-isomorphism in $\mathcal{BA}_n$.*

Proof. One sees that any object from $S_n^{\frown}$ is quasi-isomorphic to bounded complex of objects of the form $j_{n!}K$, where $K \in C_n^{\frown}$. The problem then reduces to $A = j_{n!}K$. We have a quasi-isomorphism $j_! \Sigma_{\mathbf{coass}} i_n K \to \Sigma_{\mathcal{BA}} j_{n!} K$. The composition

$$T_n j_! \Sigma_{\mathbf{coass}} i_n K \to T_n \Sigma_{\mathcal{BA}} j_{n!} K \to j_{n!} K$$

is equal to

$$T_n j_! \Sigma_{\mathbf{coass}} i_n K \to j_{n!} T_n \Sigma_{\mathbf{coass}} i_n K \to j_{n!} K,$$

and the problem is reduced to showing that

$$T_n \Sigma_{\mathbf{coass}} i_n K \to K$$

is a quasi-isomorphism, which follows from the isomorphism $T_n \Sigma_{\mathbf{coass}} i_n K \to \Sigma_{C_n} K$. □

6.5.2

Proposition 6.7. *Let $X \in S_n'$. Then the natural map $X \to I_n T_n X$ is a quasi-isomorphism in $\mathcal{BA}\vee$*

Proof. We have a quasi-isomorphism $f : X \to X'$ in $\mathcal{BA}\vee$ for some $X' \in \mathcal{BA}_n$. and the commutative diagram

$$\begin{array}{ccccc} X & \longrightarrow & I_n T_n X & \xleftarrow{\approx} & I_n T_n \Sigma X \\ \downarrow{\scriptstyle\approx} & & \downarrow & & \downarrow{\scriptstyle\approx} \\ X' & \xrightarrow{\approx} & I_n T_n X' & \xleftarrow{\approx} & I_n T_n \Sigma X', \end{array} \tag{6.1}$$

where $\approx$ means "quasi-isomorphism", therefore all the remaining arrows are quasi-isomorphisms. □

6.5.3

Proposition 6.8. *Let $A, B \in S_n^{\frown}$. Then* $\hom_{S_n^{\frown}}(A, B) \to \hom_{S'}(\Sigma I_n A, \Sigma I_n B)$ *is quasi-isomorphism.*

Proof. We have

$$\hom_{S'}(\Sigma I_n A, \Sigma I_n B) \approx \hom_{BA\vee}(\Sigma I_n A, I_n B) \cong \hom_{\mathcal{BA}_n}(T_n \Sigma I_n A, B).$$

Since we know that $T_n \Sigma I_n A \to T_n I_n A \leftarrow A$ are quasi-isomorphisms, the statement follows. □

Proposition 6.9. *Let $X \in \mathcal{BA}^\frown$ and $Y \in S'_n$. Then for any $N \geq n$ the map* $\hom(X, Y) \to \hom_{S_n^\frown}(T_n X, T_n Y)$ *is a quasi-isomorphism.*

Proof. We have a chain of quasi-isomorphisms

$$\hom(X, Y) \to \hom(X, I_N T_N Y) \to \hom(T_N X, T_N Y).$$

□

6.6 S' is saturated

6.6.1

Proposition 6.10. *Any object X in $\mathcal{BA}^\frown$ admitting the filtration as in* 6.3 *is in S'.*

Proof. Without loss of generality we can assume that the filtration on X is such that each X_i/X_{i-1} is cofibrant (i.e. the filtration is as in 6.5).

Let $A_i = \operatorname{Ker} T_i|_{X_i}$. We have $A_i \subset X_i$ and $A_{i+1} \cap X_i \subset A_i$. Let

$$Z_n = X_n / \sum_{i=0}^{n} A_i.$$

and $Z = X/\sum_{i=0}^{\infty} A_i$. It is clear that Z_n, $Z \in S$, We have obvious maps $Z_n \to Z_{n+1}$ which are inclusions and $Z = \operatorname{limdir} Z_n$. We have $Z_{n+1}/Z_n \cong T_{n+1} X_{n+1}$. Also we have an obvious projection $p : X \to Z$ such that $p(X_n) \subset Z_n$ and the associated graded map is isomorphic to the collection of maps $X_{n+1}/X_n \to T_{n+1}(X_{n+1}/X_n)$. which are quasi-isomorphisms. □

6.6.2

Proposition 6.11. *Let I be any totally ordered set and let $X \in \mathcal{BA}$ and X_i, $i \subset I$ be a filtration such that $X_i/\bigcup_{j<i} X_j$ belongs to $S_n(i)$ for some n_i. Then $X \in S'$*

Proof. Same as above. □

6.6.3

Proposition 6.12. *$S' \subset \mathcal{BA}^\frown$ is a saturated subcategory.*

Proof. Let $X_\bullet$ be a bounded complex with $X_i \in S'$. One can replace it with a termwise quasi-isomorphic bounded complex $Y_\bullet$ such that each Y_n admits a filtration 6.5. Denote this filtration by $G_0 Y_n \subset G_1 Y_n \subset G_2 Y_n \subset \cdots$. Denote

$$F_{ij} Y_k = 0, \quad k > i;$$
$$F_{ij} Y_i = G_j K_i$$
$$F_{ij} Y_k = Y_k, \quad k < i.$$

It is a filtration with respect to the lexicographic order on the set of pairs of integers $(ij), i \geq n, j \geq 0$. Now apply the previous proposition. □

6.7 t-structures

The PROP $\mathcal{BA}$ is concentrated in the zeroth degree, therefore the category $\mathcal{BA}^{\frown}$ has the standard t-structure.

6.7.1 We have $D^{\geq 0} \otimes D^{\geq 0} \subset D^{\geq 0}$.

Proof. Follows from 6.2. □

Proposition 6.13. *The t-structure on $\mathcal{BA}^{\frown}$ defines a t-structure on S'.*

Proof. It suffices to check that the operators $\tau_{\geq 0}$ and $\tau_{\leq 0}$ preserve S'. Let $H \in S'$, $G \in S$ and $f : H \to G$ a quasi-isomorphism in $\mathcal{BA}\vee$ Then we have an induced quasi-isomorphism $\tau_{\geq 0}H \to \tau_{\geq 0}G$ in $\mathcal{BA}\vee$, Where $\tau_{\geq 0}G$ is by definition a quotient of G such that $\tau_{\geq 0}G(X) = \tau_{\geq 0}(G(X))$ for any finite set X with the induced action of $\mathcal{BA}$. We see that $\tau_{\geq 0}G \in S$, therefore $\tau_{\geq 0}H \in S'$. The proof for $\tau_{\leq 0}$ is similar. □

6.7.2 Let S_0 be the category of conilpotent functors $\mathcal{BA} \to$ **vect**. The functor of taking H^0: $H^0 : T_{S'} \to S_0$ is equivalence of symmetric monoidal categories.

6.8

Proposition 6.14. *Let $F' : \mathcal{BA}^{\frown} \to BA^{\frown}$ be a weak symmetric monoidal t-functor preserving S'. Let $F : S' \to S'$ be the restriction and a symmetric monoidal isomorphism $T_F \to$* Id *is given. Then F and* Id *are homotopy equivalent.*

The proof will occupy the rest of the section.

6.8.1 Categories C^+ and C^{++} For their definitions see 3.0.11.

6.8.2 Let $C \subset \mathcal{BA}^{\frown}$ be the full subcategory of finitely generated cofibrant objects concentrated in the degree 0. It is a symmetric monoidal subcategory. Therefore, so are C^+, C^{++}. We have a symmetric monoidal compatible equivalence $\iota : C^{++} \to \mathcal{BA}^{\frown}$

6.8.3 We may assume that there exist a compatible weak monoidal exact functor $G : C^{++} \to C^{++}$ and an exact functor $G' : \mathcal{BA}\vee \to \mathcal{BA}^{\frown}$ such that there exist homotopy equivalences $G\iota \to \iota F$ and $\iota G \to G'\iota$.

6.8.4 Let $C^n \subset C^{++}$ be the full subcategory of objects X such that $\Sigma\iota X \in S_n^{\frown}$.

Lemma 6.15. *G preserves C^n for any n.*

6.8.5 Let $C^\infty = \cup_n C^n$. We have a functor $\kappa : C^{\infty++} \to C^{++}$ Let $\mathbf{SC} \subset C^{++}$ be the full subcategory formed by the image of κ. $\mathbf{SC}$ is a symmetric monoidal subcategory such that $\iota(\mathbf{SC}) \cong S'$. G preserves $\mathbf{SC}$ and our problem reduces to showing that $G|_{\mathbf{SC}}$ is homotopy equivalent to Id.

6.8.6 Category P_∞ Let P_n be the full subcategory of $S_n^{\frown}$ consisting of objects which are quasi-isomorphic to cofibrant finitely generated objects in $S_n^{\frown}$ concentrated in degree 0. We have functors $T_m : P_n \to P_m$, $n > m$ which are restrictions of $T_m : \mathcal{BA}_n \to \mathcal{BA}_m$. Let P_∞ be the category whose objects are collections $R_n \in P_n$, $n = 1, 2, \ldots$ equipped with surjective quasi-isomorphisms $p_n : T_n R_{n+1} \to R_n$. Let $R, R' \in S_\infty$. We have a natural map of complexes

$$D : \prod_k \hom_{S_k}(R_k, R'_k) \to \prod_k \hom_{S_k}(T_k R_{k+1}, R'_k) \tag{6.2}$$

defined as follows. Let $f_k : R_k \to R'_k$, $k = 1, 2, \ldots$. Then

$$D(f_k|_{k\geq 1}) = (\phi_l|_{l\geq 1}),$$

where $\phi_l : T_l R_{l+1} \to R'_l$ is given by

$$\phi_l = p_l f_{l+1} - f_l p_{l+1}.$$

Lemma 6.16. *D is surjective.*

Proof. It suffices to prove that the composition map

$$\hom(R_{k+1}, R''_{k+1}) \to \hom(T_k R_{k+1}, T_k R'_{k+1}) \xrightarrow{p_k} \hom(T_k R_{k+1}, R'_k),$$

is surjective. The second map is surjective because $T_k R'_{k+1}$ is cofibrant and T_k is surjective. The first map is isomorphic to the map

$$\hom(R_{k+1}, R'_{k+1}) \to \hom(R_{k+1}, I_{k,k+1} T_k R_{k+1})$$

induced by the canonic map $R'_{k+1} \to I_{k,k+1} T_k R_{k+1}$ which is surjective. Since R_{k+1} is cofibrant, the first map is also surjective. □

Set

$$\hom_{P_\infty}(R, R') = \operatorname{Ker} D.$$

Define the symmetric monoidal structure on P_∞ by setting $(X \otimes Y)_i = X_i \otimes Y_i$, where the tensor product in $S_i^{\frown}$ is defined in 6.4.7.

6.8.7 Categories $P_{\le 0}$ and P_0. Set

$$\hom_{P_{\le 0}}(X,Y) = \tau_{\le 0}\hom_{P_\infty}(X,Y);$$
$$\hom_{P_0}(X,Y) = H^0\hom_{P_\infty}(X,Y);$$

The categories $P_{\le 0}$ and P_0 have an induced symmetric monoidal structure and we have symmetric monoidal functors $P_0 \leftarrow P_{\le 0} \to P_\infty$ which are symmetric monoidal equivalences.

6.8.8 The functor G' extends naturally to P_∞, $P_{\le 0}$ and P_0. Namely, we set $T_{G'}(X)_i = T_i G'(X_i)$. $T_{G'}$ has a clear symmetric monoidal structure. denote by $T^\infty_{G'}$ the action of $T_{G'}$ on P_∞, by $T^{\le 0}_{G'}$ the action of $T_{G'}$ on $P_{\le 0}$, and by $T^0_{G'}$ the action of $T_{G'}$ on P_0.

6.8.9 it is immediate that $T^0_G \cong \mathrm{Id}$. Therefore T^∞_G and $T^{\le 0}_G$ are homotopy equivalent to identity.

6.8.10 The collection of functors $T'_n : C \to \mathcal{BA}^\frown \xrightarrow{T_n} S^\frown_n$ defines a functor T' : $C \to P_\infty$ by $T'(X)_n = T'_n X$.

6.8.11 Let U_∞ (resp. $U_{\le 0}$, U_0) be P^{++}_∞, (resp. $P^{++}_{\le 0}$, P^{++}_0). We have a functor $T : C^{++} \to U$ which extends T'. The functor T'_G extends to each of $U_\infty, U_{\le 0}, U_0$ and is homotopy equivalent to identity.

6.8.12

Lemma 6.17. *Let $X, Y \in$* **SC** . *Then $T : \hom(X,Y) \to \hom(TX,TY)$ is a quasi-isomorphism.*

Proof. Actually, the statement is true for any X in C^{++}. Let $X_\bullet$ be a bounded complex such that each $X^k = \oplus_i X_{ik}[n_i]$ and $X_{ik} \in C_{r_{ik}}$. Then $\hom(X,Y) = \hom(X^\bullet, Y) = \left(\prod(X_{ik}[n_i], Y), d + D_X\right)$ and

$$\hom(TX,TY) = \left(\prod \hom(TX_{ik}[n_i], TY), d + D'_X\right),$$

where D_X, D'_X are determined by the differential on $X_\bullet$. It follows from the boundedness of $X_\bullet$ that it suffices to check that $\hom(X_{ik}[n_i], Y) \to \hom(TX_{ik}[n_i], TY)$ is a quasi-isomorphism for all X_{ik}. In other words, it suffices to check the statement for $X \in C$, which we assume from now on. Similarly, Let $Y = Y_\bullet$, where $Y_k = \oplus Y_{ik}[n_{ik}]$, where $Y_ik \in C^{m_{ik}}$ Then we have $\hom(X,Y) = (\oplus_{ik}\hom(X, Y_{ik}[n_k]), d + D_Y)$ and same for $\hom(TX,TY)$. We see again that it suffices to check that the statement for $X \in C$ and $Y \in C^k$. It follows from the fact that $\hom(TX,TY)$ is quasi-isomorphic to the complex (6.2) and that $\hom(X, Y) \to \hom(T^n X, T^n Y)$ is a quasi-isomorphism for all $n \geq k$ (see 6.9). □.

6.8.13 We have a natural transformation $\mu : TG \to T_{G'}T$ which is defined as follows. $TG(X)_k = T_k G(X)$ and $T_{G'}T(X)_k = T_k G' T_k X$ and we have maps $T_k G(X) \to T_k G'(X) \to T_k G'(T_k X)$.

6.8.14

Proposition 6.18. *For any $X \in \mathbf{SC}$, μ_X is a quasi-isomorphism.*

Proof. If $X \in C_k$, then the statement follows from the quasi-isomorphism $X \to T_k X$. The statement now follows easily. □

6.8.15 Let $A \subset U$ be the full subcategory of objects quasi-isomorphic to objects from the image of T. Then $T_{G'}$ preserves A and $T_{G'}|_A$ is homotopy equivalent to G. Since $T_{G'}$ is homotopy equivalent to Id, so is $T_{G'}|_A$, whence the statement.

6.8.16 Let $F'_t : \mathcal{BA}^\frown \to \mathcal{BA}^\frown[[t]]$ be a weak symmetric monoidal functor mapping $S' \to S'[[t]]$ and let $F : S' \to S'[[t]]$ be the restriction. Assume that $F_t = F_0 \pmod t$ and that $F_0 \cong \mathrm{Id}$. Let $I : S' \to S'[[t]]$ and $i : T_{S'} \to T_{S'}[[t]]$ be natural inclusions. The following proposition is proved in the same way as the previous one.

Proposition 6.19. *Assume that there exists an isomorphism of functors $\mu : T_{F_t} \to i T_{F_0}$ such that $\mu = \mathrm{Id}(\bmod t)$. Then F_t is deformationally homotopically equivalent to $i F_0$.*

6.9 Universal conilpotent $\mathcal{BA}$-algebras

6.9.1 Let P be a PROP and let $A : \mathcal{BA} \to P\vee$ be a functor. Any such an A determines a functor $A' : \mathcal{BA} \otimes P^{op} \to$ **complexes**. Call A *conilpotent* if for any n, k and any $x \in A'([n] \otimes [k])$ there exists an m such that for any $m' > m$ and $t \in j(\mathbf{coass}([n], [m'])$ we have $(t \otimes \mathrm{Id}_{[k]})x = 0$.

6.9.2

Proposition 6.20. *If A is conilpotent, then $A\# : P\vee \to \mathcal{BA}\vee$ takes only values in S.*

Corollary 6.21. *$\Sigma(A\#) : P^\frown \to S'$ is an exact weak symmetric monoidal functor.*

6.9.3 Call a functor $F : P^\frown \to S'$ essentially strong if such is the composition $P^\frown \to S' \to \mathcal{BA}^\frown$ (see 5.2.3). We see that $\Sigma(A\#)$ is essentially strong whenever A is essentially strong. This follows from 5.2.5.

6.10 Operad B_∞

Let C be a symmetric monoidal additive category and let C' be the symmetric monoidal category of complexes in C. Let $n \in C'$ and assume $m = \oplus_{i=1}^{\infty} T^i n = Tn$ exists in C. m has a structure of cofree coassociative algebra. Call a $\mathcal{B}_\infty$-structure on n a dg-$\mathcal{BA}$-structure on m such that the induced coassociative algebra structure is the cofree one and the restriction of the differential on $n \subset Tn$ is equal to the existing differential on n.

Such a structure is uniquely determined by the corestriction of the differential $d_r : Tn \to Tn \to n$ and of the product $m_r : Tn \otimes Tn \to Tn \to n$. Certain identities should be satisfied. One sees that these identities are described by an operad denoted by B'_∞. The B'_∞-structure on the shifted space $n[1]$ is called $\mathcal{B}_\infty$-structure.

6.10.1 Let $P\mathcal{B}_\infty$ be the PROP generated by $\mathcal{B}_\infty$. Since any $\mathcal{B}_\infty$-algebra structure on n determines a BA-structure on $T(n[1])$, we have a map $H' : \mathcal{BA} \to P\mathcal{B}_\infty{}^\frown$.

6.10.2 Since the constructed bialgebra Tn is manifestly conilpotent, we have $H'\# : P\mathcal{B}_\infty \vee \to S \subset \mathcal{BA}\vee$ (see 3.7).

6.10.3 Brace structures. A $\mathcal{B}_\infty$-structure on n is called *brace structure* if

$$d_r|_{T^{\geq 2}(n[1]) \otimes T(n[1])} = 0$$

and $m_r(T^{\geq 3}(n[1])) = 0$. Thus, a brace structure on n is described by a degree zero map $m : n \otimes n \to n$ and by a collection of "braces" $b_m : n \otimes T^m n \to n$, $m \geq 1$ of degree $-m$. The operation b_k has the following notation:

$$b_k(n_0; n_1, \ldots, n_k) = n_0\{n_1, \ldots, n_k\}.$$

Brace structures are described by the operad M. We have a map $p : \mathcal{B}_\infty \to M$ and the composition $H = p^\frown \circ H' : \mathcal{BA} \to M^\frown$. We have the corresponding functor $H^\#\ P\mathcal{M}\vee \to S \subset \mathcal{BA}\vee$

6.10.4 It is known that there exists a (non-canonic) quasi-isomorphism $u : \mathbf{hoe}_2 \to M$.

6.10.5 Let n be a $\mathcal{BA}$-algebra. Define a brace structure on $B(n) = T(n[1])$ as follows. Let $m_2 : Tn[1] \otimes Tn[1]$ be the associative product and let

$$n\{n_1, n_2, \ldots, n_k\} = \Delta_k(n) \times (n_1 \otimes \cdots \otimes n_k),$$

where Δ denotes the k-fold coproduct in the bialgebra $\mathbf{1} \oplus n$ and $\times : T^k n \otimes T^k n \to T^k n$ is the component-wise product. We thus have a map $B : M \to (\mathcal{BA}^{op})^\frown$. We have a functor $B^\# : \mathcal{BA}\vee \to P_M\vee$.

Denote $\Gamma = BH : \mathcal{BA} \to \mathcal{BA}^\frown$. We have $\Gamma^\# = H^\# B^\# : \mathcal{BA}\vee \to S$.

6.10.6 Let δ_k be the k-fold coproduct on m. Observe that $\delta_2(x) = \Delta_2 x - 1 \otimes x - x \otimes 1$. Let X be a conilpotent $\mathcal{BA}$-coalgebra. Define the map $Z : X \to \Gamma^{\#}(X)$ as follows. The inclusion $i : X[1] \to T(X[1])$ defines the map $Ti : T(X[1][-1]) \to T(T(X[1])[-1])$. We have also the map $\delta = \sum_k \delta_k : X \to TX$. Set $Z = (Ti)\delta$.

6.10.7 Z is a quasi-isomorphism.

6.10.8 Thus we have a symmetric monoidal transformation $Z : \mathrm{Id}_S \to \Gamma^{\#} I$.

6.10.9 Set

$$\Delta : S' \to \mathcal{BA}^{\frown} \xrightarrow{\Gamma} S \xrightarrow{\Sigma} S'.$$

6.10.10

Proposition 6.22. *The functor Δ is a t-functor*

Proof. Indeed, for any $X \in S'$ there exists a quasi-isomorphic $X' \in S$. We have $\Delta(X) \approx \Gamma(X') \overset{Z}{\approx} X' \approx X$, where $\approx$ means "quasi-isomorphic". Therefore, Δ preserves any full subcategory closed with respect to quasi-isomorphisms, in particular $\mathcal{D}^{\geq 0}$ and $\mathcal{D}^{\leq 0}$. □

6.10.11 The transformation Z establishes equivalence $T_\Delta \cong \mathrm{Id}$. Therefore, by 6.8 Δ is equivalent to Id.

6.11

Prove that the inverse composition $L = \Sigma B^{\#} H^{\#}$ is an equivalence of symmetric monoidal categories. We have that L is homotopy equivalent to $BH^{\#}$. According to 5.2.6 it suffices to check that $BH([1])$ is quasi-isomorphic to $[1]$ (which can be done straightforwardly) and that the induced map $P_{[1]} \to P_{BH([1]}$ is a quasi-isomorphism. Note that since $[1] \approx BH([1]), P_{BH[1]} \approx P_[1]$. Since $P_{[1]} \cong PM$ and the cohomology of $PM \approx Pe_2$ are generated by $H^\bullet \hom_{PM}([2], [1]) \cong M([2]) \approx e_2([2])$, it suffices to check that the induced map

$$H^\bullet(BH) : H^\bullet \hom_{PM}([2], [1]) \to H^\bullet \hom(BH([1]) \otimes BH([1]), BH([1]))$$

is an isomorphism, which also can be done straightforwardly.

6.11.1 Corollary. *Functors $H' = \Sigma H\#$ and $B' : \Sigma B\#$ are homotopy equivalences of symmetric monoidal categories.*

6.11.2 Let $M' = M\{-1\}$ be the shift of M. Let $\mathcal{D}^{\geq 0}_{M'} \subset (PM')^{\frown}$ be the full subcategory of objects U such that $H^{<0}U(X) = 0$ for any X. We have the shift maps $\mathbf{sh} : PM' \to PM$ and $\mathbf{sh}' : PM \to PM'$. Let $H'' = H'\mathbf{sh}$ and $B'' = \mathbf{sh}'B'$.

Lemma 6.23. *$H''(\mathcal{D}^{\geq 0}_{M}) \subset H''(\mathcal{D}^{\geq 0})$ $B''(\mathcal{D}^{\geq 0}) \subset H''(\mathcal{D}^{\geq 0}_{M})$. Therefore $\mathcal{D}^{\geq 0}_{M'}$ is the category of objects quasi-isomorphic to objects from $B'(\mathcal{D}^{\geq 0})$*

Proof. Straightforward. □

Set $\mathcal{D}^{\leq 0}_{M'}$ to be the full subcategory of objects quasi-isomorphic to $B'(\mathcal{D}^{\leq 0})$. Then the pair $(\mathcal{D}^{\geq 0}_{M'}, \mathcal{D}^{\leq 0}_{M'})$ is a t-structure.

6.11.3 Fix a quasi-isomorphism $u : \mathbf{hoe}_2 \to \mathcal{M}$. Let $e_2' = e_2\{-1\}$ We have a chain of equivalences

$$Q_u : (Pe_2')^{\frown} \to (Phe_2')^{\frown} \to PM' \to S'. \tag{6.3}$$

Also we have a t-structure on Pe_2' from 3.8.9.

Proposition 6.24. *All arrows in 6.3 are t-functors.*

Corollary 6.25. *T''_H, T''_B, T_{Q_u} are equivalences of symmetric monoidal categories.*

6.12

Corollary 6.26. *The symmetric monoidal category S_0 is equivalent to the symmetric monoidal category T_M.*

6.13 Lie bialgebras and Etingof–Kazhdan quantization

6.13.1 Let $\mathcal{LBA}$ be the PROP of Lie bialgebras, **lie** be the PROP of Lie algebras and **colie** be the PROP of Lie coalgebras. We have maps $i : \mathbf{lie} \to \mathcal{LBA}$ and $j : \mathbf{colie} \to \mathcal{LBA}$.

Proposition 6.27. *The natural map $\phi : \oplus_{n \geq |Y|} \mathbf{lie}([n], X) \otimes_{S_n} \mathbf{colie}([Y], [n])$ is an isomorphism.*

6.13.2 A functor $U \in \mathcal{LBA}\vee$ is called *conilpotent* if for any finite set X and any $x \in U(X)$ there exists an N such that for all $n > N$, $j(\mathbf{colie}(X, [n]))x = 0$. Let L be the full subcategory of conilpotent functors and $L' \subset \mathcal{LBA}^{\frown}$ be the full subcategory of objects quasi-isomorphic in $\mathcal{LBA}\vee$ to conilpotent functors.

6.13.3 The category $\mathcal{LBA}^{\frown}$ has a canonic T-structure which defines a T-structure on L'.

6.13.4 Let L_0 be the category of conilpotent functors $\mathcal{LBA} \to$ **vect**. We have a symmetric monoidal equivalence $T_{L'} \cong L_0$ given by H^0.

6.13.5 Let A be a Lie bialgebra in a symmetric monoidal category C closed with respect to limits. Then $C^\bullet(A)$ is a Gerstenhaber algebra in the category of complexes in C (see 1.2.5). This construction defines a functor $B : Pe_2 \to \mathcal{LBA}^\frown$ and the corresponding strong generalized map of PROP's $B^\#$.

6.13.6 Let X be a Gerstenhaber algebra in a symmetric monoidal category C closed with respect to limits. Let $\mathbf{lie}(n)$ be the n-th space of the operad of Lie algebras and let $\mathbf{lie}^*(n)$ be its linear dual. For any $Y \in C$ Set $\mathbf{CL}(Y) = \oplus_{n=1}^{\infty} \mathbf{lie}^*(n) \otimes_{S_n} Y^{\otimes n}$. $\mathbf{CL}(Y)$ is naturally a Lie coalgebra in C. The commutative product on X defines a differential on $\mathbf{CL}(X[1])$ so that its corestriction $\mathbf{CL}(X[1]) \to \mathbf{CL}(X[1]) \to X[1]$

does not vanish only on the summand $\mathbf{lie}(2)^* \otimes_{S_2} (X[1])^{\otimes 2} \cong S^2X[2]$ and is defined by the commutative product $S^2X \to X$. The bracket on X Defines a bracket $\mathbf{CL}(X[1]) \otimes \mathbf{CL}(X[1]) \to \mathbf{CL}(X[1])$ such that the corestriction

$$\mathbf{CL}(X[1]) \otimes \mathbf{CL}(X[1]) \to \mathbf{CL}(X[1]) \to X[1]$$

does not vanish only on $X[1] \otimes X[1] \subset \mathbf{CL}(X[1]) \otimes \mathbf{CL}(X[1])$ and is defined by the bracket on X.

Thus, $\mathbf{CL}(X[1])$ is a Lie bialgebra. One sees that it is conilpotent. This construction defines a functor $B : \mathcal{LBA} \to Pe_2^\frown$ so that $G^\# : Pe_2\vee \to L$ and the composition $G^\# : Pe_2\vee \to L \to \mathcal{LBA}\vee$ is a strong generalized map of PROP's.

6.13.7 Let X be a conilpotent Lie bialgebra. We have a map $Z : X \to \mathbf{CL}(C^\bullet(X)[1])$ such that $Z : X \xrightarrow{Z'} \mathbf{CL}(X) \to \mathbf{CL}(C^\bullet(X)[1])$ and

$$Z' \in \hom(X, \mathbf{CL}(X)) = \hom(X, \oplus \sum \mathbf{lie}(n)^* \otimes X^{\otimes n})$$

is defined by the Lie coalgebra structure on X.

This construction defines a map $\mathrm{Id} \to B^\# G^\#$.

6.13.8 Functors $\Sigma G^\# : L' \to Pe_2^\frown$ and $\Sigma B^\# : Pe_2^\frown \to L'$. are t-functors. Each of them is a homotopic symmetric monoidal equivalence.

The proof of there statements goes along the same line as the corresponding proof for $H' : PM^\frown \to S'$ and $B' : S' \to PM^\frown$ (6.10.9-6.11)

6.13.9 Let $G' = T_{\Sigma G^\#}$ and $B' = T_{\Sigma B^\#}$. They establish a symmetric monoidal equivalence between L_0 and $T_{Pe_2^\frown}$.

6.13.10 We thus have symmetric monoidal equivalences $\mathbf{Q} : L_0 \to T_{Pe_2^\frown} \to S_0$ and $\mathbf{DQ} : S_0 \to T_{Pe_2^\frown} \to L_0$ so that $\mathbf{Q}$ and $\mathbf{DQ}$ are mutually inverse strong generalized maps of PROP's (this follows from the fact that both **Q** and **DQ** are induced by t-functors which are compositions of strong generalized maps of PROP's.

6.13.11 Let C be an additive symmetric monoidal category and let $C' \subset C^{op}\vee$ be the full subcategory of functors $C^{op} \to \mathbf{vect}$. We have the Yoneda embedding $C \to C'$ Let $X : \mathcal{LBA} \to C'$ be a conilpotent algebra. We can rewrite $X : \mathcal{LBA} \otimes C \to \mathbf{vect}$ or $X : C^{op} \to \mathcal{LBA}$. Since X is conilpotent, $X : C^{op} \to L_0$. Set $Q(X) = \mathbf{Q} \circ X : C^{op} \to S_0$. $Q(X)$ is a conilpotent bialgebra in C'. (It is a strong $\mathcal{BA}$-algebra because Q is the restriction onto the core of an essentially strong generalized map of PROP's.) Thus Q is a map from the category of conilpotent Lie bialgebras in C' to the category of conilpotent Hopf algebras in C'

6.13.12 Similarly, the composition with **DQ** defines the quasi-inverse functor DQ from the category of conilpotent Hopf algebras in C' to the category of conilpotent Lie bialgebras in C'. (Again DQ(X) is always a strong $\mathcal{LBA}$-algebra, because DQ is the restriction onto the core of an essentially strong generalized map of PROP's).

6.13.13 The two previous subsections imply that the pair of functors Q and DQ establishes an equivalence between the categories of conilpotent Lie bialgebras in C' and conilpotent Hopf algebras in C'.

6.14 Action of GT

6.14.1 Recall that we have fixed a quasi-isomorphism $u : \mathbf{hoe}_2 \to \mathcal{M}$. Therefore, we have a T-equivalence $Q_u : P\mathbf{hoe}_2^{\frown} \to S'$. In the same way, we have a quasi-isomorphism $v : \mathbf{ho}\mathcal{M} \to e_2$ which is homotopy inverse to u. We can construct a map $Q_v : S' \to PM \to P\mathbf{ho}\mathcal{M} \to P\mathbf{hoe}_2$ so that $Q_u Q_v$ and $Q_v Q_u$ are homotopy equivalent to identity.

6.14.2 Let $\mathbf{hoe}_2$ be a resolution of e_2 and let $m : \mathbf{hoe}_2 \to \mathbf{hoe}_2[[t]]$ be an action of a formal one-parametric group on $\mathbf{hoe}_2$ such that its derivative at $t = 0$ produces a non-trivial element in the cohomology of the deformation complex of e_2. Then we have the map $Q_{um} : P\mathbf{hoe}_2^{\frown} \to S'[[t]]$.

Theorem 6.28. *There is no equivalence between $T_{Q_{um}}$ and iT_{Q_u} which reduces to identity modulo t.*

Proof. Assume there is. Then $T_{Q_u Q_v}$ and $T_{Q_{um} Q_v}$ are also equivalent in such a way that the reduction of the equivalence modulo t is identity. Hence, in virtue of Proposition 6.19, $Q_u Q_v$ and $Q_{um} Q_v$ are deformationally homotopy equivalent and so are Q_u and Q_{um}. Let $m_* : \mathbf{hoe}_2^{\frown} \to \mathbf{hoe}_2^{\frown}[[t]]$ be the induced map. Then $Q_{um} = Q_u m_*$, therefore m_* is deformationally homotopy equivalent to i. Let $e \in \mathbf{hoe}_2^{\frown}$ be such that

$e(X) = \hom([1], x)$. Note that $m_* e = e$. The full operad $\mathcal{O}_e$ is equivalent to the operad of $\mathbf{hoe}_2$ algebras and the induced map $\mathcal{O}_e \to \mathcal{O}_e[[t]]$ is just m. Since m_* is deformationally homotopy equivalent to i and i is deformationally trivial, so does m according to 4. This contradicts to the assumption. □

6.14.3 In [9] it was proven that the Lie algebra $\mathfrak{grt}$ of the graded Grothendieck–Teichmüller group acts on $\mathbf{hoe}_2$ by derivations so that the induced map $\mathfrak{grt} \to H^0$ **def** $\mathbf{hoe}_2$ is injective. Therefore, the conclusion of Theorem 6.28 is true when m is induced by a one parametric subgroup of the graded Grothendieck–Teichmüller group.

References

[1] Beilinson, A., Bernstein, J., Deligne, P., Faisceaux Pervers (French) [Perverse sheaves], Analysis and topology on singular spaces, I, Luminy 1981, *Astérisque* 100, Soc. Math. France, Paris 1982, 5–171.

[2] Cartan, H., Eilenberg, S., *H*omological Algebra, Princeton University Press, Princeton, NJ, 1956.

[3] Drinfeld, V., On quasitriangular quasi-Hopf algebras and a group closely related to Gal($\bar{\mathbb{Q}}/\mathbb{Q}$), (Russian) *Algebra i Analiz* 2 (1990), no. 4, 149–181; translation in *Leningrad Math. J.* 2 (1991), no. 4, 829–860.

[4] Etingof, P., Kazhdan, D., Quantization of Lie bialgebras, I, *Selecta Math.* 2 (1996), no. 1, 1–41.

[5] Enriquez, B., Etingof, P., On the invertibility of quantization functors of Lie bialgebras, preprint.

[6] Kontsevich, M., Operads and motives in deformational quantization, *Lett. Math. Phys* 48 (1999), no. 1, 35–72.

[7] Kontsevich, M., Soibelman, Y., Deformation of algerbas over operads and Deligne's conjecture, in *Quantization, Deformations, and Symmetries* (Dito, G., Sternheimer, D., eds.), Conférence Moshé Flato 1999, Math. Phys. Stud. 22, Kluwer Academic Publishers, Dordrecht, Boston, London 2000, 255–307.

[8] Tamarkin, D., Formality of the chain operad of small squares, `math.QA/9809164`.

[9] Tamarkin, D., Action of the Grothendieck-Teichmüller group on the operad of Gerstenhaber algebras, preprint

Bei Fragen zur Produktsicherheit wenden Sie sich bitte an:
If you have any questions regarding product safety,
please contact:

Walter de Gruyter GmbH
Genthiner Straße 13
10785 Berlin
productsafety@degruyterbrill.com